# Mobile Computing and Wireless Communications

# Mobile Computing and Wireless Communications

Edited by Timothy Kolaya

CLANRYE
INTERNATIONAL
www.clanryeinternational.com

Clanrye International,
750 Third Avenue, 9th Floor,
New York, NY 10017, USA

ISBN: 978-1-63240-596-8

**Cataloging-in-Publication Data**

Mobile computing and wireless communications / edited by Timothy Kolaya.
    p. cm.
Includes bibliographical references and index.
ISBN 978-1-63240-596-8
1. Mobile computing. 2. Wireless communication systems. 3. Mobile communication systems.
4. Electronic data processing--Distributed processing. I. Kolaya, Timothy.
QA76.59 .M63 2017
004--dc23

For information on all Clanrye International publications
visit our website at www.clanryeinternational.com

CLANRYE
INTERNATIONAL

Printed in the United States of America.

# Contents

# Preface

Over the recent decade, advancements and applications have progressed exponentially. This has led to the increased interest in this field and projects are being conducted to enhance knowledge. The main objective of this book is to present some of the critical challenges and provide insights into possible solutions. This book will answer the varied questions that arise in the field and also provide an increased scope for furthering studies.

Mobile computing is a form of human-computer interaction facilitated by wireless communications. Some of the significant aspects of this field include mobile communications, mobile software, data transmission, etc. There has been rapid progress in this field and its applications are finding their way across multiple industries. This book includes contributions of experts and scientists which will provide innovative insights into this field. A number of latest researches have been included to keep the readers up-to-date with the global concepts in this area of study. It will be of great help to students and teachers in the fields of telecommunications, wireless engineering and software engineering.

I hope that this book, with its visionary approach, will be a valuable addition and will promote interest among readers. Each of the authors has provided their extraordinary competence in their specific fields by providing different perspectives as they come from diverse nations and regions. I thank them for their contributions.

**Editor**

# On the embedded warning system based on wireless sensor networks for marine monitoring and defending

**YUE Xiangyu**

School of Management and Engineering, Nanjing University; Nanjing Jiangsu; 210008; PR China

**Email address:**

yuexiangyu168@gmail.com

**Abstract:** The warning system for ocean monitoring and defending is a remote video ocean monitoring system engineering using the information gathering technologies, such as ultrasonic, radar, laser, infrared, machine vision, interactive, and so on, and other integrated technologies, such as wireless sensor networks, modern communication, computer information processing, solar energy. The engineering, based on information collecting technology and wireless sensor networks and combined with the technologies such as Beidou navigation and positioning, images and meteorological sensor information, geographic information, remote sensing information, forms a marine resource digital information management system that provides a full range of electronic monitoring and protection for the oceans. The wireless sensor network is a self-organizing network composed of a large number of sensor nodes, which sets such three technologies as the sensors, micro-electromechanical systems and network in one. The network's aim is to sense, collect and process the information of the objects in the network coverage and send it to the data processing center to provide the basis for ocean monitoring and managing and protecting the marine safety. With the continuous development of micro-electromechanical system MEMS, wireless communication technology and electronic technology, the practical field of wireless sensor networks is becoming wider and wider.

**Keywords:** Embedded System, Wireless Sensor Networks, Beidou Positioning System, Wireless Communication, Detection, Warning

With the growing increase of China's comprehensive national strength as well as the continued exploration and development of marine resources, marine monitoring and early warning system are being gradually demanded and valued by people. However, due to the huge ocean area, it has more difficulty to control. This article designed an embedded marine monitoring and early warning system based on wireless sensor networks which can play a good role in marine monitoring and marine security protection.

## 1. Current Research Status of Wireless Sensor Network at Home and Abroad

Wireless sensor network has important scientific value and broad application prospects which also causes a great concern of many countries' military department, industry and academia. Representatives of which is the "Smart Dust" laboratory jointly established by the University of California at Berkeley and Intel Corporation. Its goal is to make the U.S. military's surveillance of the activities of the enemy without being noticed. U.S. Army proposed the "Smart Sensor Network Communication" program in 2001. It connected the unattended type of ammunition, sensors and robotic systems in Future Combat Systems into a network so that exponentially increased the ability of a single sensor, while improving the system's efficiency and robustness. Intel Corporation also issued a "new computing development plan based on micro-sensor networks "in 2002, dedicated to the application of sensor networks in the fields of environmental monitoring , forest fire fighting , Haiti sector etc. Research institutions in many other countries such as Britain , Germany, Canada, Japan have also joined the research of wireless sensor network, major companies includes Philips, France Telecom, Siemens, NEC, Omron , Sky2 and so on.

Domestic research on wireless sensor network technology is still in its infancy. Compared with foreign countries, there is still a long way to go. In the theoretical research, Tsinghua University and other universities and the CAS etc. all have had some preliminary research on network security technology, energy technology and other core processing technology, routing algorithms and communication protocols and so on. In practical applications, the program that combines the wireless communications technology and sensor technology appeared even later. For instance, wireless sensor network node hardware and software platforms for monitoring the temperature, humidity and soil pH inside the greenhouse of vegetables, developed by the CAS Institute of Computing Ningbo segments which based on ZigBee protocol and a dedicated low-power processor chips and intelligent health monitoring system for major engineering structures developed by academicians Ou Jinping etc. of Harbin Industrial University. In addition, some of the other domestic universities and research institutions have also made some achievements on wireless sensor network technology. But there is less innovative research on the system and there are many problems, so the level of research is relatively backward. Therefore, future research and application in this area should be strengthened to narrow the gap between foreign countries of technologies in this field.

## 2. Key Technologies of Wireless Sensor Networks in Marine Monitoring and Defending early Warning System

The wireless sensor network in ocean monitoring and defending early warning system is a wireless infrastructure network. Compared with the wireless ad hoc networks, it has many defects in the aspects of resources and so on. To compensate for these shortcomings, we have taken the following key measures in the research and design to the wireless sensor network:

### 2.1. Increase Self-Management Skills and Self-Organizing Capacity of the Wireless Sensor Network

The application of the wireless sensor networks in ocean monitoring and defending early warning system will be impacted and restricted by the surrounding environment. The main structure wireless sensor networks used is a wireless ad hoc network structure, and its sensor nodes are placed in the monitoring area. So it must be able to automatically launch and must take the initiative in the signal, contacting with an adjacent node when having information, and then feedback the acquired position and work status of the adjacent nodes to the work station. Work station will then plan the network topology structure according to the recorded information.

### 2.2. Strengthen Energy Management in Wireless Sensor Networks

As the network nodes of the wireless sensor in marine monitoring and defending early warning system is limited by the battery to supply energy, the use of energy must be accurately calculated which could make them reasonably switch between dormancy and working conditions, thus to achieve the purposes of saving energy as much as possible, reducing space and a reduction of cost.

### 2.3. Optimize Sensor Network's Data Management, Processing and Query

The wireless sensor network in ocean monitoring and defending early warning system is data-centric, so data storage, management, processing and query are its core technologies. This is mainly manifested in the following aspects:

(1) In the application of wireless sensor networks in ocean monitoring and defending early warning system, monitoring data and information is of the most importance. However, the working condition of each specific node is not the main part.

(2) Because marine monitoring and defending early warning system will monitor too large scope and too many nodes, so the data collection is very large and complex, and thus it has a feature of redundancy .

(3) For the purpose of data processing by data-centric wireless sensor network, sensory data management becomes one of the core technologies of ocean monitoring and defending early warning system, including sensory data storage, query, analysis, mining , understanding, decision analysis and many other aspects.

(4) Due to the large quantity of wireless sensor network nodes in marine monitoring and defending early warning system, when network bandwidth is not very satisfactory, energy issues are of the most importance.

(5) Since the attribute addressing manner of the wireless sensor network in ocean monitoring and defending early warning system is node allocation address, there is no need to assign each node a unique address.

### 2.4. Strengthen the Node Localization of the Wireless Sensor Network

When the ocean monitoring and defending early warning system is applying to the wireless sensor network, most critically, the observer should make the right data processing to the monitored areas based on collected data information, and the data information collected in each node must have a comprehensive consideration on the information in the coordinate system of measured position.

Only in this way can improve the routing efficiency, therefore to grasp the load balancing and topology structure of network better. Meanwhile, during the formation of the algorithm, it must be carefully considered about the issues like how the network of each node generates its own spatial coordinates and so on. We use Beidou navigation and positioning technologies, the world's best satellite navigation and positioning system, to position each node.

## 2.5. Construct Protocol Stack for Sensor Network

The protocol stack for wireless sensor network of ocean monitoring and defending early warning system includes five layers of protocols: the physical layer, the network layer, the data link layer, the transport layer and the application layer. This five-layer respectively correspond to each protocol layer of the Internet protocol stack. Additionally, the protocol stack for wireless sensor network also includes the following three management platforms: energy management platform, task management platform and mobile management platform.

In the protocol stack for wireless sensor network of the marine monitoring and defending early warning system, location and time synchronization sub-layer are the most important. It will not only provide information support to other wireless sensor network protocol layers, but also make cooperative positioning and time synchronous negotiation relying on the data transmission channel. In each protocol layer, QoS management designs queue management, priority mechanism or bandwidth reservation mechanism etc., and gives special treatment to the specific application data. Topology control achieves topology generation through the link layer, physical layer and routing layer, which also provides basic information support to optimize the MAC protocol process, and reduce the energy consumption of network, improving the efficiency of agreement. Wireless sensor network management then requires each protocol layers to not only conveniently embed a variety of information interfaces but also be able to regularly collect traffic information and operating status of the protocol as well as coordinate all protocol components in the control network, which can make it operate normally.

## 3. Implementations of the Monitoring Node for Wireless Sensor Network in Ocean Monitoring and Defending early Warning System

The monitoring node for wireless sensor network in ocean monitoring and defending early warning system is mainly responsible for the monitoring and data collection of crashers throughout the ocean. Its core is the low-power, 32-bit chip STM32F103 ZET6 based on the ARM Cortex-M3 kernel, which combined with wireless transceiver chip CC2420. The mode of surveillance and information gathering technology includes ultrasonic wave, radar, laser, infrared ray, machine vision, and interaction and so on. The information processing, however, mainly analyzes and processes the collected information in the ARM chip. And information determination is a dynamic and real-time measurement and identification of risk or security status based on the timing, the weather, and the information such as the distance, relative acceleration, relative speed etc. between the detection node and monitoring object. When the information is identified and judged as dangerous, the main responsibility for warning information is to calculate the risk level and dangerous direction, and give appropriate sound and light alarm as well as transmit information through a wireless sensor network, notifying the control room located onshore. The monitoring node is battery-powered. According to the performance requirements of marine monitoring, ultrasonic sensors, infrared sensors, radars, cameras etc. can be used to gather information. Additionally, if it is aiming to detect partial marine ecological information in the use of this system in the meanwhile, salinity sensors, temperature and humidity sensors can be also installed on it. For example, using temperature and humidity sensor SHTll, it employs advanced CMOSensTM digital sensor technology. The stability and reliability of which is relatively high. The interface mode for the sensor is two-wire digital mode, which can be directly connected to the microcontroller so that its peripheral circuit has been greatly simplified. The package is SMD chip which greatly reduces the area of PCB.

## 3.1. Structure of the Monitoring Node for Wireless Sensor Network

The main function of the monitoring node for wireless sensor network in ocean monitoring and defending early warning system is to monitor and collect data of marine suspicious object and then transfer it to the base station node. It consists of four parts-the sensor module, the processing module, the transmission module and the power supply module. The sensor module uses an interactive electronic whiteboard including electromagnetic induction, infrared sensing and ultrasonic sensing, which connects the processing module by simulating 112C interface. The processing module controls the routing protocols and the node's power management etc. between other modules and the wireless sensor network. The transmission module mainly consists of a low-power wireless communication chip CC2420 and its peripheral circuits. The software runs the appropriate communication protocols. The energy supply module uses two 5 AA solar rechargeable batteries.

## 3.2. Hardware Design of the Monitoring Node for Wireless Sensor Network

### 3.2.1. Processing Circuit Module

STM32F103ZET6, selected for the monitoring node for wireless sensor network in ocean monitoring and defending early warning system, is a low-power 32-bit microcontroller based on the ARM Cortex-M3 kernel. The internal structure of this type of chips is Harvard with the advantages of low power consumptions and abundant resources etc. The choice of storage is a 512KB on-chip ROM and a 64KB SRAM. In addition, this chip has 112 IO ports, three 12-bit ADC channels, a 12-bit DAC, two basic timers, four general purpose timers and two senior timers, which can work normally in a variety of different modes. It also has an SDIO interface, an FSMC interfaces, three SPI, two I2C bus interfaces as well as multiple serial ports and CAN bus interfaces, which can be designed by two programming approaches-ISP and JTAG. It supports three kinds of low-power mode, resulting in a best balance point among low power consumption, short startup time and available wakeup sources.

### 3.2.2. Transmission Circuit Module

The selection of the RF transceiver of the monitoring node for wireless sensor network in ocean monitoring and defending early warning system is CC2420, coming from the SmartRF03 technology of Chipcon, which is made by 0.18um CMOS process. It has many advantages such as minimal external components, very low power consumptions, relatively stable performance, low working voltage(2.1 ~ 3.6v), small volume etc. Therefore it is quite suitable for integration. Using this chip in wireless sensor network in ocean monitoring and defending early warning system, data transfer rates can be up to 25OKbPs, resulting in achieving rapid multipoint networking. CC2420 sets the operation mode of 4-wire SPI (SI, S0, SCLK, CSn) bus chip, realizing functions such as read (write) functions of the cached data and status register etc. The transmit/receive buffer is set by controlling the state of FIFO's and FIFOP's pin interface.

### 3.2.3. Sensor Circuit Module

The sensor module can integrate a variety of sensors into one board, which makes it with a high degree of integration. It integrates all functions like infrared sensing, ultrasonic sensing, temperature and humidity sensing, signal conversion, A/D conversion etc. into one single chip, resulting in free and flexible mode choosing and high energy saving ability. Especially, after monitoring and communication, the sensor module can be automatically transferred into the low-power mode which greatly reduces the energy consumption.

## Brief Introduction to the Author

Yue Xiangyu (1992- ), who is a male, Han nationality, was born in Weifang, Shandong, with BA, Nanjing University, mainly engaged in electrical information and automation, aesthetics and other aspects of learning and research, specializing in modern industrial embedded control systems, networked control systems, intelligent control. He presided over one national college student science and technology innovation project, taking part in the research on one "Eleventh Five-Year Plan" education and science key project of the Education Ministry, and one soft science research project of Shandong Province. He has 12 science and technology papers published in Chinese and oversea academic journals. He has won the 1st scholarship of China, the 1st scholarship of people, the 1st and the 2nd prizes of China Education Robotics Competition, the 1st prize of China Mathematical Modeling Contest, the top award of scientific research achievement of Nanjing University, two China science and technology patents and the honorary title, "three-good-student" of Jiangsu Province. In addition, as a representative of Nanjing University, he went to the National University of Singapore to participate in an academic exchange program. Address: School of Management and Engineering, Xianlin Campus of Nanjing University, 163, Xianlin Avenue, Qixia District, Nanjing, Jiangsu Province, PR China; Zip: 210008.

## References

[1]  Ye Tao, Conghua Lan (2011), "Research on the Forest Disaster Warning System Based on Wireless Sensor Networks", Computer Knowledge and Technology 7.

[2]  Xiaoqiang Yi, Jing Zhao (2011), "Research on Anti-Rear-End Collision Warning System", Technological Development of Enterprise 12, pp.8-9.

[3]  Xuejia Cai, Xu Li, Feng Deng(2011), "Research on Embedded Remote Control System Based on Wireless Sensor Networks", Modern Electronic Technology 16, pp.96-98.

# AquaMesh - design and implementation of smart wireless mesh sensor networks for aquaculture

**Adinya John Odey, Li Daoliang**

College of Information and Electrical Engineering, China Agricultural University, Beijing, PRC

**Email address:**

Johnodey@yahoo.com(A. J. Odey), li_daoliang@yahoo.com(Li Daoliang)

**Abstract:** In this paper we design and implemented a dynamic and smart wireless mesh sensor network for aquaculture and water quality management applications. This system utilizes the Waspmote embedded systems platform developed by Libelium, mesh networking transceivers from Digi International and smart sensors from UNISM to implement a novel smart Wireless Mesh Sensor network –Aquamesh with multiple gateways of different technologies (Zigbee, GPRS and WIFI). The system is designed to continuously monitor aqua-environmental parameters and then initiate an alert or early warning to system user when certain thresholds are exceeded. The data generated from this system is stored locally on the gateway or sent to a remote web server. Data on the local database or remote web server can be accessed with smart mobile phones or personal computers. The experimental results show that the system presented in this paper is feasible to implement and present results consistent with traditional aqua-quality monitoring systems. This system will find application in the monitoring of marine and wetlands environments like fish ponds, coastal water pollution monitoring systems, effluent and sewage treatment plants, offshore oil and gas drilling facilities.

**Keywords:** Aqua Mesh, Wireless Mesh Sensor Networks, Aquaculture, Dynamic and Smart Gateway

## 1. Introduction

Recent advances in semiconductor technologies have presented a new paradigm within computer science: the networking of small-sized sensors which are capable of sensing, processing, and communicating. These Small and smart devices equipped with a processing unit, storage capacity, and small radios offers wireless communication and has presented new application opportunities[1]. Wireless Sensor Networks (WSN) are becoming ubiquitous and permeating every aspect of our everyday life ranging from unobtrusive applications in military and intelligence control to new areas of habitat and environment monitoring, disaster control, health care and home automation.

Wireless mesh sensor networks combine the advantages of wireless mesh networks and wireless sensor networks. As the name implies, wireless mesh sensor networks are;

- Comprised of wireless Sensor nodes. A node in this type of network consists of a sensor or an actuator that is connected to a bi-directional radio transceiver. Data and control signals are communicated wirelessly in this network and nodes can easily be battery operated.
- Arranged in a networking topology called "mesh net-

work". Mesh networking is a type of network where each node in the network can communicate with more than one other node thus enabling better overall connectivity.

State of the art WMSN shares some of the following characteristics; self healing, self organizing and performs multi-hops from data source to destination. They posses decisive advantages, compared with other technologies previously used to monitor environments via the collection of physical data. The combination of these advantages enables the monitoring of phenomena of high variability, both in time and space and also makes it possible to retrieve data in real-time from locations that are difficult to access, either temporarily or permanently, thus addressing stringent responsiveness and accessibility requirements. Their capacity to organize spontaneously in a network makes them easy to deploy, expand and maintain, as well as resilient to the failure of individual measurement points.

Wireless Sensor Network (WMSN) technologies have enabled many interesting applications in pervasive and ubiquitous computing [2]. For example using wireless sensor networks in the field of Agriculture has witness a tremendous increase as practices and technologies to increase world food production becomes the focus of various countries as world population continues to increase. WSN in the agricultural industry find usage in wide and varied areas of applications like Precision Agriculture, vineyard monitory, irrigations and control systems, aquaculture and water quality monitoring.

Aquaculture involves cultivating freshwater and saltwater populations such as fishes, crustaceans, molluscs and aquatic plants under controlled conditions [3]. The positive impacts of aquaculture are well documented. For instance, proponents of aquaculture argue that aquaculture reduces the world dependence on wild stocks of fish, provides new jobs, and helps to feed the worlds growing population [4]. The reported output from global aquaculture operations would supply one half of the fish and shellfish that is directly consumed by humans [5]. In 2004, Aquaculture contributed more than one third of the world total production of fishes [6] and there have been consistent growth rate in production recorded over the last 30 years by different countries.[7]. For example Puerto Montt in Chile has become famous for farming and exporting Salmond and it is reported that about 90% of all the shrimps and crustaceans consumed in the United states is farmed and imported [8]. While agricultural revolution in Aquaculture is an especially important economic activity in China accounting for 70% of world production [9,10], it is also becoming one of the fastest growing areas of food production in the United states.

In modern aquaculture management, water quality monitoring plays an important role. Appropriate control of the water quality to keep the concentration of the water environment parameters in the optimal range can enhance the growth rate of aquatic organisms, affect dietary utilization and reduce the occurrence of large-scale diseases incidences. Without gathering information of physical and chemical parameters of water quality together with the related ecological factors it is almost impossible to perform the appropriate water quality control at the right time in the right place.

To realize real-time data collection in a secure, robust, manageable and low-cost manner, without long-distance cable connections, is the motivation for this paper. This paper examines how best to develop a wireless mesh sensor network to provide these and other services to aquaculture farming which primarily benefits the community and enhance return on investment to the farmer. We present the research, design and development of a wireless mesh sensor network called "Aquamesh" to be deployed in an aqua cultural environment and identify some of the benefits this type of network could offer.

## 2. Systems Design

For aquaculture monitoring, we are interested in continuous monitoring of multiple environmental parameters in an aqua-culture environment through the use of wireless mesh sensor network. The parameters of interest in paper are PH, Dissolved Oxygen (DO), Electrical Conductivity (EC) and Temperature. The systems architecture and network topology is shown in Figure 1.

### 2.1. Data Generation

Our system adopts a hybrid data generation strategy, combining all the data needs and generation requirements of time, event and query driven approach.

The system is expected to generate data values of PH, Electrical Conductivity (EC), Dissolved Oxygen (DO) and Temperature periodically, forwarding the results to the gateway for storage in the database or to the internet server when available. This allows for the best energy consumption profile, because radio resources are used only when it is considered necessary by the application.

Also if the parameters exceeds or fall below the thresholds of 6.0 or 8.5 for PH, 20mg/l for DO, 200 µs/cm for EC and 340C for temperature, short message service (SMS) alert are sent to the farmer or administrators to enable instant remedial actions to be initiated. This allows capture on the flight important events, whose freshness is important for the proper operation of the system.

The design also allows for users generated queries to be performed on the sensors at any time to acquire needed data. These queries can be generated through any of the WIFI, GPRS/SMS or ZigBee gateway.

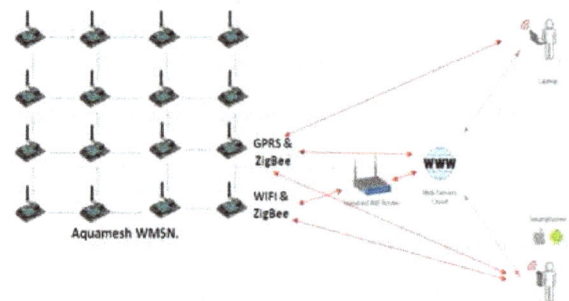

*Figure 1. Systems architecture of Aquamesh.*

### 2.2. Data Transport

We envisioned a wireless meshed sensor network where data generated in the field can be accessed and processed in the comfort of a remote office. We consider a direct connection through a single gateway station unfeasible due to the remote locations of most farms. On the other hand, a pure complete mesh network, where full connectivity is required, would not be the case in this design because of the complexities and resources utilization of such a topology. In this system, a partial wireless mesh network topology is utilized to achieve data transport from the end nodes through intermediate nodes to the gateways and finally the remote user's location.

To accomplish high data throughput and availability, prompt access to data from remote location, to ensure maximum reliability and avoid data loss, the system is designed with multiple gateways using a mix of technologies

(GPRS, WIFI, ZigBee) to ensure sensed data is communicated to the end user.

# 3. Systems Implementations

Aquamesh is composed of (a) Wireless Mesh sensor nodes, (b) a GPRS/SMS gateway, (c) a WIFI gateway, and a network implementation strategy using (d) mesh topology to transport data generated from (e) smart sensor probes to (f) remote web server (g) smart mobile phones or a personal computer fitted with (h) ZigBee dongle gateway.

## 3.1. Wireless Mesh Sensor Node

The wireless mesh sensor node implemented is a Waspmote embedded systems board integrated with Atmega 1281 micro controller unit (MCU) [11], a wireless mesh network transceiver radio module (DigiMesh 2.4 transceiver radio chip) [12], smart sensors [13, 14, 15] for PH, EC, DO and temperature connected through a Waspmote [16] Prototyping board [17] and a battery power source complemented with an energy harvesting solar module [18].

Waspmote does not support RS485 smart sensors either directly or through the Prototyping board. To implement our Aquamesh WMSN, an RS485 to RS232 UART conversion was done using CAU-RTU-0-01 chip module. Smart Digital Sensor probes CAU-PH-3000, CAU-DO-1100, and CAU-EC-4000 for PH, DO, and EC respectively developed by UNISM were used to acquire data and communicate on request from the MCU using the Unibus Protocol. These sensors are connected to the Waspmote node through the RS485 interface of the CAU-RTU-0-01 chip .The prototyping board provides +3.2V power which powers all the Sensors. As part of the design to enable low power consumption and individual control of the sensors, the software ensures the sensors are powered serially one after another. This allows only one sensor to be active at a time.

The architecture of Aquamesh is designed to have dynamic and multiple gateways implementing a mix of wireless technologies as shown in figures 1 to ensure reliability and access to data generated at any time.   Figure 3 shows the Aquamesh node with Digimesh transceiver module, Sensor boards and smart sensors integrated in Waspmote board. The technologies used for the gateway are GPRS, WIFI and Zigbee as shown in Figure 3, Figures 4 and Figures 5.

**Figure 2.** *Aquamesh WMSN node   .*

**Figure 3.** *GPRS/SMS gateway nodes*

**Figure 4.** *WIFI gateway nodes*

**Figure 5.** *Zigbee gateway*

## 3.2. Power Source

The design for Aquamesh WMSN requires the infrastructural mesh nodes to be powered continuously. In this system the radio devices account for most of the power consumption.

The current design utilizes a battery pack of three rechargeable nickel metal hydride (NiMH) AA type batteries which produces over 4.2 volts and has a capacity of 2.5Ah. In order for the mesh node to be continuously powered, it requires a supplement power source derived from renewable sources of solar. This supplement power source can be used to charge the batteries and power the mesh node. Figure 6 shows the solar panel and casing for the implemented Aquamesh node.

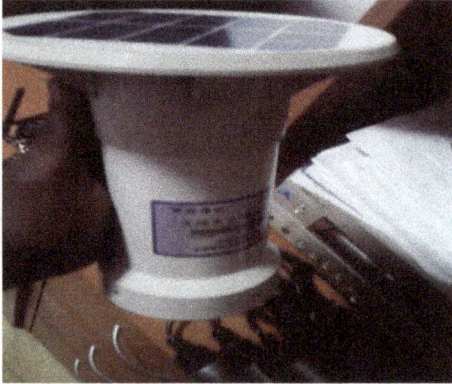

*Figure 6. Solar Panel and final housing for Aquamesh.*

### 3.3. Software

Aquamesh software implementations cut across several platforms. From C and Java programming languages in the embedded Waspmote hardware to internet explorers and Android platforms for implementations of the mobile interface, the programming implementation is flexible and powerful enough to allow fast deployment.

### 3.3.1. WaspMote IDE

The Waspmote-IDE [199] (Integrated Development Environment) is used to program Waspmote. It is based on open source Arduino platform compiler, following the same style of libraries and operation. It is the interface for programming as well as controlling the Waspmote.

### 3.3.2. Web Interface

Part of the system design is to store and display data generated from Aquamesh on a remote web server where it will be accessed with an ordinary internet explorer. To validate this functionality, the data generated from the network between 16th February, 2013 and 17th March, 2013 was sent once a day to the cloud web server hosted by Digi International [18]. The system support the use of SMS messages as well as analysis of data; the status of the network, graphs and Network diagram as well as records of the data generated can be accessed from the webpage. Data update to the web server is received through the GPRS/SMS or WIFI gateways.

### 3.3.3. Android Mobile Phone

The designs of this system also include the use of smart phones to send control commands, access and analyze data. The Smartphone is expected to pair with the WIFI gateway in an adhoc network topology or as a client through an access point (AP).

## 4. Testing and Results

To validate the designs and implementations of Aquamesh Wireless Mesh Sensor Networks, six nodes were deployed in an aquaculture fish pond at the Laboratory of Internet of Things a China Agricultural University from February 16th, 2013 to March, 14th, 2013. Three of the nodes served as data acquisition nodes each fitted with three smart sensor probes described in section 4. The three other nodes serve as the gateway nodes with the GPRS/SMS node made the primary gate way over WIFI and ZigBee gates.

*Figure 7. Experimental Deployment of Aquamesh in Fish Pond*

The pond equipped with aerator machines and had thirty Koi fishes breeding. The Sensors were set to acquire data every six hours and the results stored at the local gateway and a copy forwarded to the web server once every day.

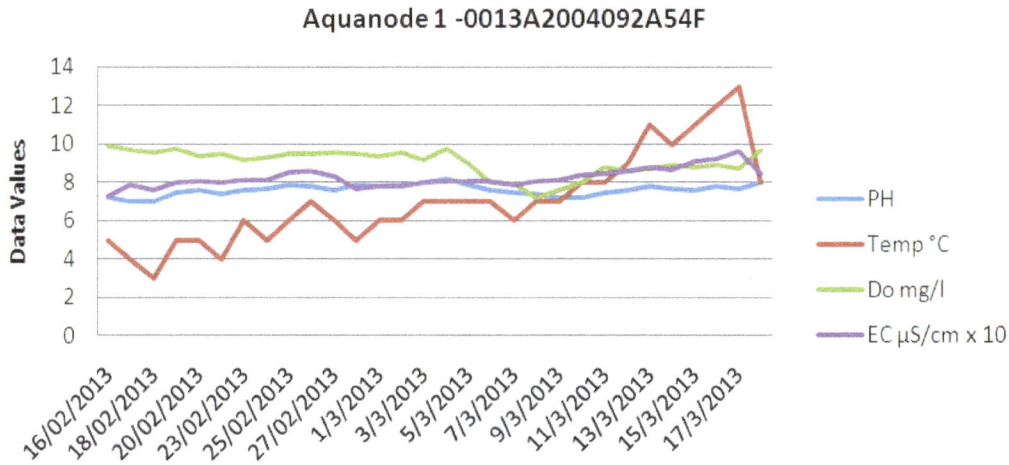

**Figure 8.** *Test Sensors readings from Aquamesh node 1*

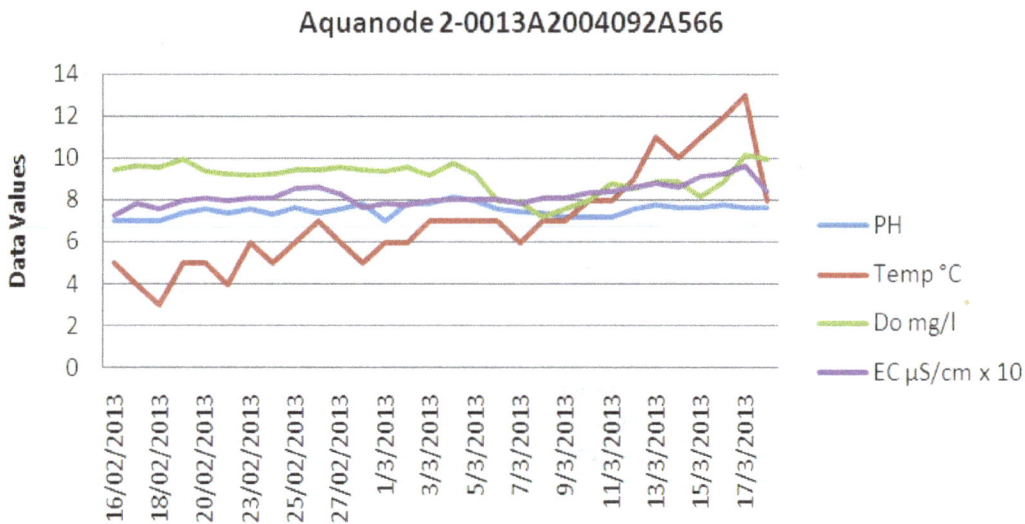

**Figure 9.** *Test Sensors readings from Aquamesh node 2*

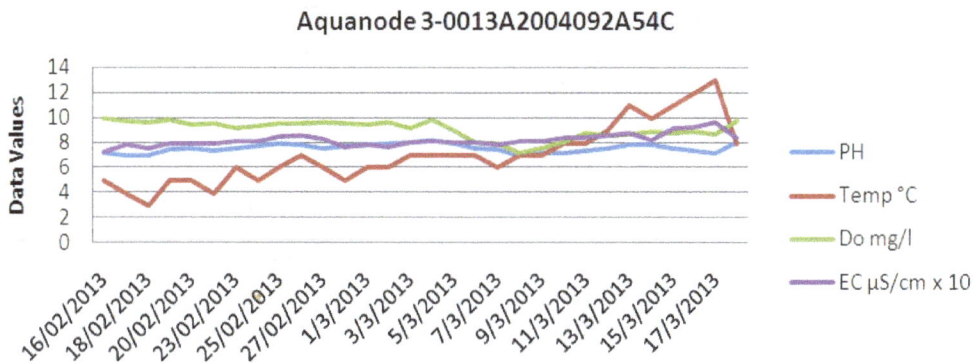

**Figure 10.** *Test Sensors readings from Aquamesh node 3*

## 4.1. Data Acquisition

Figure 5.4- Figure 5.6 shows the sensor readings of DO, pH, EC and Temperature from the three nodes. The graph shows the pH, EC, DO concentration and water temperature trends from February 16th to March 14. We can see the periodic change of the water temperature every day, but DO concentration only fluctuated around 9 mg/l before March 5th. The DO concentration of March 7th, March 8th and March 9 is lower than the other days because the aerator machine was manually turned off. After March 10th, the Aerator machine returned to work normal and DO concentration return to previous range. Also an SMS alert for DO was sent out from the three nodes on the 8th of March since the threshold for alarm was set for 7.3mg/l. It is also observed that the EC values tend to rise with increase in temperature. The reason for this trend is outside the scope of this work. The data for pH sensors are also shown. The values from the three nodes fall within the range of 6.7 to 8.7 and this shows a trend consistent with pH range in aquaculture environments. The gateways functionalities were validated by changing the Hierarchy of the three nodes alternatively. The data transported through the ZigBee gateway was captured using the WaspMote IDE programming interface while data sent through the WIFI and GPRS/SMS were captured using the remote web server of Digi Cloud. Results captured through the three gateway nodes shows consistencies in accuracy; however the ZigBee gateway seems to handle data transfer faster than the other technologies. These could be as a result of routing paths used by the different technologies to send data to the server.

### 4.2. Node and Route Discovery Test

This test is to verify the basic operation to initialize the mesh nodes with a hop count and routing information. The nodes were placed in discovery mode. A discovery command was sent from the PC through the Zigbee Gateway to the network using X-CTU from Digi International. This command was relayed as a broadcast by the gateway node. Node 2 was positioned to be the only node in RF range of the gateway. Node 2 then relayed this discovery broadcast to Node 1 which in turn relayed it to Nodes 3.

This test is configured so that all nodes would report their hop count and routing data to the gateway node after 2 seconds. This test was verified using X-CTU to monitor the data packets received by the PC. All nodes were correctly initialized.

Also if the source node doesn't have a route to the requested destination, the packet is queued to await a route discovery (RD) process. This process is also used when a route fails. A route fails when the source node uses up its network retries without ever receiving an ACK. Figure 10 shows the network route map.

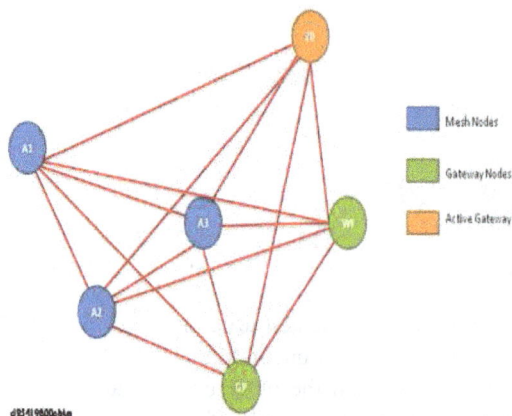

***Figure 11.*** *Network Route Map*

## 5. Conclusion

In this paper, we presented the design and implementations of a smart, wireless mesh sensor network suitable for use in data acquisition in Aqua culture. The system implemented was deployed to continuously monitor water chemistry in a fish pond. Aquamesh is able to initiate an alert when certain thresholds are exceeded and the data generated was successfully sent to a remote web server for storage and processing or directly to human in the loop for remedial actions. The data generated show consistency with data generated using other methods of aqua culture measurements.

One of the key design features in this paper is the use of multiple, dynamic gateway nodes. These gateway nodes were implemented using a mixture of wireless technologies like GPRS/SMS, WIFI, and Zigbee. The system is designed to consume little power, generate data and transport data more efficiently to remote locations. The system presents the advantages of mesh networking and the simplicity of using a modular approach to systems designs.

## References

[1]   Marcel Busse."Algorithms for Energy Efficiency in Wireless Sensor Networks". PhD Dissertation, 2007. Accessed on 26/07/2012. Available at http://d-nb.info/987608029/34J. Clerk Maxwell, A Treatise on Electricity and Magnetism, 3rd ed., vol. 2. Oxford: Clarendon, 1892, pp.68–73.

[2]   Holger Karl, and Andreas Willig," Protocols and Architectures for Wireless Sensor Networks. John Wiley & Sons, 2005, pp.15-329. doi:10.1002/0470095121

[3]   Aquaculture's growth continuing: improved management techniques can reduce environmental effects of the practice.(UPDATE)." Resource: Engineering & Technology for a Sustainable World 16.5 (2009): 20-22. Gale Expanded Academic ASAP. Web. 1 October 2009. http://find.galegroup.com/gtx/start.do?prodId=EAIM.

[4]   Half Of Fish Consumed Globally Is Now Raised On Farms, Study Finds Science Daily, September 8, 2009.

[5]   FAO (2006) The State of World Fisheries and Aquaculture (SOPHIA)

[6]   Blumenthal, Les (August 2, 2010). "Company says FDA is nearing decision on genetically engineered Atlantic salmon". Washington Post. Retrieved August 2010.

[7]   The State of World Fisheries and Aquaculture (SOFIA) 2004

[8]   Wired 12.05: "The Bluewater Revolution." http://www.wired.com/wired/archive /12.05/fish_pr.html. Accessed August 12th, 2010.

[9]   Eilperin, Juliet (2005-01-24). "Fish Farming's Bounty Isn't Without Barbs". The Washington Post.

[10]  Waspmote Technical Guide http://www.libelium.com/uploads/2013/02/waspmote-technical_guide_eng.pdf. Accessed on 2013/03/15.

[11]  Digi International "XBee/XBee-PRO®DigiMesh 2.4 RF

Modules data sheet".http://www.digi.com/products/wireless-wired-embedded-solutions/zigbee-rf-modules/zigbee-mesh-module/xbee-digimesh-2-4#docs. Accessed on 2013/03/15.

[12] Yaoguang Wei, Qisheng Ding, Daoliang Li, Haijiang Tai, and Jianqin Wang "Design of an Intelligent Electrical Conductivity Sensor for Aquaculture Sensor Lett. 2011, Vol. 9, No. 3 1546-198X/2011/9/1044/005 doi:10.1166 /sl.2011.1395

[13] Qisheng Ding, Haijiang Tai, Daokun Ma, Daoliang Li and Linlin Zhao. "Develop ment of a Smart Dissolved Oxygen Sensor Based on IEEE1451.2" Sensor Lett. 2011, Vol. 9, No. 3 1546-198X/2011/9/1049/006 doi:10.1166/sl.2011.1397

[14] Haijiang Tai, Qisheng Ding , Lihua Zeng, Shuangyin Liu , Daoliang Li An intelligent ammonia sensor based on multi-parameter for aquaculture" Sensor Lett. 2011, Vol. 9, No. 3 1546-198X/2011/9/1049/006 doi:10.1166/sl.2011.1397

[15] Waspmote Prototyping Board http://www.libelium.com/uploads/2013/02/prototyping-sensor-board_2.0. Accessed on 2013/03/15.

[16] Akyildiz, I.F., Melodia, T., and. Chowdhury, K., *A survey on wireless multimedia sensor networks.* Computer Networks,vol.51, issue4: p. 921-960, 2007.

[17] "X-CTU Configuration & Test Utility Software". http://ftp1.digi.com/support/documentation/90001003_A.pdf. Accessed on February 2nd, 2013.

[18] Etherios, "Device cloud". https://my.idigi. com/home.do#. Accessed on May 2nd, 2013.

# Comparative performance of forwarding protocols based on detection probability and search angle in a Multi-hop CDMA wireless sensor networks

**Uma Datta[1], Sumit Kundu[2]**

[1]Electronics & Instrumentation Dept., CSIR-CMERI, Durgapur, India
[2]Dept. of ECE, National Institute of Technology Durgapur, Durgapur, India

**Email address:**

uma_datta58@yahoo.in (U. Datta), sumitkundu@yahoo.com (S. Kundu)

**Abstract:** Energy conservation is one of the most important issues in wireless ad hoc and sensor networks, where nodes are likely to rely on limited battery power. Many new algorithms have been proposed for the problem of routing data in sensor networks. In this paper, we propose a new forwarding technique where nodes are placed in a random fashion maintaining a minimum distance between any two nodes and study its performance on a multi-hop CDMA wireless sensor network (WSN). a new routing protocol where selection of intermediate node is based on a metric combining detection probability and maximum forwarding distance towards the sink is proposed to reduce retransmissions in a wireless channel impaired by path loss and shadow fading. We provide a detail description of the routing scheme based on the proposed concept and report on its energy and latency performance using end-to-end ARQ between source and final destination. Performance of this scheme is compared with an existing search angle based protocol, called nearest neighbor based forwarding, where an intermediate node in the route selects the nearest node within a sector angle, considered as search angle, towards the direction of the destination as the next hop. Lifetime of network in both cases is compared. Further, all parameters are estimated by incorporating error control scheme as applicable to WSN and compared with simple ARQ scheme. A solution for packet size optimization is introduced such that the effect of multi hop routing by varying different parameters are captured. Optimization solution is formalized by using different objective functions, i.e., packet throughput, energy efficiency, and resource utilization.

**Keywords:** Multi-Hop Communication, Forwarding Protocol, Search Angle, Route Diversity, Packet Size Optimization, Node Lifetime

## 1. Introduction

In wireless sensor networks energy conservation is one of the key technical challenges. It is necessary to devise networking schemes which make judicious use of limited energy resources without compromising the network connectivity and the ability to deliver data reliably to the intended destination. Many new algorithms have been proposed for the problem of routing data in sensor networks. Almost all of the routing protocols can be classified as data-centric, hierarchical or location-based although there are few distinct ones based on network flow or QoS awareness [1]. Data-centric protocols are query-based and depend on the naming of desired data, which helps in eliminating many redundant transmissions.

Hierarchical protocols aim at clustering the nodes so that cluster heads can do some aggregation and reduction of data in order to save energy. Location-based protocols utilize the position information to relay the data to the desired regions rather than the whole network. The last category includes routing approaches that are based on general network-flow modeling and protocols that strive for meeting some QoS requirements along with the routing function. Several routing protocols are described in the literature, based on location information of sensor / relay nodes [2, 3, 4, 5]. A routing scheme where each intermediate node in a multi-hop route selects the nearest node within a sector of angle ($\theta$) toward the direction of the destination as the next hop, is considered in [2]. However, node isolation may occur in case of low search angle in such scheme. Geographic–information-based forwarding (GIF) is an efficient scheme for finding the appropriate relay node for next hop utilizing the location information while avoiding a large number of control packets during route discovery. The routing scheme where selection of next relay node is based on maximum advanced distance, to minimize the number of hops from the source to the destination node,  is described in [3], ignoring the unreliability of wireless channel. Consequently, performance may degrade substantially over a bad channel and consumes more energy for successful reception of a packet due to increase in number of retransmissions. An effective approach for the reduction of unnecessary retransmission due to propagation impairment is to choose the next hop relay node that is in good channel condition. An efficient advancement metric (EAM) is proposed in [4], considering the forwarding distance and the probability of successful packet transmission in a wireless channel situation within a specified region, defined by $(R_f, \phi/2)$ to $(R_f, -\phi/2)$, as shown in Fig.1. However, in a wireless channel impaired by path loss and shadow fading, a new routing protocol where selection of intermediate node is based on a metric combining detection probability following [6], and maximum forwarding distance towards the final destination may provide another solution where reduction of unnecessary retransmission due to propagation impairment may be possible.

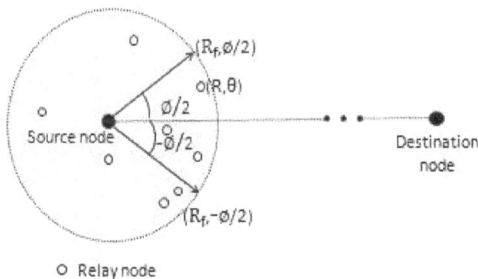

*Figure 1.* *Scenario of a source node and the potential relay nodes in Channel Aware Geographic Informed Forwarding (CAGIF) protocol [4].*

Given that the sensors have limited energy, buffer space, and other resources, different MAC protocols are being developed by several researchers. To handle a large number of nodes, where a number of nodes simultaneously and asynchronously access a channel, CDMA is a good choice as a MAC protocol [7, 8, 9]. CDMA has been widely used in cellular networks where power control is required to combat the near-far problem. As WSN are operated without any central infrastructure and all nodes are battery operated with simple transceivers, it is difficult to achieve perfect power control. CDMA has been advocated for WSN in [7, 8], where distribution of interference power in randomly distributed nodes is discussed. BER and energy consumption in CDMA WSN multi-hop communication with fixed hop length is studied in [10] using infinite automatic repeat request (ARQ) with CRC. Above analyses of multi hop communication do not include any routing

topology for the selection of neighboring nodes until it reaches the destination.

Framework as proposed in [2, 3, 4], where the positions of the nodes are random, are more realistic sensor network scenarios. In this paper, considering random placement of nodes, we are proposing one new routing protocol where selection of intermediate node is based on a metric combining probability of detection of neighbors considering a maximum forwarding range in shadowed environment [6] and maximum advanced distance towards the final destination. We are analyzing the performances of CDMA based WSNs using the proposed routing protocol and compare with the routing protocol as proposed in [2] considering the same wireless channel environment.

ARQ and forward error correction (FEC) are the key error control strategies in wireless sensor networks. Hybrid ARQ (HARQ) scheme, incorporating both retransmission and FEC, is another approach to minimize energy[11,12]. Proper combination of ARQ and FEC using different retransmission strategies needs investigation for minimization of energy. For example, hop by hop error detection and correction scheme using error control code results in higher energy consumption at every node, especially due to significant energy consumption for decoding which reduces network life time in turn. However, if decoding is considered only at the sink which is not energy constrained, power consumptions at intermediate nodes may be reduced significantly. This technique may be investigated in designing an efficient wireless sensor network.

Further, the determination of optimal packet size is important parameters for energy constrained and delay sensitive WSN. A cross layer solution for packet size optimization is presented in [13], where cross layer effects of multi hop routing, the broadcast nature of wireless channel using carrier sense mechanism, and the effects of error control techniques are captured. Such analysis may be extended in context of designing an efficient CDMA based multi hop WSN.

In the present paper, our contributions are as follows:

1. A new forwarding protocol in a random WSN, keeping a minimum distance between any two nodes as in [8], is proposed and simulated on MATLAB; which captures preselected probability of detection and maximum forwarding distance towards the final destination and, considering path loss and log normal shadow fading.

2. We apply this protocol on a CDMA based WSN for performance evaluation. An analytical framework has been developed to model the MAI and NI at each hop to evaluate link BERs followed by average route BER between the sink and the source.

3. An appropriate model has been used to evaluate the energy consumption and delay for successful delivery of a packet from a source node to sink in multi hop using end-to-end ARQ and HARQ type I (HARQ-I) scheme using BCH coding, where decoding is done only at sink.

4. Further, multi-hop performance is also analyzed for

the model where selection of next hop neighboring node is based on the nearest node within a sector of angle (θ) towards the direction of the destination as proposed in [2] considering same wireless channel condition having same reference distance between source and destination as in the proposed protocol.

5. Impact of different network parameters like node density, search angle, probability of detection on energy consumption, delay are analyzed and compared using the schemes.

6. Network lifetime, which is defined as the time to the first node failure in the path [2], is evaluated in each case and compared.

7. Packet size optimization is formalized by using three different objective functions, i.e., packet throughput, energy efficiency, and resource utilization.

The remainder of the paper is organized as follows: The system model for the existing search angle based and the proposed protocol is presented in section 2. Section 3 presents our analytical approach to evaluate the end-to-end BER, energy consumption, delay, and node/network lifetime. Cross layer solution for packet optimization is discussed in section 4. Results based on our developed frame work are presented in Section 5. Finally we conclude in Section 6.

2. Network Model and Problem Description

In this section, we describe the wireless sensor network model using two forwarding protocols and the basic assumptions considered in the paper.

A. Routing Protocol Based on Search Angle

We consider a routing scheme following [2], where each intermediate node in a multi-hop route selects the nearest node within a sector of angle '$\theta$' towards the direction of the destination as the next hop as shown in Fig.2. Let '$W$' be a random variable denoting the distance to the nearest neighbor (as indicated by $r_R$ in earlier chapters) in a two dimensional Poisson node distribution.

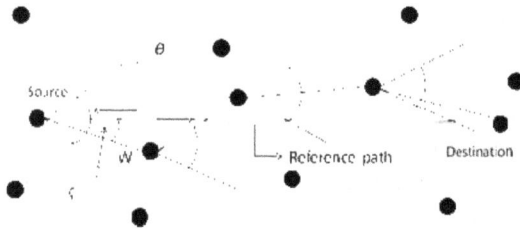

**Figure 2.** *Multi-hop route between the source and the destination in a random topology following nearest neighbor based routing protocol [2].*

For a fixed node spatial density with large '$N$', the CDF of the distance to the nearest neighbor within a sector angle of '$\theta$' in a torus, as expressed in [2] is appropriately modified in context of our network scenario, as shown in Fig.3 where we incorporate a minimum distance ($r_0$) parameter as:

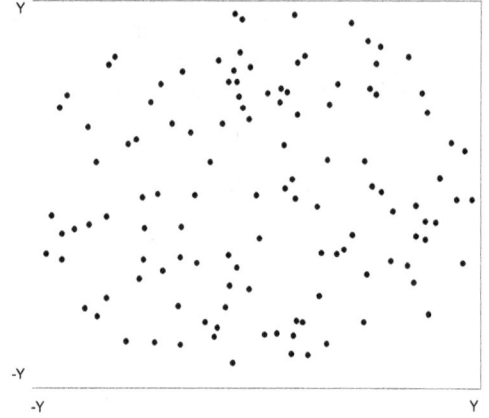

**Figure 3.** *Network with random distribution of sensor nodes.*

$$F_W^{(\theta)}(w) = 1 \qquad w > Y$$
$$= 1 - e^{\rho_s \theta (w - r_0)^2 / 2} \qquad r_0 \le w \le Y \qquad (1)$$
$$= 0 \qquad \text{otherwise.}$$

The number of hops depends on the node density of the network, search angle '$\theta$' and distance between the source and destination. In our analysis we consider an average number of hops to evaluate the performance of WSN. The average number of hops on a route depends on the value of search angle (θ) and is estimated as [Panichpapiboon, 2006; Panichpapiboon, 2004]:

$$\bar{n}_{rand} = \frac{\bar{Z}}{E[W \cdot \cos\varsigma]} \qquad (2)$$

where $\bar{Z}$ is the average length of the reference path between a source and destination node. Here $E[W \cdot \cos\varsigma]$ is the average projected hop length and $\bar{Z}$ can be represented by

$$\bar{Z} = \frac{Y}{3}[\sqrt{2} + \ln(1 + \sqrt{2})] \qquad (3)$$

In our analysis, we assume that each hop deviates with respect to the reference path, by an angle $\varsigma$, where $\varsigma$ is uniformly distributed in the interval between ($\theta/2, -\theta/2$) as shown in Fig.1. Following [2], average projected hop length $E[W \cos\varsigma]$ is expressed by:

$$E[W \cos\varsigma] = E[W] \cdot E[\cos\varsigma] \qquad (4)$$

For large network area, following [14], it can be shown that in our case, $E[W] \approx \sqrt{\pi/(2\rho_S\theta)} + r_0$ , and $E[\cos\varsigma] \approx (2/\theta) \cdot \sin(\theta/2)$

Thus, we obtain the average number of hops as:

$$\bar{n}_{rand} = \frac{\frac{Y}{3}[\sqrt{2} + \ln(1 + \sqrt{2})]}{(\sqrt{\pi/(2\rho_S\theta)} + r_0) \cdot (2/\theta) \cdot \sin(\theta/2)} \qquad (5)$$

**B. Forwarding Protocol Based on Probability of Detection:**

We propose an algorithm for the selection of next hop neighbors as intermediate nodes for the multi-hop path between source node and the final destination i.e. sink node. We consider a network architecture as shown in Fig.2, where nodes are randomly distributed keeping minimum distance between any two nodes, equal to $r_0$ . We are focusing on a circular region of radius $\overline{Z}$ , as stated by (3), to keep the reference distance between the source and sink almost same (equal to $\overline{Z}$ ) in both cases, i.e. the proposed scheme and nearest neighbor based scheme, for performance comparison. Sink is considered to be at the centre and sources are at the periphery. Other nodes are considered only as intermediate relay nodes. Channel conditions, i.e. path loss and shadowing are assumed to be fixed at a particular level throughout the network. The proposed algorithm of the routing protocol, considering path loss and shadowing in the wireless channel, is simulated on MATLAB platform through following steps:

i) Locations of all nodes are assumed to be known. Node at the centre is considered as sink and nodes at the periphery region are considered as sources. Other nodes are considered only as intermediate relay nodes. Once a source node has a packet to send, selection of intermediate relay nodes to send the packet from source to the destination i.e. the sink is governed by the following steps:

ii) Channel conditions, i.e. path loss ( $\beta$ ), and shadowing ( $\sigma$ ) are assumed to be fixed at a particular level. Considering a maximum forwarding distance and standard deviation of shadowing, average sensing distance, $\overline{R}$ , is evaluated using Table I as in [Tsai, 2008]. Next, following [Tsai, 2008], probability of detection of all other nodes in the network from the designated source node is estimated as:

$$P_{\det}\left(r_R(i,n)\right) = Q(\frac{10 \cdot \beta \cdot \log_{10}(r_R(i,n)/\overline{R})}{\sigma}) \quad (6)$$

where $r_R(i,n)$ is the distance between the transmitter and the ' $n$ -th' node for $i$ -th hop. Initially for the source node, $i$ =1.

iii) Amongst all nodes, the relaying node, i.e. the receiver for the present hop, is selected as the one who simultaneously satisfies the following conditions:

Having probability of detection greater than or equal to a preselected value, $(P_{\det})_{sel}$ , i.e. $P_{\det}\left(r_R(i,n)\right) >= (P_{\det})_{sel}$ at that channel condition, closest to the destination, i.e. the sink, and towards the sink.

iv) The selected relay node of the present hop will become the transmitter for the next hop. The relay node for the next hop, i.e. $i$ =2, is found following same approaches of steps (ii) and (iii). This is repeated until the packet reaches the final destination through multi-hop transmission.

v) All hop distances for the selected relay nodes, i.e. $r_R(i,m)$ , where $m$ is the selected node in $i$ -th hop are

calculated from known locations, with $i$ =1, 2, ----H, for H number of hops.

Selection of intermediate nodes based on MATLAB simulation is shown in Fig.4.

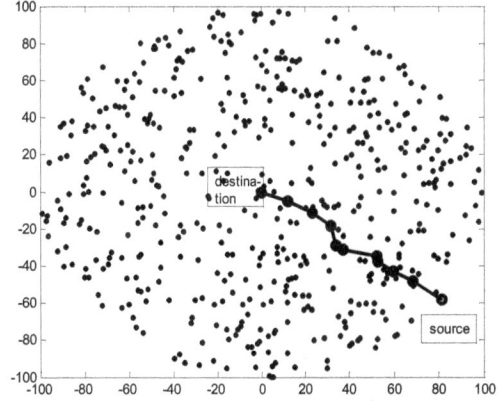

***Figure 4.*** *Possible multi-hop route between source and destination in a random topology following the proposed protocol.*

**C. MAC Protocol and Transmission Scheme**

Here we consider CDMA-based MAC protocol. In this present work we assume two routing protocols separately, where a node can identify its intended neighbor, governed by the routing protocols as described above, ensures the data flow from source to sink in multihop. For example in Fig.5, destination node, i.e. sink, D is receiving information from the source nodes S, S1, S2, S3 etc., via multi-hop communication using digital relays, as employed in any regenerative systems. Relays regenerate the received signal and then transmit the same with power control to the next hop. As we are considering a CDMA sensor network, any node can transmit to its nearest next neighbor node at any time, so that source information finally reaches at the destination D. At a particular instant, considering two protocols separately, we assume that nodes (f,h,i), (d,e,h,i), (j,k,l), (a,b,c) are used as intermediate nodes to route the information from source nodes S, S1, S2 ,S3 respectively to the sink, i.e. the destination D. Here we consider that each transceiver is having a receive threshold $P_r$ . Each transmitter adjusts its transmission power so as to achieve a given level of received power ( $P_r$ ) at its intended receiver. Accordingly the transmit power depends upon the distance between the transmitter and the receiver pair ( $r_R(i)$ ), and the statistics of shadowing. Several concurrent nodes those are sending their information to any intermediate relay node on the path, e.g. 'h' within the area $\pi r_R^2$ would cause MAI at 'h'. The concurrent transmitted signal power generated by the source/relay nodes situated within the interference range $r_I$ ( $r_I = 2r_R$ ) of 'h', which are sending information to their respective destination node, might be sensed by 'h'. This would cause NI to 'h'. During propagation of signal from source to sink via different nodes, the desired signal at each receiver node is accompanied with MAI and NI. We obtain the interference power distribution at each receiver node following assumptions and definitions [8, 10]:

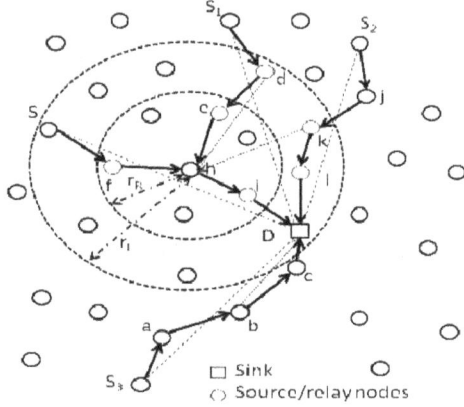

*Figure 5. Signal flow at a particular instant in multi-hop communication*

# 3. Analysis of Route BER, Energy Consumption, Delay, and Lifetime

We assume an average H number of hops between a source node and the destination node, or sink, i.e. $H = \overline{n}_{rand}$ , as determined by (5). Channel conditions are assumed to be same for all the hops from source to sink.

## 3.1. Estimation of Interference Power

Following [10], the mean value of the collected interference power $\eta'$ from an interfering node (i.e. d is transmitting signal to it's intended receiver e) to any desired receiver (i.e. h):

$$\eta' = \eta . e^{\left(m_{S_{dh}} + \frac{\sigma_{S_{dh}}^2}{2}\right)} . e^{-\left(m_{S_{de}} + \frac{\sigma_{S_{de}}^2}{2}\right)} . e^{\left(m_{R_{de}} + \frac{\sigma_{R_{de}}^2}{2}\right)} \quad (7)$$

$\eta$ can be found out as in [8] as:

$$\eta = \frac{4P_R \left(r_R^{\alpha+2} - r_0^{\alpha+2}\right)\left(r_I^{\alpha-2} - r_0^{\alpha-2}\right)}{\left(\alpha^2 - 4\right)r_0^{\alpha-2} r_I^{\alpha-2}\left(r_I^2 - r_0^2\right)\left(r_R^2 - r_0^2\right)} = \gamma P_R \quad (8)$$

where $m_{S_{de}}$ , $m_{R_{de}}$ are the mean and $\sigma_{S_{de}}$ , $\sigma_{R_{de}}$ are the standard deviation of shadowing ( $S_{de}$ ) and pce ( $R_{de}$ ) respectively for an arbitrary path $de$ . Further $\alpha$ is the path loss exponent $2 < \alpha < 6$ and $m_{S_{dh}}$ , $\sigma_{S_{dh}}$ are the mean and standard deviation of shadowing of the path $d_{dh}$ between the nodes $d$ and $h$ .

## 3.2. Number of Potential Interfering Neighbors

Under the Poisson approximation, for very large number of nodes ( $N$ ) and very small area of interference region compared to the total area (A), the expected numbers of nodes within the receiving and interference range of the receiver are [8]

$$c_1 = \rho_S \pi (r_I^2 - r_R^2) \quad (9)$$

$$c_2 = \rho_S \pi r_R^2 \quad (10)$$

where $\rho_S$ is the node density, $r_I$ is the interference distance equal to $2r_R$ . Activity factors determine the number of active nodes at any instant in the two layers contributing MAI and NI, which are fractions of $c_1$ and $c_2$ as given in (9) and (10).

## 3.3. Estimation of Average Route BER, Energy Consumption and Latency

With direct sequence BPSK of spreading bandwidth $BW$ , and for constant received signal power levels with power control error, following [10, 16], the mean probability of error at any hop can be approximated by:

$$P_e = Q\left(\sqrt{2\frac{E_b}{\chi}}\right) = Q\left(\sqrt{\frac{2P_R e^R / R_b}{I / BW}}\right)$$

$$= Q\left(\sqrt{\frac{2P_R e^R / R_b}{P_R e^\phi / BW}}\right) = Q(e^\psi) \quad (11)$$

We estimate average BER for each hop individually as each hop, having average distance of $r_{Ri}$ for $i$ -th hop, will have different number of interferers. Next we estimate the average route BER considering an average number of ' $H = \overline{n}_{rand}$ ' hops. The route BER for $H$ hops, without any error correction mechanism applied at the intermediate relay nodes, is expressed by the relation [14],

$$\overline{(P_e)}_H = 1 - \prod_{i=1}^{H} \left(1 - \overline{P_e}^{(i)}\right) \quad (12)$$

where $\overline{P_e}^{(i)}$ is the mean probability of error at $i$ -th hop.

Further $n_f$ bits/packet is considered in forward transmission of information and $n_b$ bits/packet for NACK /ACK with an assumption of instantaneous error free reception of NACK/ACK. Assuming perfect error detection of a CRC code and infinite retransmissions, ARQ and HARQ-I mechanisms are used for error correction. In the present schemes, the packet is checked only at destination (D) for error control; retransmissions of the packet are requested to source (S), with a NACK coming back from D to S via the same multi-hop path till the packet is received correctly. Considering the sink is not an energy constrained node, two schemes of error control are proposed:

Scheme I: Error control is based on simple ARQ, i.e. the packet from source consists of $n_f$ bits message including CRC, and overhead ($\beta$ bits) is transmitted and is checked at sink only for correctness. It sends a retransmission request to the source for an entirely new retransmission in case of receipt of erroneous packet.

Scheme II: Error control is based on hybrid ARQ type I

(HARQ-I), where the packet from source consisting of $\ell_{BCH}$ bits including $n_f$ bits message, and encoding bits for one bit error correction capability using BCH coding is transmitted. At sink, the packet is decoded. It discards erroneous packets (when errors remain after BCH decoding), sends a retransmission request to the source for an entirely new retransmission. Retransmissions take place at the same code rate until the packet is correctly decoded.

Scheme I

Average end-to-end packet error level for H hops is

$$\overline{(P_f)}_H = 1 - (1 - (\overline{P_e})_H)^{n_f} \qquad (13)$$

where $(\overline{P_e})_H$ is given in (12). Let successful delivery of a packet from source to destination is occurred at $n$ th number of retransmissions. Average number of retransmissions for successful delivery of a packet [17]:

$$(\overline{N})_{ARQ} = 1 / \left(1 - (\overline{P_f})_H\right) \qquad (14)$$

The energy $E_b$ required to communicate one bit of information from source to sink through H-hop communication i.e. end-to-end delivery [18]:

$$E_b = \sum_{i=1}^{H} (P_{ti} + P_r) / R_b \qquad (15)$$

where $R_b$ is the data rate, $P_{ti}$ is the mean of transmitted power for i-th hop of length $r_{Ri}$, and is represented by [10]:

$$P_{ti} = P_R \, r_{Ri}^{\alpha} \, e^{-\left(m_{S_{de}} + \frac{\sigma_{S_{de}}^2}{2}\right)} . e^{\left(m_{R_{de}} + \frac{\sigma_{R_{de}}^2}{2}\right)} \qquad (16)$$

We have included the energy consumption due to start-up transients of transceivers while calculating the energy consumption for data communication. It is observed that start-up energy consumed in the transmitter / receiver varies approximately from 10 micro Joule to 45 micro Joule [19], [20]. Assuming average start-up energy from sleep mode to either transmit/receive mode is equal to 10 micro Joule, the energy consumed per packet from source to sink, i.e. single loop transmission of information from source to sink via H hops, with ACK/NACK from sink to source via multi-hop is:

$$(E_{pkt})^{ARQ} = E_b n_f + E_b n_b + 10 \times 10^{-3}.H \qquad (17)$$

We assume that each node will be wake up, either in transmit or receive mode, from sleep state and will remain in the active state, without going to sleep state, until the information is received correctly at sink. Since on the average, each packet requires $(\overline{N})_{ARQ}$ number of retransmissions from source to destination for successful delivery, average energy consumed by a packet through

multi-hop communication of H hops is:

$$E_{av}^{ARQ} = (\overline{N})_{ARQ}.(E_b n_f + E_b n_b) + 10 \times 10^{-3}.H \qquad (18)$$

Energy consumed only by the message in single transmission, i.e. the effective energy

$$E_{eff}^{ARQ} = ((\sum_{i=1}^{H} (P_{ti} + P_r) \cdot (n_f - \beta)) / R_b + 10 \times 10^{-3}.H \qquad (19)$$

where $\beta$ is the number of overhead bits.

Average packet delay for successful transmission of packet is obtained as [21]:

$$D_{av}^{ARQ} = (\overline{N})_{ARQ}.n_f / R_b \qquad (20)$$

Now we describe the scheme incorporating HARQ-I using BCH coding.

Scheme II

The packet error rate (PER) for the scheme as in [11] is:

$$(PER_{HARQ-I})_{BCH} = 1 - \sum_{i=0}^{t} \binom{\ell_{BCH}}{i} . ((\overline{P_e})_H)^i . (1 - (\overline{P_e})_H)^{\ell - i} \qquad (21)$$

$\ell_{BCH} = (\partial + n_f)$ is the total number of bits to be transmitted and is the sum of length of frame check sequence $\partial$, and $n_f$ bits as described above.

Average number of retransmissions for successful reception of a packet [17]:

$$(\overline{N})_{BCH} = 1 / (1 - (PER_{HARQI})_{BCH}) \qquad (22)$$

In the present case, we assume negligible energy for BCH encoding at source node, and energy consumed at sink is ignored since the sink is not an energy constrained node. With these assumptions, the energy consumed per packet by the source and relay nodes in single loop transmission of information via H hops including start up energy of each node, with ACK/NACK via multi-hop is:

$$(E_{pkt})_{BCH} = E_b \cdot \ell_{BCH} + E_b \cdot n_b + 10 \times 10^{-3}.H \qquad (23)$$

where $E_b$ is expressed in (15). Average energy required for successful delivery of packet

$$E_{av}^{BCH} = (\overline{N})_{BCH}.(E_b \cdot \ell_{BCH} + E_b \cdot n_b) + 10 \times 10^{-3}.H \qquad (24)$$

Average packet delay for successful transmission of packet is obtained as [21]:

$$D_{av}^{bch} = (\overline{N})_{BCH}.\ell_{BCH} / R_b \qquad (25)$$

### 3.4. Network Lifetime

Here, following [2], we consider the time to the first node failure as the network lifetime (i.e., a worst-case approach). We assume that every node has an initial finite battery energy denoted by $E_{batt}$ and packets are transmitted with average rate $\lambda_t$ without queuing. The average energy depleted per second due to transmission and reception is simply $\lambda_t \cdot E_{packet}$, where $E_{packet} = E_{av}/H$ is the average energy consumed per node while delivering a packet successfully from source to sink in each scheme. Finally, the total time it takes to completely exhaust the initial battery energy of any node can be written as:

$$\tau = E_{batt}/(\lambda_t \cdot E_{packet}) \qquad (26)$$

This simple analysis does not take into account the energy consumed when a node is processing packets. Thus, the lifetime of a node will be shorter than what is predicted by our analysis.

## 4. Packet Size Optimization

Following [13], optimization framework based on end-to-end performance metrics using infinite ARQ between source and sink, as described in scheme I, is formulated. Optimization solution is formulated by using three different objective functions, which highlight different aspect of communication in WSN and can be selected according to the requirements. The two objective functions are defined as [13]:

1. Packet throughput: This function considers the end-to-end packet success rate and the average delay for successful reception of a packet of payload $l_d = (n_f - \beta)$ through multi hop communication.

$$U_{pktput} = (n_f - \beta) \cdot \left(1 - (\overline{P_f})_H\right)/D_{av}^{ARQ} \qquad (27)$$

Maximizing this function, by setting $\dfrac{d}{dl_d}(U_{pktput}) = 0$, results in optimal packet size $L_{opt}^{pktput}$ that achieve high packet throughput for a particular forwarding protocol and channel condition. After simplification, $L_{opt}^{pktput}$ is expressed by:

$$L_{opt}^{pktput} = \frac{\sqrt{\beta^2 - \dfrac{4\beta}{\ln\left(1 - (\overline{P_e})_H\right)}} - \beta}{2} \qquad (28)$$

2. Resource utilization: This function considers both energy consumption and delay for successful reception of a packet, and is expressed by:

$$U_{res} = E_{av}^{ARQ} \cdot D_{av}^{ARQ}/l_d \cdot \left(1 - (\overline{P_f})_H\right) \qquad (29)$$

where $l_d = (n_f - \beta)$ is the message length. Minimizing this function, by setting $\dfrac{d}{dl_d}(U_{res}) = 0$, results in optimal

packet size ($L_{opt}^{res}$), that balances the tradeoff between energy consumption and latency, especially useful for delay sensitive WSN. After simplification, $L_{opt}^{res}$ is obtained as:

$$L_{opt}^{res} = \frac{\sqrt{\beta^2 + \dfrac{4\beta}{\ln\left(1 - (\overline{P_e})_H\right)}} - \beta}{2} \qquad (30)$$

Energy efficiency is another objective function, which may be considered for optimization. Considering successful reception of a packet from source to sink, energy efficiency is expressed by:

$$\xi = E_{eff}^{ARQ}/E_{av}^{ARQ} \qquad (31)$$

Maximizing this function results in optimal packet size, that achieves high energy efficiency.

However, considering startup energy, closed form solution of optimized packet length using energy efficiency $\xi$ is not straight forward to obtain. We evaluate it with the help of simulation under such startup energy included case.

## 5. Results

Same channel condition and average reference distance between source and sink are considered for both schemes. Parameters used in present analysis, based on semi-analytic method, are given in Table 1.

*Table 1. Parameters used in the analysis.*

| Parameter | Value |
|---|---|
| Value of Y | 100 m |
| Min. distance between two nodes ($r_0$) | 1 m |
| Processing Gain ( pg ) | 128 |
| Constant receive power ($P_R$) | $1.0 \times 10^{-07}$ mW |
| Path loss parameter ($\alpha$) | 3 |
| Transmission rate ($R_b$) | 20.0kbps |
| NACK/ACK ($n_b$) | 2 bits |
| Standard deviation of pce ($\sigma_R$) | 1dB |
| Standard deviation of Shadowing ($\sigma_s$) | 3 dB |
| Band width | 5 MHz |
| Overhead ($\beta$) | 8 bits |
| Start up energy/node | 10micro Joule |
| Maximum forwarding distance | 30 m |

Mean of all shadowing and pce components are considered to be zero. We assume that 50% of total nodes within receiving distance ($r_{Ri}$) of sink are active for MAI while 25% of nodes between receiving and interfering distance of sink are active for NI. For other intermediate relay nodes, 25% of total nodes within $r_{Ri}$ are active for MAI while same percentage of nodes between $r_{Ri}$ and $r_{Ii}$ are active for NI at any instant. All parameters at each hop are calculated considering distance between two consecutive nodes as $r_{Ri}$ meter, where $r_{Ri}$ s' are calculated by using (1) and (5) for search angle based protocol. The procedure adopted for the evaluation of link BER, followed by route BER and average route BER for the estimation of different QoS parameters, using the two forwarding protocols, are described below:

(A): Forwarding protocol based on search angle with nearest neighbor:

1. Average number of hops between source and destination $\overline{n}_{rand}$ is calculated by using (5) for a particular search angle ($\theta$), and reference distance as expressed by (3), considering Y=100m.

2. Average hop distances ($r_{Ri}$) are estimated using (1), by generating $\overline{n}_{rand}$ number of random variables.

3. Link BER at each hop is evaluated using (11), followed by route BER for that $\theta$ using (12). Energy consumption at each link for transfer of fixed length information for a single realization of source to destination path is estimated.

4. Average route BER and route energy consumption for sending a fixed length information from source and receiving at sink through multi hop for a particular $\theta$ is obtained by computing arithmetic average of large number of realizations of route BER and route energy consumption.

5. Average delay for sending information from source to sink, node lifetime, and objective functions, like packet throughput, energy efficiency, and resource utilization are evaluated by using the value of average route BER and average route energy consumption using end to end ARQ.

6. Parameters like node lifetime, delay for successful reception of information from source to sink are also estimated using end to end HARQ-I using BCH coding.

(B): Forwarding protocol based on probability of detection with maximum advancement:

Same channel condition and average reference distance between source and sink as previous protocol are considered. Under the proposed protocol, all link distances i.e. $r_{Ri}$ s' are estimated after selection of intermediate relay nodes by the developed algorithm as described in section 2. Next the QoS parameters, e.g. node lifetime, delay are evaluated by similar method as in scheme A.

Fig.6 compares the variation of average route BER with node densities under two different protocols. Impact of search angles ($\theta$) of nearest neighbor based protocol, and probabilities of detection with maximum sensing distance of 30m on proposed protocol are shown considering different correlation amongst interferers only. In case of nearest neighbor based protocol, increase in node density or search angle improves the link BER as described earlier. This in turn results in reduction in average route BER with node density or $\theta$ (curves i and iii). In case of channel sensitive forwarding protocol, higher probability of detection reduces distance of next hop neighbor for a fixed channel condition, which results in improved link BER followed by lower route BER (curves v and vi). With increase in node density, keeping probability of detection fixed at a level, number of interferers increases, which in turn increases the average route BER. Thus the proposed

channel sensitive forwarding protocol may outperform the other in some cases and choice of search angle plays an important role. Increase in correlation amongst interferers results in increase in link BER followed by increase in average route BER under both the protocols (curve ii, iii and curve iv, vi).

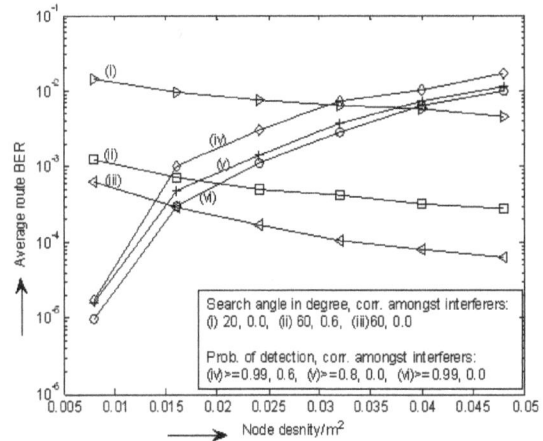

**Figure 6.** *Variation of average route BER with node density and correlation amongst interferers under two different forwarding protocols.*

Following the definition of network lifetime as defined earlier, average energy consumed by each node for successful transmission of a packet directly governs the network lifetime. Fig.7 compares the variation of average energy consumed by each participating relay node for successful transmission of a typical packet of length 63 bits with node densities under several values of search angle ($\theta$), probability of detection for two different forwarding protocols. In case of nearest neighbor based forwarding protocol, with increase in node density as well as increase in search angle, link distance decreases, thus transmit power at each node decreases which leads to the decrease in average energy consumed by each node for successful transmission of a packet (curve i, ii; Fig.7) followed by increase in node lifetime with increase in search angle or node density as shown in (curve iii, iv; Fig.8). In case of channel sensitive forwarding protocol, for a fixed channel condition of shadowing and path loss, variation of link distance is insignificant with increase in node density, as it is dictated by the probability of detection. Thus variation of average energy consumption per node and subsequently variation of node lifetime with node density is insignificant (curve iii, iv of Fig.7 and curve i, ii of Fig.8). High detection probability ($P_{det} \geq 0.99$) reduces the link distances. At lower value of detection probability ($P_{det} \geq 0.8$), link distance increases which causes higher transmit energy at each node and significant increase in number of retransmissions at higher node densities. Consequently, average energy consumption per node increases and average node lifetime decreases with increase in node density (curve iii, iv of Fig.7 and curve i, ii of Fig.8 respectively).

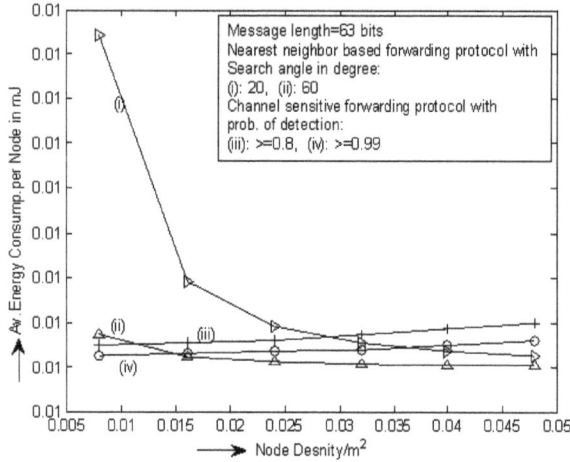

*Figure 7. Variation of average energy consumption/node for successful transmission of packet with node density under two different forwarding protocols.*

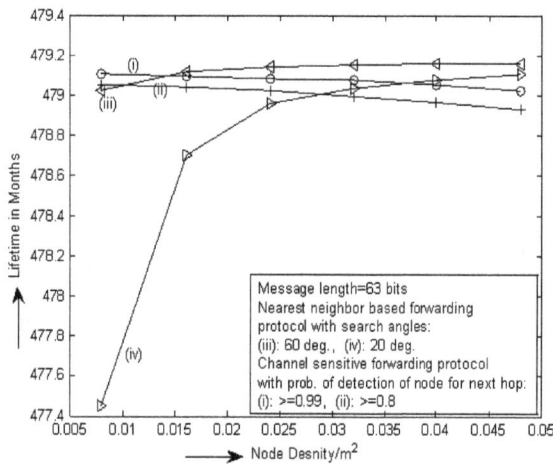

*Figure 8. Variation of average node lifetime with node density for two different forwarding protocols under E2E ARQ transmission.*

Fig.9 compares the variation of end to end (E2E) delay/latency with node densities under two forwarding protocols for successful transmission of a packet of length 63 bits using ARQ. Using nearest neighbor based forwarding protocol, average route BER improves with increase in node density as well as increase in search angle. Thus average delay of the network decreases with increase in $\theta$ or node density due to reduction in number of retransmissions (curves i, iv) under such protocol. However, in case of channel sensitive protocol, with increase in node density at a fixed probability of detection, as well as decrease in probability of detection keeping node density fixed; number of interferers increases, which increases delay (curves ii, iii) due to increased number of retransmissions.

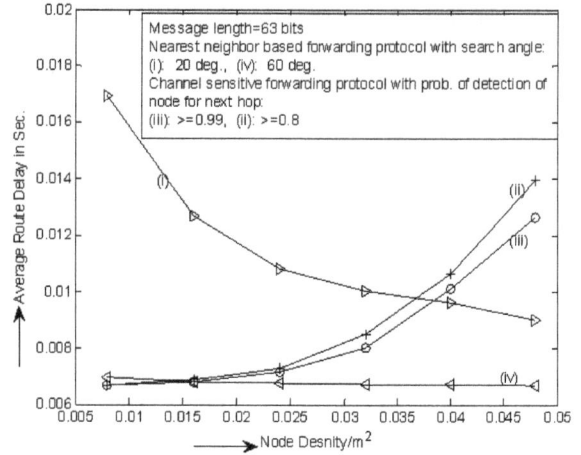

*Figure 9. Variation of average route delay with node density for two different forwarding protocols.*

Fig.10 depicts and compares the variation of packet throughput ($U_{pktput}$), as derived in (28) using end to end ARQ, with packet lengths under two different forwarding protocols for a fixed node density 0.016/m². At low packet length region, low packet throughput is due to overhead, which is comparable with the packet length. At high packet length region, packet throughput decreases due to degradation in PER. This decrease is significant in case of low search angle, i.e. 20 degree, (curve vi) or low probability of detection, i.e. $P_{det} \geq 0.6$, (curve v). It is due to higher number of interferers, associated with higher hop length which occurs under the two cases as mentioned above, i.e. low $\theta$, as well as low $P_{det}$. With increase in search angle or detection probability, route PER for a fixed packet length improves due to improved route BER. Subsequently, delay for successful delivery of packet decreases and packet throughput increases as in Fig.10. The optimized packet lengths ($L_{opt}^{pktput}$), as observed from the curves, for different search angles and detection probabilities, match approximately with those obtained directly by using (28), as shown in table 2.

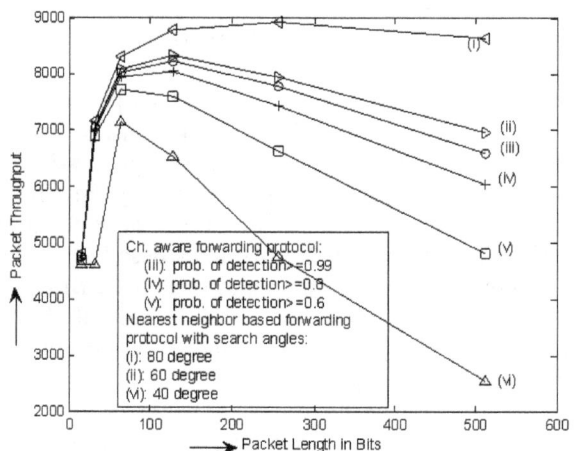

*Figure 10. Variation of packet throughput from source to sink with packet lengths for two different forwarding protocols.*

*Table 2. Optimized packet length using packet throughput for different search angles, probabilities of detection; node density $0.016/m^2$*

| Routing protocol based on search angle | | | Routing protocol based on probability of detection | | |
|---|---|---|---|---|---|
| Search angle | $40^0$ | $60^0$ | Detection prob. | >=0.99 | >=0.8 |
| Optimized packet length (bits/pkt) | 69.59 | 141.69 | Optimized packet length (bits/pkt) | 164.09 | 127.8 |

Fig.11 shows the variation of resource utilization ($U_{res}$), as in (30), with packet length under two different forwarding protocols for a fixed node density $0.016/m^2$, considering start up energy at each node. At low packet length region, $U_{res}$ is significantly high due to reasonable size of overhead, which is comparable with the message length. At high packet length region, due to increase in number of retransmission, $U_{res}$ increases slowly. In case of low search angle, the rate of increase is appreciably high (curve i) due to high link distance and associated high BER. It is observed that there exists an optimum packet length, depending on the network condition, which yields minimum resource utilization beyond which resource utilization increases due to significant increase in route PER. The optimized packet lengths ($L_{opt}^{res}$), as observed from the curves via semi-analytic approach match with those obtained numerically by using (30), as shown in table 3.

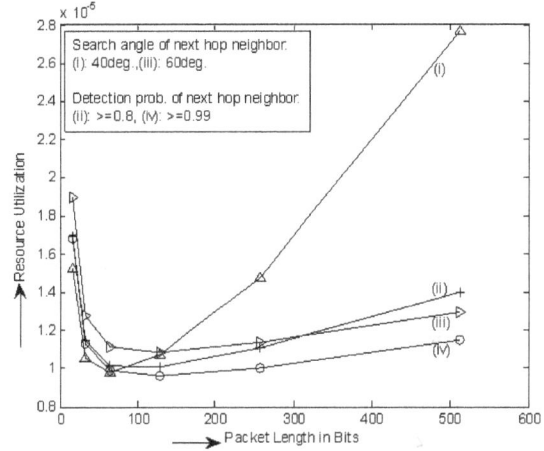

*Figure 11. Variation of resource utilization with packet lengths for two different forwarding protocols.*

*Table 3. Optimized packet length using resource utilization for different search angles, probabilities of detection; node density $0.016/m^2$*

| Routing protocol based on search angle | | | Routing protocol based on probability of detection | | |
|---|---|---|---|---|---|
| Search angle | $40^0$ | $60^0$ | Detection prob. | >=0.99 | >=0.8 |
| Optimized packet length (bits/pkt) | 65.82 | 137.80 | Optimized packet length (bits/pkt) | 160.18 | 123.92 |

Fig. 12 depicts the variation of energy efficiency, as expressed in (31), with packet length for a fixed node density $0.016/m^2$, under several values of search angle and detection probability using the two forwarding protocols. With high search angle or high $P_{det}$, insignificant decrease in energy efficiency with packet length is observed (curve i and ii). It is due to the lesser energy consumption for communication as compared with the energy consumption in case of higher hop length associated with lower $P_{det}$ and lower search angle (curve iii, iv).

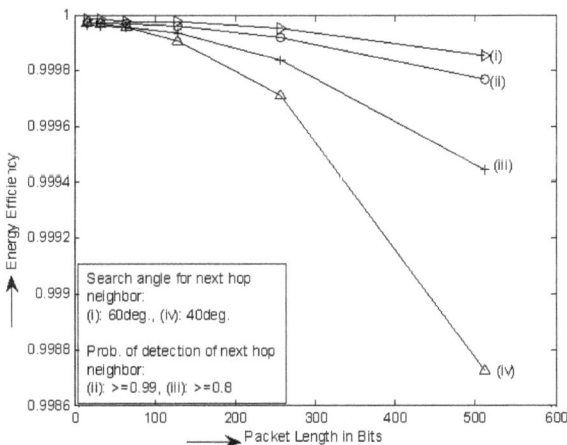

*Figure 12. Variation of energy efficiency with packet lengths for two different forwarding protocols.*

Fig.13 compares the variation of node lifetime with node densities for several values of search angles ($\theta$) and detection probability using the two forwarding protocols. Cases with and without incorporating error control and correction schemes using BCH coding (63, 57, 1) at source and sink are considered. With E2E ARQ, the message length is 57 bits. Using channel sensitive protocol, and considering $P_{det} \geq 0.99$, it is seen that there is no significant difference in lifetime of each source/ relay nodes of the sensor network system under ARQ with respect to the scheme using BCH (curve i, ii). It is due to insignificant difference in number of retransmissions for link distances associated with $P_{det} \geq 0.99$. Using nearest neighbor based forwarding protocol, at low search angle, e.g. 40 degree, where link distance is comparatively high, lifetime is higher using HARQ-I with BCH coding as compared with only infinite ARQ from sink to source (curve v, vi), in spite of transmission of extra bits after encoding. This is mainly due to significant reduction in PER, which subsequently results in significantly less number of retransmissions as compared with the scheme without error correcting capability at sink. Considering nearest neighbor based protocol, with increase in node densities or ($\theta$), average hop distance decreases. In this case incorporation of error correction capability is not significant as compared to only ARQ (curve iii, iv).

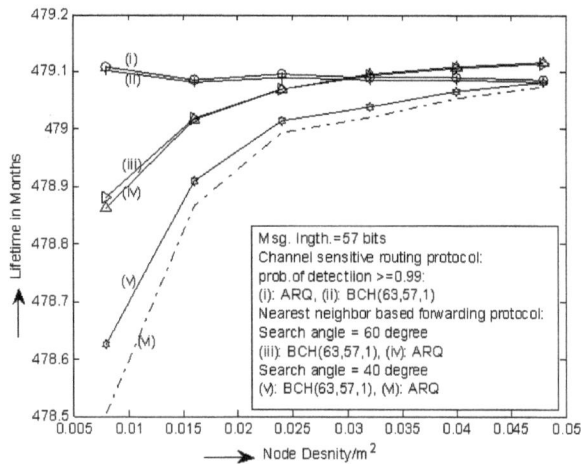

*Figure 13. Variation of average lifetime with node densities for different error control schemes under two forwarding protocols.*

Fig.14 compares the variation of average route delay with node densities under two different forwarding protocols. Cases based on infinite ARQ of 57 bit packet and a BCH (63,57,1) coded packet based HARQ-I are considered for both the forwarding protocols. In case of channel aware protocol with $P_{det} \geq 0.99$, ARQ outperforms the other scheme at low node density, i.e. low interference region. However at high node density region, BCH coding shows better performance. It is seen that using nearest neighbor based forwarding protocol with search angles 40 degree, where average hop distance is high, BCH coding outperforms ARQ scheme over all node densities. Further, with increase in search angle ($\theta$) as well as node density, average route BER improves under nearest node based routing. In this condition, incorporation of HARQ-I at sink may lead to increase in delay due to transmission of extra overhead (curve iii and iv).

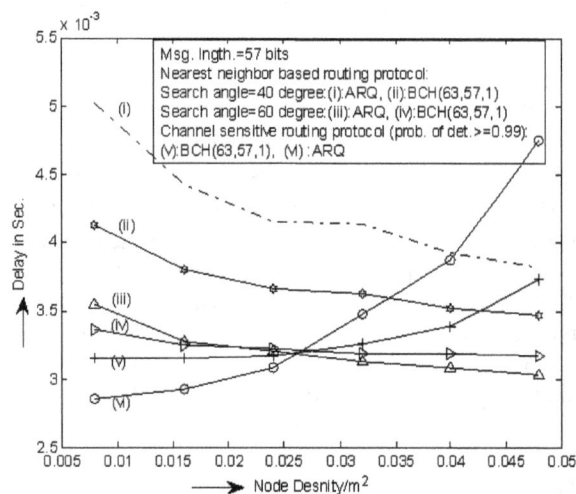

*Figure 14. Variation of average delay for successful transmission of a packet with node densities for different error control schemes under two forwarding protocols.*

## 6. Conclusions

In this paper, using a semi-analytical model, the end-to-end performance in terms of average Route BER, average node lifetime, delay in successful reception of packetized data from source to sink via multi hop is estimated using a new channel aware forwarding protocol, where selection of intermediate relay nodes for multi hop operation is based on the probability of detection combined with maximum advanced distance with respect to the destination. The performance is compared with the nearest neighbor based forwarding protocol, where intermediate relay nodes is selected as the nearest node within a sector of angle ($\theta$) towards the direction of destination. Variation of energy consumption/node and node/network lifetime with node density is lower in case of channel aware protocol than nearest neighbor based protocol. Packet throughput, energy efficiency, and resource utilization metric are also compared for the two protocols. Optimum packet length which yields best packet throughput, energy efficiency, and resource utilization are indicated under two protocols. Best packet throughput is obtained with lower link distance. Best resource utilization within a range of packet length is observed with nearest neighbor based protocol with high search angle and channel aware based protocol with high probability of detection, i.e. with lower link distance. Considering start up energy, high energy efficiency is observed at low packet length region, where packet error rate is minimum, which results decrease in average number of retransmissions followed by lower average energy consumption for successful transmission of packet. Incorporation of HARQ-I between source and sink provide better performance in terms of increased lifetime and minimum delay in successful transmission of packetized data. Performance using nearest neighbor based protocol with low search angle, e.g. 200, degrades significantly. This study will be useful in selecting efficient forwarding protocol, which is an important step in designing energy efficient along with delay critical CDMA WSN.

## References

[1]  Akkaya K., and Younis M., "A Survey on Routing Protocols for Wireless Sensor Networks", Journal on Ad Hoc Networks, Elsevier publication, vol.3, issue 3, May 2005, pp.325-349.

[2]  Panichpapiboon, S., Ferrari, G., Tonguz, O.K., "Optimal Transmit Power in Wireless Sensor Network", IEEE Transactions on Mobile Computing, Vol.5, No. 10, October 2006, pp.-1432-1447.

[3]  Zorzi and R. Rao, "Geographic random forwarding (GeRaF) for ad hoc and sensor networks: Multihop performance," IEEE Trans. Mobile Comput., vol. 2, no. 4, pp. 337–348, Oct.–Dec. 2003.

[4]  L. Zhang and Y. Zhang, "Energy-Efficient Cross-Layer Protocol of Channel-Aware Geographic-Informed Forwarding in Wireless Sensor Networks," IEEE Transactions On Vehicular Technology, vol. 58, no. 6, July 2009, pp. 3040-3052.

[5]  Panigrahi B., De S., Panda B.S., and Lan Sun Luk J.D., "Energy-Efficient Greedy Forwarding Protocol for Wireless Sensor Networks", Proceedings of IEEE Vehicular Technology Conference(vtc 2010- spring), pp.1-5.

[6]  Tsai Yuh-Ren, "Sensing Coverage for Randomly Distributed Wireless Sensor Networks in Shadowed Environments", *IEEE Transactions on Vehicular Technology*, January 2008, Vol.57, No.1, pp.556-564

[7]  Hyunduk Kang, Heonjin Hong, Seokjin Sung, and Kiseon Kim, "Interference and Sink Capacity of Wireless CDMA Sensor Networks with Layered Architecture", ETRI Journal, Volume 30, Number 1, February 2008, pp.13-20.

[8]  De, S., QIAO, C.,. Pados, D. A., Chatterjee, M. and Philip, S. J.,"An Integrated Cross-layer Study of Wireless CDMA Sensor Networks," *IEEE Journal on Selected Areas in Communications*, September, Vol.22, No.7, pp. 1271-1285.

[9]  Muqattash, A. and Krunz , M., " CDMA based MAC protocol for wireless ad hoc networks", *Proceedings of ACM MobiHoc*, June, Annapolis, MD, USA, Vol 1, pp 153-164.

[10]  Datta, U., Sahu, P. K., Kundu, C. and Kundu, S., "Energy Level Performance of Multihop Wireless Sensor Networks with Correlated Interferers," Proceedings of the Fifth IEEE International Conference on Industrial and Information Systems (ICIIS-2010), July 29- August 01, 2010, Mangalore, India, pp 35-40.

[11]  Zhen, T., Dongfemg, Y. and Quanquan, L.(2008) "Energy Efficiency Analysis of Error Control Schemes in Wireless Sensor networks", Proceedings in the International Wireless Communication and Mobile Computing Conference, WCMC-2008, pp.401-405.

[12]  Mehmet C. Varun, Ian F. Akyiliz, (2009) " Error Control in Wireless Sensor Networks: A Cross Layer Analysis", IEEE/

ACM Transactions on Networking, Vol.17, No. 4, August 2009, pp.1186-1199.

[13]  Mehmet C. Vuran, Ian F. Akyildiz, " Cross-layer Packet Size Optimization for Wireless Terrestrial, Underwater, and Underground Sensor Networks", Proceedings of IEEE INFOCOM 2008, PP.780-788.

[14]  Tonguz, O. K. and Ferrari, G. "Adhoc Wireless Networks", A Communication – Theoretic Perspective, Wiley publication, John Wiley and sons, 2004, England.

[15]  Datta, U.and Kundu, S., "Packet Size Optimization for Multi HopCDMA Wireless Sensor Networks with Nearest Neighbors Based Routing", Proceedings of the Third International Conference on Emerging Applications of Information Technology (EAIT 2012), Nov. 29-Dec. 01, 2012, at Indian Statistical Institute, Kolkata.

[16]  Jerez, R., Ruiz-Garcia, M. and Diaz-Estrella, A. (2000) "Effects of Mutipath Fading on BER Statistics in Cellular CDMA Networks with Fast Power Control", IEEE Communications Letters, November, Vol.4, No.11, pg.-349-35.

[17]  Kleinschmidt, J. H., Borelli, W. C. and Pellenz, M. E. (2007) "An Analytical Model for Energy Efficiency of Error Control Schemes in Sensor Networks" *IEEE Communications Society, ICC 2007 proceedings*, pp.3895-3900. Sakhir, M., Ahmed, I., Peng, M. and Wang, W. (2008) "Power Optimal Connectivity and Capacity in Wireless

[18]  Sensor Network", *2008 International Conference on Computer Science and Software Engineering*, pp.967-970.

[19]  "ASH Transceiver Designer's Guide" RFM-TR1000 Transceiver, November 2001.

[20]  "ATMEL Transceiver Designer's Guide" AT86RF212 Transceiver.

[21]  Kim, Joon Bae and Honig, Michael L. (2000) "Resource Allocation for Multiple Classes of DS-CDMA Traffic", *IEEE Transaction on Vehicular Technology*, Vol.49, No.2, March, pp.506-519.

# 4

# Node selection in Peer-to-Peer content sharing service in mobile cellular networks with Reduction Bandwidth

S. Uvaraj[1], S. Suresh[2], E. Mohan[3]

[1]Arulmigu Meenakshi Amman College of Engineering, Kanchipuram, India
[2]Sri Venkateswara College of Engineering, Chennai, India
[3]Pallavan College of Engineering, Kanchipuram, India

**Email address:**

ujrj@rediffmail.com(S. Uvaraj), ss12oct92@gmail.com(S. Suresh), emohan1971@gmail.com(E. Mohan)

**Abstract:** The peer-to-peer service has entered the public limelight over the last few years. Several research projects are underway on peer-to-peer technologies, but no definitive conclusion is currently available. Comparing to traditional server-client technology on the internet, the P2P technology has capabilities to realize highly scalable, extensible and efficient distributed applications. At the same time mobile networks such as WAP, wireless LAN and Bluetooth have been developing at breakneck speed. Demand for using peer-to-peer applications over PDAs and cellular phones is increasing. The purpose of this study is to explore a mobile peer-to-peer network architecture where a variety of devices communications each other over a variety of networks. In this paper, we propose the architecture well-adapted to mobile devices and mobile network. In P2P file sharing systems over mobile cellular networks, the bottleneck of file transfer speed is usually the downlink bandwidth of the receiver rather than the uplink bandwidth of the senders. In this paper we consider the impact of downlink bandwidth limitation on file transfer speed and propose two novel peer selection algorithms named DBaT-B and DBaT-N, which are designed for two different cases of the requesting peer's demand respectively. Our algorithms take the requesting peer's downlink bandwidth as the target of the sum of the selected peers' uplink bandwidth. To ensure load balance on cells, they will first choose a cell with the lowest traffic load before choosing each peer.

**Keywords:** Network Security, P2P, Mobile Cellular Networks, Peer Selection, Load Balance, Reduction Bandwidth

## 1. Introduction

In the cooperated network, the server-client technology has been used as the traditional way to handle network resources and provide Internet services. It has advantages to regulate Internet services by only maintain limited number of central servers. As a result, peer-to-peer technology has become popular and has been used in networks which manage vast amounts of data daily, and balance the load over a large number of servers.

At the same time, mobile Internet services have become very popular. In the past four years, the market of mobile Internet services has considerably grown successful in Japan where imode is the most famous example. The mobile environment is different from the fixed Internet in that it is an extremely constrained environment, in terms of both communication and terminal capabilities. This should be taken into account when developing systems which will work in a mobile environment. Additionally, various wireless have been emerging of network environments such as IMT-2000 (International Mobile Telecommunications-2000, for example FOMA)[10], Wireless LAN and Bluetooth, and users can select them to satisfy their network demands. In the near future, an environment where many sensors, users and different kind of objects exist, move and communicate with one another, called "ubiquitous communication environment", will appear.

In this paper we study the problems of peer selection in P2P file sharing systems over mobile cellular networks with consideration of downlink bandwidth limitation. Our motivation is that, since the file transfer speed is limited by the requesting peer's downlink bandwidth, some other performance indicator such as load balance on cells should be focused on. So our goal is to achieve load balance on cells under the precondition that the requesting peer's demand is satisfied. In P2P file sharing systems the requesting peer's demand can be divided into two cases, one is that the requesting peer demands a lower bound of the sum of the selected peers' uplink bandwidth (as in some P2P media sharing systems [5]), and the other is that the requesting peer demands a certain number of selected peers (denoted as *Numwant* in BitTorrent Tracker protocol [6]). We consider the two cases both and propose two algorithms for the two cases respectively. The first one is named DBaT-B (Downlink Bandwidth as Target, Bandwidth satisfied), and the second one is named DBaT-N (Downlink Bandwidth as Target, Number satisfied). Major features of our algorithms can be described as follows. First, they take the requesting peer's downlink bandwidth as the target of the sum of the selected peers' uplink bandwidth. Second, they choose a cell with the lowest traffic load before choosing each peer. Difference of the two algorithms lies in using different criteria in each peer selection round to satisfy the different demand. Moreover, we also provide a Fuzzy Cognitive Map (FCM) [7] that can be used in our algorithms to estimate peers' service ability according to multiple influential factors. Simulation results show that in respective cases DBaT-B and DBaT-N algorithms can both achieve favorable load balance on cells in mobile cellular networks while ensuring good file transfer speed, and compared with other two traditional peer selection algorithms our algorithms are nice improvements.

The principal goal of our work is thus to design a mobile peer-to-peer architecture and a general peer-to-peer platform that enhances communication capabilities for mobile clients, by utilizing network resources efficiently and supporting mobility in an integrated and practical way.

The rest of this paper is organized as follows. Section 2 gives an overview of our peer-to-peer architecture, describes each key element of our architecture and mobile device adaptation. Section 3 describes the Algorithm design. We introduce DBAT-B and DBAT-N algorithms in detail. Section 4 we present the FCM that can be used in our algorithms. In Section 5 we evaluate the performance of our algorithms through simulations. Finally we give out conclusions for this paper.

# 2. Architecture

## 2.1. Architecture Overview

The proposed mobile peer-to-peer architecture is shown in Fig. 1. All of the peer-to-peer communication entities that have a common set of interest and obey a common set of policies construct one peer-to-peer community. This architecture consists of the following basic components:

*Peer-to-peer node*: The peer-to-peer node is an independent communication entity in the peer-to-peer network.

*Mobile proxy*: Theoretically, all the mobile devices (e.g. WAP or i-mode terminals) can be independent nodes in the peer-to peer architecture. However some of them are functionally limited and cannot act as autonomous nodes. The mobile proxy is a function in a node, which acts as a proxy

for the mobile devices with constrained capability, so that these mobile devices can join the peer-to-peer architecture.

*Pure peer-to-peer architecture*: There are only peer-to-peer nodes in the pure peer-to-peer architecture. The proposed pure peer-to-peer architecture is shown in Fig. 2 (a). The connection between peer-to-peer nodes is established on their mutual trust. Each peer-to-peer node is an independent entity and can participate in and leave the peer-to-peer network at its convenience. Messages are sent from a peer-to-peer node to another one directly or by passing them via some intermediary peer-to-peer nodes.

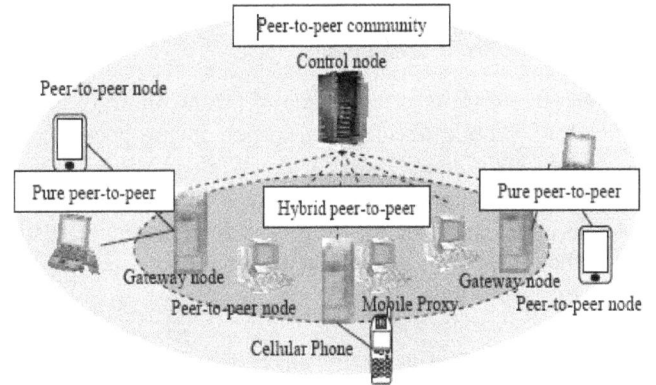

**Fig 1.** *The mobile peer-to-peer architecture*

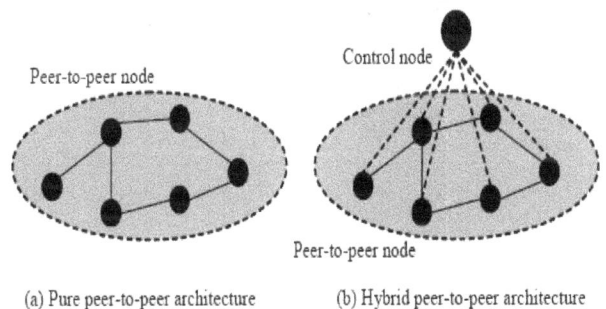

(a) Pure peer-to-peer architecture    (b) Hybrid peer-to-peer architecture

**Fig 2.** *Pure peer-to-peer and hybrid peer-to-peer*

*Hybrid peer-to-peer architecture*: The hybrid peer-to-peer architecture resolves the disadvantages of the pure peer-to-peer architecture such as inefficient routing, splits of network and lack of security, by introducing a control node. The proposed hybrid peer-to-peer architecture is shown Fig. 2 (b). In our architecture, the control node provides the functions for providing routing information to a destination node, discovering the first peer-to-peer node, recovering from the splitting of the peer-to-peer network, improving the network topology and security such as authentication, in order to improve the inefficiency of the pure peer-to-peer architecture. To realize the hybrid peer-to-peer architecture, the control node and the gateway node are defined.

*Control node*: Control node is an administrative entity which manages a peer-to-peer community in the peer-to-peer network. It provides several functions independent of particular applications such as name resolution, route information provision, the first per-to-peer discovery, network topology optimization, node authentication and multicast group management.

*Gateway node*: Gateway node is a connection entity linking between a pure peer-to-peer network and a hybrid peer-to-peer network. It provides for nodes in pure peer-to-peer network with several proxy functions such as routing information provision, node authentication, and multicast group management. A control node receives a request from a peer-to-peer node and provides it with routing information, topology optimization function and security function. A gateway node collects topology information on a pure peer-to-peer network and reports it to the control node.

### 2.2. Mobile Device Adaptation

Another distinct characteristic of this architecture is that it allows mobile devices to take part in the peer-to-peer network via a mobile proxy node. While a mobile device, such as a cellular phone, may have enough capabilities to act as an independent peer-to-peer node in the future, it still has the following limitations at this time:

• Limited storage
• Small runtime heap
• Modest processor performance
• Constrained electrical power

Thus, a current mobile device can't fully perform the role of a peer-to-peer node that offers services to other peer-to-peer nodes in a peer-to-peer network. In order to incorporate a mobile device into a peer-to-peer network, some functions must be done by other nodes on behalf of a mobile device. Through the cooperation of a mobile proxy, a mobile device can virtually act as a peer-to-peer node and can perform the necessary functions in the peer-to-peer architecture. From the point of view of the peer-to-peer architecture, mobile devices are modeled in the three ways.

In Fig.3(a), mobile devices share the same proxy node. From the point of view of the network, the mobile proxy acts as one peer-to-peer node. In Fig. 3(b), a mobile device has its own node name and acts as a separate peer-to-peer node in the peer-to-peer architecture.

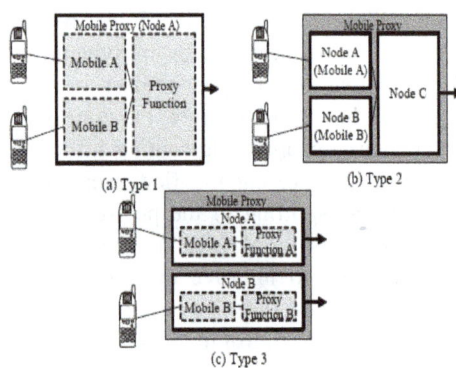

*Fig 3. Mobile Device Adaptation*

For realizing this type of mobile proxy, some proxy functions should be implemented on Node C such as transforming a message received from a mobile device into a message of a peer-to-peer protocol. In Fig. 3(c), a pair of a mobile device and a proxy function constructs a peer-to-peer node in the peer-to-peer architecture. In this case, a mobile device has its own node name, and acts as a separate node through a mobile proxy. The mobile proxy does not act as an independent node. It will be decided by requirements of peer- to- peer applications, as to which type of mobile proxy model will be preferable.

## 3. Dbat Algorithms

In this section we depict the details about DBaT-B and DBaT-N algorithms, which can be denoted as DBaT algorithms as a whole. Due to the existence of downlink bandwidth limitation, it is unnecessary to always choose peers with high uplink bandwidth. So DBaT algorithms take the requesting peer's downlink bandwidth as the target of the sum of the selected peers' uplink bandwidth. Besides, to ensure load balance on cells, DBaT algorithms will first choose a cell with the lowest traffic load before choosing a peer.

In DBaT algorithms service ability is used as one of the criteria for peer selection. As we have mentioned before, in mobile environments estimation of peers' service ability is complicated since it is influenced by multiple factors. In this section we just focus on the details of DBaT algorithms. A method about how to estimate peers' service ability in P2P file sharing systems over mobile cellular networks will be provided in next section. Here we list the common notations used in this section and their meanings:

> ▬ Br: Upper bound of the requesting peer's downlink bandwidth.
> ▬ Lest: Estimated traffic load on a cell after a peer in this cell is selected. Initial value of lest is just the initial traffic load on this cell before peer selection.
> ▬ Bpa: Available uplink bandwidth of a candidate peer.
> ▬ Bpe: Estimated uplink bandwidth of a peer after it is selected.

The traffic load on a cell is defined as the ratio of the current used radio bandwidth over the total radio bandwidth of the base station in the cell. For example, assuming that the current used radio bandwidth of a 3G base station is 1.2Mbps, the traffic load on this cell is 0.6 since the maximum radio bandwidth provided by this 3G base station is 2Mbps. Obviously the value of traffic load on a cell ranges from 0 to 1.

### 3.1. Dbat-B Algorithm

DBaT-B algorithm is designed for the case that the requesting peer demands a lower bound (denoted as $Bd$) of the sum of the selected peers' uplink bandwidth. The relationship between $Bd$ and $Br$ can be discussed as follows.

On one hand, if $Bd$ is higher than $Br$, it makes no sense due to the performance limitation imposed by $Br$. On the other hand, if $Bd$ is lower than

$Br$, we can easily take $Bd$ as $Br$ and DBaT-B algorithm still works. Here we list the notations used in this subsection and their meanings:

- $Bs$: Sum of the estimated uplink bandwidth of the selected peers. Initial value of $Bs$ is 0.
- $\Delta B$: Difference between $Br$ and $Bs$, more specifically, $\Delta B = Bs - Br$.

In more detail, DBaT-B algorithm works in the following steps:

Step 1. Choose a cell:
1-1. Sort the cells according to the traffic load;
1-2. Choose a cell with the lowest traffic load, go to Step 2.
Step 2. Check all the peers in this cell with $Bpa$ and $|\Delta B|$:
2-1. Compare the values of $Bpa$ of all the candidate peers in this cell with $|\Delta B|$;
2-2. If there is no peer satisfying $Bpa > |\Delta B|$, go to Step 3, otherwise go to Step 4.
Step 3. Choose a peer in this cell according to service ability:
3-1. Choose a peer with the highest service ability in this cell;
3-2. Calculate $Lest$ for this cell according to (1), calculate $Bpe$ according to (2);
3-3. Recalculate $Bs$ and $\Delta B$, go to Step 1.
Step 4. Check all the peers in this cell with $Bpe$ and $|\Delta B|$:
4-1. Calculate the value of Lest and Bpe for each candidate peer that satisfies $Bpa > |\Delta B|$ in this cell;
4-2. Compare the values of Bpe of all the peers that satisfy $Bpa > |\Delta B|$ in this cell with $|\Delta B|$;
4-3. If there is no peer satisfying $Bpe > |\Delta B|$, go to Step 3, otherwise go to Step 5.
Step 5. Choose a peer in this cell according to service ability, $Bpe$ and $|\Delta B|$:
5-1. Choose a peer with the highest value of estimated service ability from the peers that satisfy $Bpe > |\Delta B|$ in this cell;
5-2. Recalculate Bs and $\Delta B$, end the peer selection process.

### 3.2. Dbat -N Algorithm

DBaT-N algorithm is designed for the case that the requesting peer demands a certain number of selected peers. In traditional fixed network environments, file transfer speed usually increases with more serving peers. However, in mobile cellular network environments, file transfer speed may not increase as the number of serving peers increases because it would be bottlenecked by the downlink bandwidth limitation. So the motivation of DBaT-N is to choose a certain number of selected peers whose sum of uplink bandwidth is fitly over Br. Here we list the notations used in this subsection and their meanings:

- n: Number of selected peers demanded by the requesting peer.
- Bref: Referenced value of uplink bandwidth for peer selection, more specifically, $Bref = Br/n$
- B': Cumulative referenced value of uplink bandwidth for peer selection.
- $\Delta b$: Difference value of uplink bandwidth for peer selection. Initial value of $\Delta b$ is set to 0.
- S: Number of candidate peers in a cell.
- K: Number of peers in the candidate set in a cell.

In each peer selection round, $Bref$ can be seen as a fixed target of the selected peer's uplink bandwidth, $B'$ can be seen as the actually used target of the selected peer's uplink bandwidth since it represents the adjustment of $Bref$ after last peer selection round, and $\Delta b$ is used to record the value difference between $B'$ and a candidate peer's estimated uplink bandwidth. So, the basic idea of DBaT-N can be described as follows. In each peer selection round DBaT-N chooses a peer with uplink bandwidth close to $B'$, and in the last round it chooses a peer that makes the sum of selected peers' uplink bandwidth fitly over $Br$ according to $\Delta b$.

In more detail, DBaT-N algorithm works in the following steps:

Step 1. Choose a cell:
1-1. Sort the cells according to the traffic load;
1-2. Choose a cell with the lowest traffic load, go to Step 2.
Step 2. Calculate $\Delta b$ for each peer in this cell:
2-1. Calculate the value of $Lest$ and $Bpe$ for each candidate peer in this cell;
2-2. Calculate $B' = Bref - \Delta b$, then calculate $\Delta b = Bpe - B'$ for each candidate peer in this cell;
2-3. Check the value of $n$, if $n > 1$, go to Step 3, if $n = 1$, go to Step 4.
Step 3. Choose a peer according to service ability, $K$
3-1. Choose a peer with the highest estimated service ability from the $K$ peers that have the lowest values of $|\Delta b|$ in this cell;
3-2. Record the value of $Bpe$ and $Lest$ for this peer, record the value of $\Delta b$, let $n = n - 1$, go to Step 1.
Step 4. Choose a peer according to service ability, $K$ and $\Delta b$:
4-1. Choose a peer with the highest estimated service ability from the $K$ peers that have the lowest values of $\Delta b$ and satisfy $\Delta b > 0$ in this cell;
4-2. Record the value of $Bpe$ and $Lest$ for this peer, record the value of $\Delta b$, end the peer selection process.

In Step 4, it is possible that the number of peers satisfying $\Delta b > 0$ (denoted as $k$ here) is lower than $K$. In this case, the algorithm will take $k$ as $K$ in Step 4. Moreover, it is also possible that the value of $k$ is 0.

## 4. Fcm for Estimation of Peers' Service Ability

Service ability in environments of mobile cellular networks:

• *Uplink Bandwidth.* A peer's uplink bandwidth has direct and great impact on its service ability.

• *Delay.* The delay between a peer and the requesting client will affect the peer's service ability to some extent.

• *Packet Loss Probability.* According to our simulations, Packet loss probability on the link between a peer and the requesting client impacts the peer's service ability greatly.

• *SINR.* SINR (Signal to Interference and Noise Ratio) value of a radio link can indicate the radio link quality which has a direct impact on the packet loss probability of the link.

• *Energy Level.* Since the battery energy is relatively limited on mobile hosts, the energy level has a direct impact on *peer churn probability* which greatly affects the peer's service ability. Moreover, a peer with higher energy level is expected to stay longer.

• *Lingering Time.* As in we assume that if a peer stays longer its *peer churn probability* is lower.

• *Moving Speed.* In mobile environments, a higher moving speed usually means a lower radio link quality, and thus means a higher packet loss probability. Moreover, a peer with higher moving speed often has higher *cell handover probability* which affects its service ability to some extent.

### 4.1. Simple P2P Protocol for Cellular Phones

We have designed the P2P protocol for cellular phones with a simple text format over HTTP, since they can only support HTTP and cannot process the protocol based on XML. The P2P protocol for cellular phones is provided by mobile proxy. Each mobile proxy acts as a virtual peer-to-peer node for a cellular phone and converts the XML based P2P protocol to the simple P2P protocol for the cellular phone. An example of a P2P message for cellular phone is shown in Figure 4. A message included in the HTTP body, is written in a simple text format and is composed of two parts. The first part of the message corresponds to the P2P core protocol, and the second part corresponds to the application protocol.

```
P2PFRM 92
Source: Node ID of source node
MulticastGroupID: ID of multicast group
ApplicationURI:
http://www.mml.yrp.nttdocomo.co.jp/ED/2003/03/p2p_in
stantmsg_app
InstantMessage
UserMessage: Hello All!
FRMEND
```

**Fig. 4.** *A sample of P2P message for cellular Phone*

## 5. Performance Evaluation

In this section we evaluate the performance of DBaT algorithms on our simulation platform built upon OMNet++. Network topology used in the simulations is shown. where there are 20 cells in total. As shown in Fig. 4, we adopt a BitTorrent-like [6] architecture for the P2P systems used in our simulations.

### 5.1. Simulation Settings

In our simulations, to simulate a relatively realistic and moderate initial status, the initial traffic load on each cell is randomly generated in the range of [0.25, 0.75]. The number of requesting peer is 10, and each requesting peer has 100 candidate peers. So there are totally 1000 candidate peers that are randomly distributed in 20 cells. Some parameters of each candidate peer are set as follows. The delay between a candidate peer and the requesting peer is randomly generated in the range from 0 to 500ms. The SINR value of a radio link is randomly generated in the range from 0 to 100dB. The energy level is randomly generated in the range from 0 to 5. The lingering time is generated by a random number in combination with the value of the energy level, and has a range from 0 to 5 hours. The packet loss probability of each radio link is generated by a random number in combination with the SINR value and the value of the moving speed, and has a range of [0.001, 0.01].

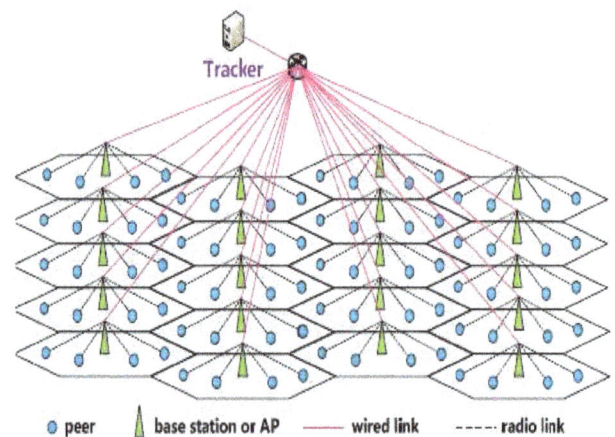

**Fig 5.** *FCM for P2P systems over mobile cellular networks*

Moreover, during the data transmission process, each serving peer has a probability of peer churn, which is generated by the value of the energy level in combination with the value of the lingering time. After a peer churn happens, the churned peer will leave the system, and the requesting peer will ask the tracker to return a new serving peer to get the remaining data.

Furthermore, each serving peer also has a probability of moving to a neighboring cell, which is called cell handover probability and is generated by the value of the moving speed in combination with a random number. After a serving peer move to a neighboring cell, its parameters including delay, SINR value and packet loss probability will be regenerated. We evaluate the performance of DBaT algorithms using a BT-like file sharing process. First, the 10 requesting peers join in the system and send requests for files to the Tracker simultaneously. Second, the Tracker performs the same peer selection algorithm for the 10 requesting peers and returns them a peer list respectively. Finally, the 10 requesting peers receive a file from their own serving peers respectively. We record the average transfer time of the 10 files and the Standard Deviation (SD) of the traffic load on the 20 cells during the file sharing process.

### 5.2. Case I: A Lower Bound of the Sum of the Selected Peers' Uplink Bandwidth is Demanded

In this subsection we consider the case that the requesting peer demands a lower bound of the sum of the selected peers' uplink bandwidth. We compare the performance of DBaT-B algorithm with other two algorithms. One is called HSA-B Highest Service Ability, Bandwidth satisfied), and the other is called RS-B (Random Selection, Bandwidth satisfied). HSA-B is in fact the same with the peer selection algorithm that always chooses peers with the highest service ability [5], and RS-B is in fact the traditional peer selection algorithm used in BitTorrent that chooses peers randomly [6].

Fig. 6 shows the simulation results under the scenario of 3G networks when the downlink bandwidth of each requesting peer ranges from 200Kbps to 1Mbps. The results can be observed and analyzed as follows.

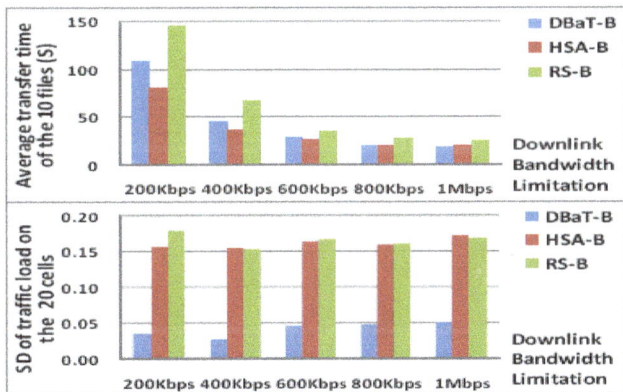

*Fig 6. Simulation results under the 3G scenario in case I*

First, with any value of the downlink bandwidth, the SD value of DBaT-B is much (at least 66%) lower than those of HSA-B and RS-B. This indicates that DBaT-B achieves much better load balance on the 20 cells than the other two algorithms. Second, with any value of the downlink bandwidth, the average file transfer time of RS-B is obviously (at least 33%) higher than that of DBaT-B or HSA-B. This indicates that the service ability of the serving peers chosen by RS-B is still lower than those chosen by DBaT-B or HSA-B, although the sum of their uplink bandwidth is over the downlink bandwidth of each requesting peer. Third, as the downlink bandwidth increases, the difference between

the average file transfer time of DBaT-B and that of HSA-B gets smaller. This can be explained as follows. When the downlink bandwidth is small, the number of serving peers in the system will not be great.

From the results and analysis in subsection 5.2 we can draw conclusions including:

1) Downlink bandwidth limitation has great impact on file transfer speed;

2) Our DBaT algorithms can achieve much better load balance on cells than traditional HSA and RS algorithms; 3) Our DBaT algorithms can achieve file transfer speed similar with that of HSA, especially when the traffic load on cells is relatively high.

## 6. Conclusion

In this paper we studied the problems of peer selection in P2P file sharing service over mobile cellular networks with consideration of downlink bandwidth limitation. Our major contribution was to propose two peer selection algorithms (named DBaT-B and DBaT-N) that can achieve load balance on cells under the precondition that the requesting peer's demand is satisfied. The two algorithms were designed for two different cases of the requesting peer's demand respectively. Our algorithms take the requesting peer's downlink bandwidth as the target of the sum of the selected peers' uplink bandwidth. Compared with the traditional HSA and RS algorithms, our DBaT algorithms can achieve much better load balance on cells, and they can also achieve file sharing speed better than that of RS, or similar with that of HSA, especially when the traffic load on cells is relatively high. So our algorithms can be seen as nice improvements. Peer-to-peer security will be an important issue for such an environment. We are considering the incorporation of peer-to-peer security into our mobile peer-to-peer architecture. Additionally, we continue to develop new mobile peer-to-peer applications and will evaluate efficiency and performance of our peer-to-peer protocols.

## References

[1]   Y. Li, Y. C. Wu, J. Y. Zhang, *et al.*, "A P2P Based Distributed Services Network for Next Generation Mobile Internet Communications," in Proc.WWW'09, pp. 1177-1178, 2009.

[2]   L. Popova, T. Herpel, W. Gerstacker, "Cooperative mobile-to-mobile file dissemination in cellular networks within a unified radio interface," Computer Networks, vol. 52(6), pp. 1153-1165, 24 April 2008.

[3]   H. Xie, R. Yang, A. Krishnamurthy, "P4P: Provider Portal for Applications," in Proc. ACM SIGCOMM'08, vol. 38, pp. 351-362, 2008.

[4]   [4]W. Li, S. Z. Chen, T. Yu, "UTAPS: An Underlying Topology Aware Peer Selection Algorithm in BitTorrent," in Proc. IEEE AINA'08, pp. 539-545, 2008.

[5]   T. S. Eugene Ng, *et al.*, "Measurement-Based Optimization

Techniques for Bandwidth-Demanding Peer-to-Peer Systems," in Proc. IEEE INFOCOM'03, pp. 539-545, Orlando, Florida, USA, 2003.

[6]   BitTorrent,Inc. 2008. URL: http://www.bittorrent.com/

[7]   B. Kosko, "Fuzzy Cognitive Maps," International Journal of Man-Machine Studies, vol. 24, pp. 65-75, 1986.

[8]   X. Li, H. Ji, F. R. Yu, R. M. Zheng, "A FCM-Based Peer Grouping Scheme for Node Failure Recovery in Wireless P2P File Sharing," in Proc. IEEE ICC'09, pp. 1-5, 2009.

[9]   OMNeT++, URL: http://www.omnetpp.org/2011

[10]  NTT DoCoMo, Inc., "FOMA (Freedom Of Mobile multimedia Access)", http://foma.nttdocomo.co.jp/English/.

# 5

# Internet authentication and billing (hotspot) system using MikroTik router operating system

**Adam Mohammed Saliu[1], Mohammed Idris Kolo[1], Mohammed Kudu Muhammad[2], Lukman Abiodun Nafiu[3]**

[1]Department of Computer Science, Federal University of Technology, Minna, Nigeria
[2]Academic Planning Unit, Federal University of Technology, Minna, Nigeria
[3]Department of Mathematics and Statistics, Federal University of Technology, Minna, Nigeria

**E-mail address:**
firdaousmama@gmail.com(L. A. Nafiu)

**Abstract:** This paper is based on using MikroTik Router Operating System (OS) to build an authentication and billing System. This kind of system is always used to create security, billing and administration of users on a network connected to the Internet. It is used to restrict usage and bandwidth allocations to users. A case study of Federal University of Technology, Minna website was considered and it was found that additional stages and radius server services should be incorporated with the antenna design in its hotspot server.

**Keywords:** Internet, Hotspot, MikroTik, Bandwidth and Wireless

## 1. Introduction

The advent of information technology has brought about globalization. The world as we know is fast becoming a global village, where transactions can easily be done from any part of the world with the help of the computer. This achievement owes its credit to the advent of networking. Wireless networks represent one of the most significant inventions in human history with wireless networking as in [1].

According to [2], it is opined that wireless network guarantees access to files regardless of where house and computer facilitates with synchronization of data on a laptop. Using wireless networks makes it so much simpler to send files between computers as compared to the traditional emails or CDs. Moreover, if printer is connected, documents can be written anywhere and finally printed through a printer connected to another computer. In other words, a printer plugged into one of the computers on the network is available to all the computers saving the expense of having multiple printers.

The work in [3] is based on being able to eliminate wires from running all over house is probably the biggest reason why wireless networking plays significant advantage over wired networking. Wired networking costs more and the packed wires make house looking ugly and not safe.

This paper involves the construction of an authentication and billing (Hotspot) server for Federal University of Technology, Minna, Nigeria with the aid of MikroTik Router OS (a Linux-based operating system). Figures 1 and 2 give an overview of operation of a hotspot network and a generalized web login of MikroTik hotspot respectively. This offers a server design that meets the needs of the public in achieving the main aim of true mobility in surfing the internet with imperial management and full administrative control of the network which is one of the distinguishing features of servers as illustrated in figure 3.

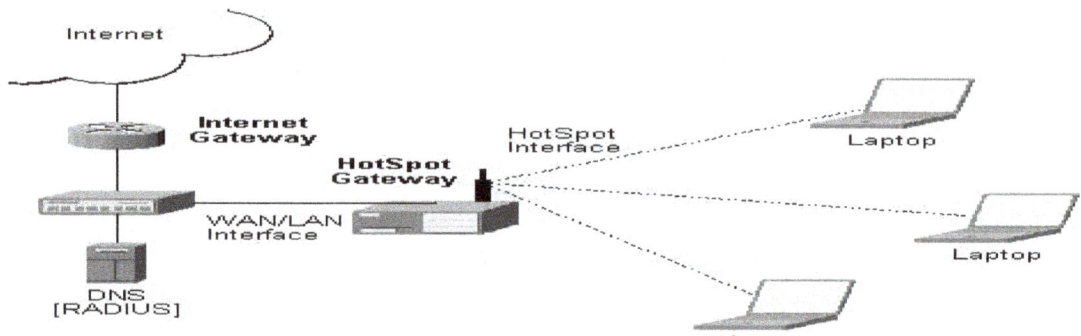

**Figure 1.** *An Overview of Operation of a Hotspot Network*

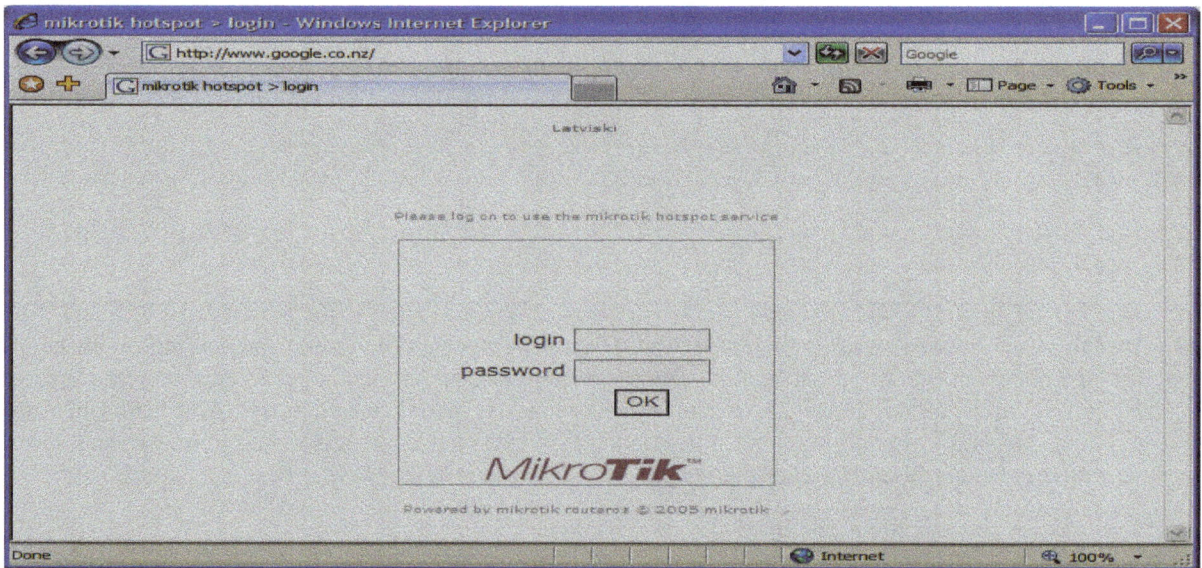

**Figure 2.** *A Generalized Web Login of MikroTik Hotspot*

**Figure 3.** *A Server Computer*

To begin browsing, a client must go through a registration process with the provider and then enter a user-name and password in a browser Log-in window that appears on the attempt to open a webpage. Hotspot technology proposes providers to establish and administer a user database, which can be useful for such enterprises as airports, hotels or universities that offer wireless or Ethernet Internet connectivity to employees, students, guests or other groups of users as in [4].

A Wireless Network Interface Controller (WNIC) as given in figure 6 is a network card connected to a radio-based computer network, unlike a regular network interface controller (NIC) which is connected to a wire-based network such as Ethernet. A WNIC is an essential component for wireless desktop computer. This card uses

an antenna to communicate through microwaves. A WNIC in a desktop computer usually is connected using the PCI bus (Peripheral Component Interconnect), an industry-standard bus for attaching peripherals to computers. Other connectivity options are USB as given in figures 4 and 5, and card as given in figure 6 based on [5].

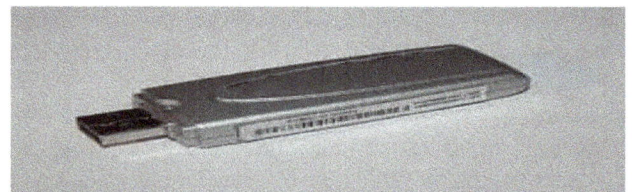

**Figure 4.** *A USB Wireless Network Interface Device*

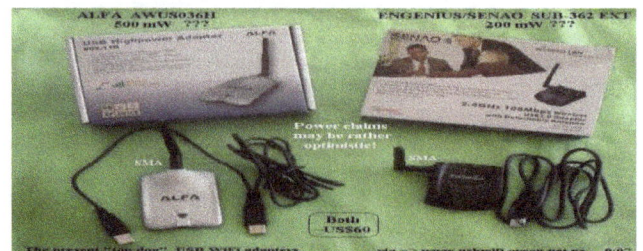

**Figure 5.** *A USB Wireless Network Interface Device*

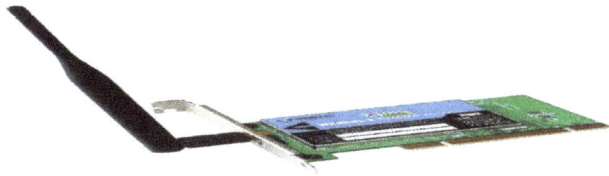

*Figure 6. Wireless Network Interface Card*

## 2. Materials and Methods

MikroTik Limited known internationally as MikroTik is a Latvian manufacturer of computer networking equipment. The main product of MikroTik is a Linux-based operating system known as MikroTik Router OS. It allows users to turn a selected PC-based machine into a software router, allowing features such as firewall rules, VPN Server and Client, bandwidth shaper Quality of Service, wireless access point and other commonly used features for routing and connecting networks together. The system is also able to serve as a captive-portal based hotspot system (wiki.mikrotik.com).

For this paper, we used Router OS version 2.9.7. The Router OS, combined with its hardware product line, known as MikroTik Router BOARD is marketed at small to medium sized wireless internet service providers, typically providing broadband wireless access in remote areas. We also made use of a MiniPCI to PCI adapter which is an advanced adapter for 3.3v MiniPCI to PCI standard sockets. The adapter is provided in PCI card form with an integrated MiniPCI slot that allows using MiniPCI cards in standard PCI sockets. The MiniPCI adapter includes two LEDs connected to the MiniPCI.

In line with [6], Mini PCI was added to PCI version 2.2 for use in laptops. It uses a 32-bit, 33-MHz bus with powered connections (3.3 V only, 5V is limited to 100mA) and support for bus mastering and DMA. The standard size for Mini PCI cards is approximately 1/4 of their full-sized counterparts as found in [5]. As there is limited external access to the card compared to desktop PCI cards, there are limitations on the functions they may perform.

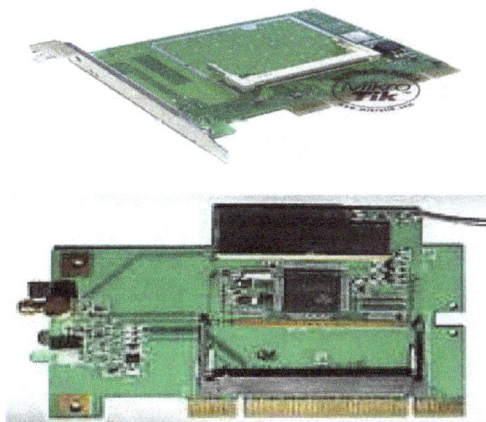

*Figure 7. Mikrotik Router Board 11 and MiniPCI-to-PCI Converter Type III*

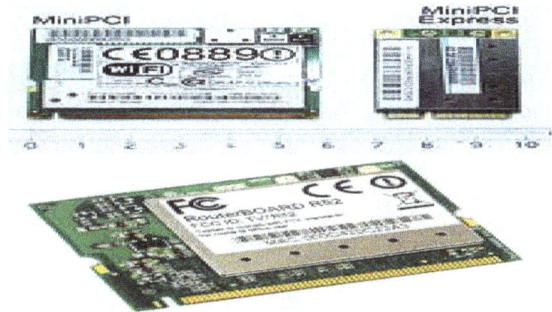

*Figure 8. MikroTik MiniPCI and MiniPCI Express cards in comparison*

Many Mini PCI devices were developed such as Wi-Fi, Fast Ethernet, Bluetooth, modems, sound cards, cryptographic accelerators, SCSI, IDE/ATA, SATA controllers and combination cards. Mini PCI cards can be used with regular PCI-equipped hardware using Mini PCI-to-PCI converters as given in figures 7 and 8 based on [7].

In this paper, construction was carried out in FIVE stages as follows based on [8]:

### 2.1. Hardware Configuration Stage

The hardware units used in this paper are:

(1) A desktop unit consisting of 512RAM, CDROM drive, Hard disk drive of 20Gigabytes, Motherboard with processor of 2.0Gigahertz speed

(2) A laptop with wireless ready and internet ready

(3) MikroTik router board 11 with 400mw MikroTik MiniPCIcard and 2dbi antenna

(4) MikroTik Router's CD for installation

(5) Microsoft FrontPage CD 2003 package for the web hotspot interface design

### 2.2. Software Installation Stage using MikroTik Router OS CD

The desktop unit is being assembled with keyboard, monitor, mouse and UPS.

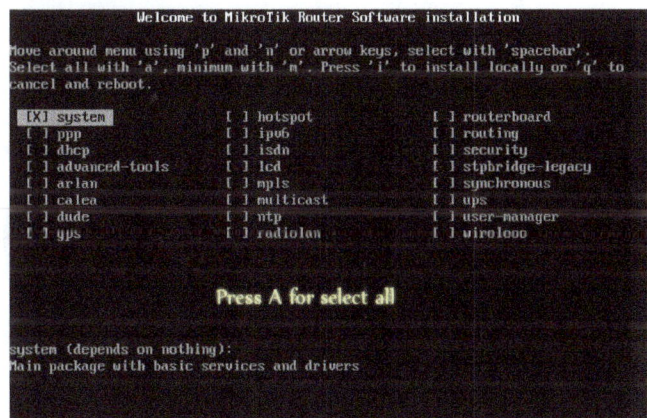

*Figure 9. MikroTik Router Software Installation*

Press "a" key for select all and then press "i". Then, answer "n" to first question and "y" to second as can be seen in figure 9.

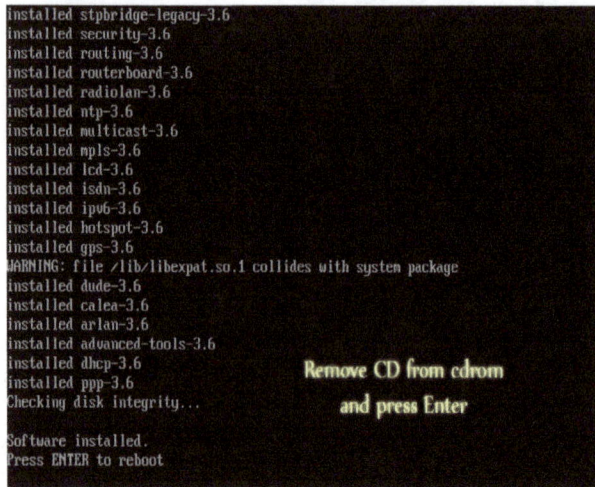

**Figure 10.** *Procedure to Remove CD from CDROM*

Installation removes CD from CD-ROM and press "Enter". After rebooting, type "admin" as login and no password required as contained in figure 10.

### 2.3. Network Planning and Implementation

Network planning involves planning of the network IP addresses and subnet address. The IP address for the LAN, as for this paper is concerned with 192.168.8.54, while the subnet address is 255.255.255.0 and gateway is 192.168.8.1. It has two types of network structures: one is a wireless while the other is a LAN network. The addresses listed above are for the LAN network which supplies the Hotspot server with internet. The next planning is for the wireless in which one gets the privilege of choosing the set of IP addresses to use for the wireless network interface. For Federal University of Technology, Minna website as found in [9], we applied 10.5.50.1 as the gateway for the wireless interface while the subnet address remains 255.255.255.0.The DNS still remains 192.168.8.1.

### 2.4. Configuration and Test Running

The following are the initial steps to take after installation is done successfully. The newly installed router OS needs to be initially configured with an IP address through the command line interface prior being able to continue the configuration through the Winbox or web interface. Once the server has been restarted, it boots up with username and password. IP address for the unit can be either statically or dynamically assigned. In this work, we used static IP address. The step to sign IP starts when the server prompts the following MikroTik RouterOS™ Welcome Screen on the command prompt menu.

### 2.5. Hotspot Web Browser Login Window Design

Press Enter to setup dhcp server. Type "a" to configure IP address and gateway.

Then, key in 192.168.8.54/255.255.255.0 (IP/subnet) and press enter key. Next press g to add default gateway to the

server. Default gateway will be 192.168.8.1. The same process will be used for wireless LAN interface IP configuration but in this case it is represented by WLAN. IP address is 10.5.50.1/255.255.255.0 (IP/Subnet).

**Figure 11.** *Web interface of MikroTik RouterOS.*

Once the web interface of MikroTik Router OS in figure 11 is downloaded, we run it to access the router by entering the LAN IP address, username and password as given in figure 12.

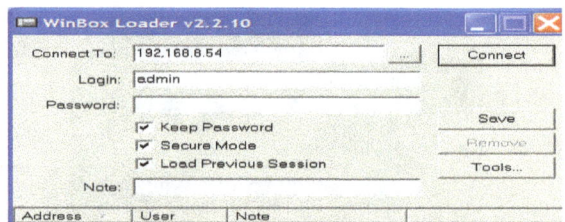

**Figure 12.** *WINBOX Login Interface*

On logging in, the following menu given in figure 13 appears

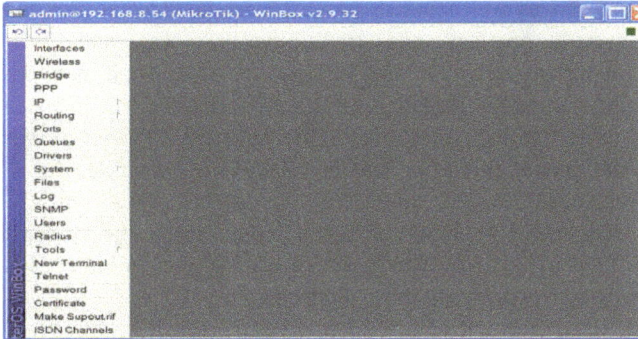

*Figure 13. WINBOX full interface*

Next is the Hotpot setup given in figure 14. On the left hand menu, click on the + sign to add a hotspot interface.

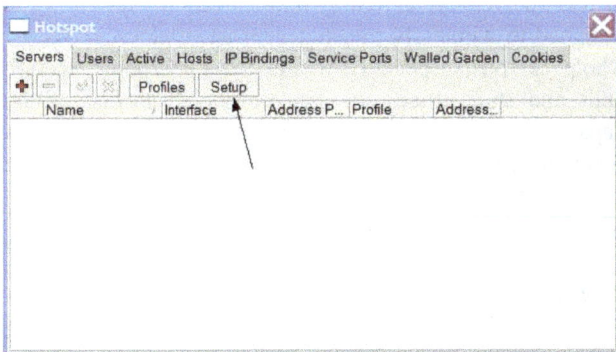

*Figure 14. Hotspot setup interface*

Click on Setup to select the hotspot interface WLAN1

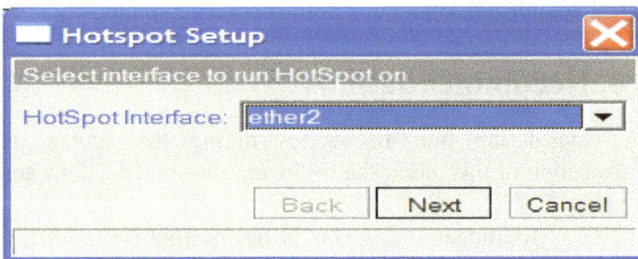

*Figure 15. Hotspot Interface Setup1*

Select the desire IP address for the Hotspot as given in figure 16

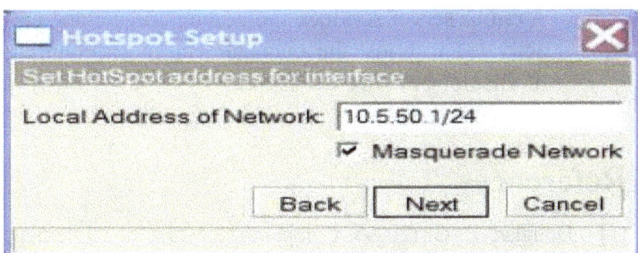

*Figure 16. Hotspot Interface Setup 2*

Select the IP address range as contained in figure 17

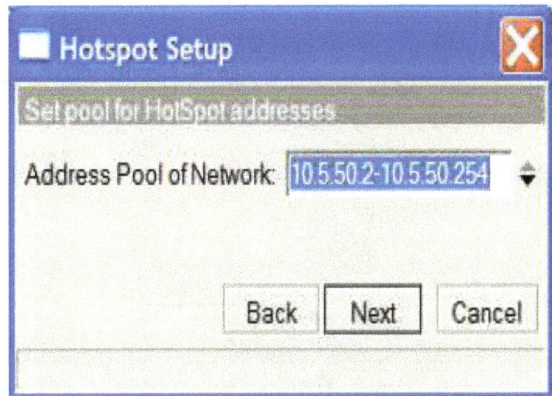

*Figure 17. Hotspot Interface Setup 3*

Then enter SSL certificate for the MikroTik as contained in figure 18

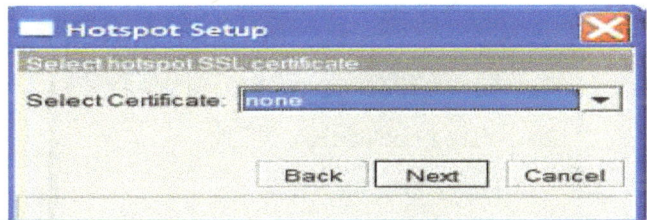

*Figure 18. Hotspot Interface Setup 4*

Most Hotspot providers will not add their SMTP server to avoid clients registering for short period and using their servers for spam as in figure 20.

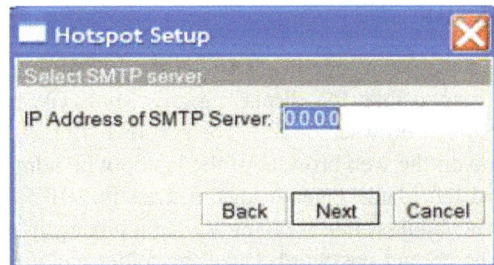

*Figure 19. Hotspot Interface Setup 5*

Enter the DNS server address for the MikroTik as contained in figure 20

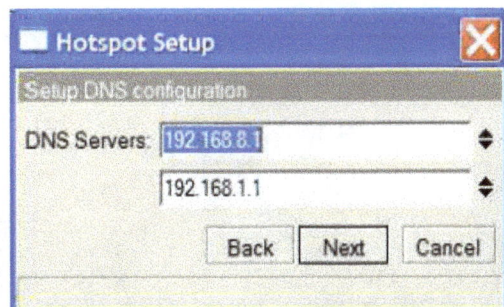

*Figure 20. Hotspot Interface Setup 6*

Enter the local DNS name for the MikroTik as given in figure 21

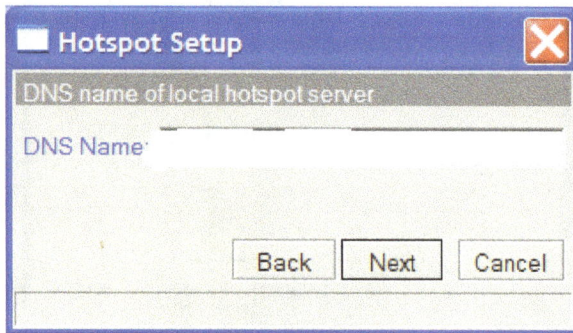

*Figure 21. Hotspot Setup 7*

Enter an admin hotspot user for local Winbox and Internet account as contained in figure 22. After this step, hotspot setup is completed successfully.

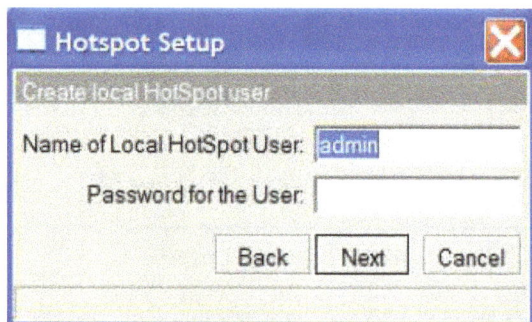

*Figure 22. Hotspot Setup 8*

## 3. Results

To edit the login, one needs to get the HTML page file that is responsible for displaying the MikroTik default hotspot web browser login. To get this file, fill the username on the web browser of the laptop with admin and password left blank. Once it logs in, type ftp://10.5.0.1 on the internet address bar. It prompts a login box again asking for username and password. Open the folder and copy the file "login.html" to the desktop of the laptop. Minimize the browser and open the file "login.html" with already installed Microsoft FrontPage in the laptop. With this web design skills, we came up with the design given in figure 23.

*Figure 23. A Customized Design Page for FUT, Minna Internet Service*

## 4. Discussion

The MikroTik Router OS version 2.9.7 used for this paper was well studied, proper understanding on how the system operates and troubleshooting was done. All important network terminologies and strategic network infrastructural planning were considered and put in place before project execution. The MikroTik hotspot server designed in this work has only one wireless interface due to consideration of where it would be placed in the department so that the signal would reach the students (final users). However, a distance of 30 meters radius was observed on testing its coverage area.

User account and session management was effectively managed securely, thanks to the undoubted efficiency of MikroTik Router OS. The routing speed and packets transfer rate was very fast due to the configuration of the desktop computer's processor and RAM installed in the computer.

The university Internet authentication system was designed to wirelessly transmit internet services across the whole campus, especially within the classrooms, lecture halls and staff offices. It also offers free surfing of the department's rich mini portal website to the students. It also serves as a virtual e-notice board to students and lecturers for disseminating vital information across the campus using wireless Wi-Fi technology.

## 5. Conclusions

The performance of this work after testing met the design specifications. In view of all that we have experienced, learnt and put into practice, we deemed it necessary to enumerate the importance of WI-FI technological development to the national building.

## 6. Recommendations

Considering our discoveries during the course of execution of this work, the following recommendations are made:

i.   Additional stages should be incorporated with the antenna design in the hotspot server of Federal University of Technology, Minna to boost its coverage on the campus by amplifying the signal the wireless PCI card is producing with an outdoor antenna and amplifier setup.

ii.  A radius server services should be implemented and activated to boost the management and accounting of this hotspot wireless server.

## References

[1]   Douglas, E. C. (2009). Computer Networks and Internets. Cisco Research Inc. United States: Pearson Prentice Hall Publisher.

[2]  Nafiu, L. A. (2011). Unpublished Lecture Notes on Net-Centric Computing. Federal University of Technology, Minna, Nigeria.

[3]  Jamrich, J. P. & Dan, O. (2001). Computer Concepts Illustrated Introductory Enhanced. United State: Gex Publishing Service.

[4]  Christian, M. and Arnaud, A. (2004). Media Freedom Internet Cookbook. United States: OSCE Publishers.

[5]  Wendell, O. (2004). Computer Networking: First Step.

United States: Cisco Press Publishers

[6]  MikroTik Reference Manual available at www.wiki.mikrotik.com

[7]  MikroTik Limited (2007). MikroTik Router OS Reference Manual.

[8]  Encyclopedia of Internet available at www.en.wikipedia.com

[9]  Federal University of Technology, Minna website available at www.futminna.edu.ng

# Effect of temperature variation on microstrip patch antenna and temperature compensation technique

**Sarita Maurya[1], R. L. Yadava[1], R. K. Yadav[2]**

[1]Department of Electronics & Communication Engineering, Galgotia's College of Engineering and Technology, Greater Noida, India
[2]Department of Electronics & Communication Engineering, I.T.S Engineering College, Greater Noida, India

**Email address:**

sarita8815@gmail.com(S. Maurya), rly1972@gmail.com(R. L. Yadava), ravipusad@gmail.com(R. K. Yadav)

**Abstract:** This paper describes the effect of temperature variation on microstrip patch antenna for different substrate materials. Eight materials are chosen as substrate and the effect of temperature variation is studied on each substrate material. A technique of temperature compensation has also been developed with substrate height variation. It is also seen that the change in resonance frequency due to variation of temperature can be compensated by varying the height of the substrate. The proposed antenna is designed and simulated by using HFSS software.

**Keywords:** Microstrip Patch Antenna, Substrate Material, Temperature Variations, Compensation

## 1. Introduction

The microstrip patch antenna has number of advantages over conventional antennas, such as low profile, light weight and low production cost. For better antenna performance, a thick dielectric substrate having a low dielectric constant is more desirable since this provides better efficiency, larger bandwidth, and better radiation [1]. It is well known that antenna is a very important component of a communication system.

One of the most important requirements for an antenna is to provide the stability of antenna parameters under meteorological factors alteration, in particular, under temperature conditions change. During a year the environment temperature depending on the geographical position can vary in the range from -50°C to +50°C. Under the influence of solar radiation or other factors the top limit for the antenna heating temperature can reach a much larger values [2]. In some applications, a microstrip antenna is required to operate in an environment that is close to what is defined as a room or standard conditions. However, antennas often have to work in harsh environments characterized by temperature variations. In this case, the substrate properties suffer from some variations [3]. Antenna ground plane performance depends on its temperature, humidity and conductivity. Antenna temperature and the temperature of its environment correlate to radiation resistance. According to

"*Antenna and Wave Propagation*", the noise temperature of a lossless antenna is equal to the sky temperature and not the physical temperature. Higher temperatures equal a higher radiation resistance. This increases the signal loss of the antenna and interferes with the performance of the ground plane. The effect of that variation on the overall performance of a microstrip conformal antenna is very important to study under a wide range of temperature. For a microstrip antenna fixed on a projectile that fly at a long distance, the temperature will be an issue for the performance of that antenna. The temperature affects the dielectric constant of the substrate and also affects expansion of the material which increase or decrease the volume of the dielectric with increasing or decreasing the temperature. As the temperature increases, the effective dielectric constant is also increases for different materials used. On the other hand, the resonance frequency decreases with increasing temperature, while VSWR and return loss decreases as the temperature increases [4]. In this paper, a microstrip patch antenna operating at 3 GHz frequency is designed simulated and the effect of temperature variation on eight substrate materials (GaAs, FR4, Quartz, Polyimide and polyethylene, Rogers, Neltec, Teflon) is analyzed. A method of temperature compensation is presented with the increased height of the substrate. The effects of temperature changes on the performances; resonance frequency, input impedance, voltage standing wave ratio, and return loss of microstrip patch antenna have also been presented.

## 2. Temperature Sensitivity of RMSA

The resonant frequency of a MSA is sensitive to temperature variations. There are two major factors affecting the resonant frequency of a microstrip antenna exposed to large temperature variations [5]. The metallic expansion or contraction of the radiating patch due to a change in temperature affects the resonant frequency. With an increase in temperature, the metallic patch expands, making the effective resonant dimension longer and, therefore, decreasing the operating frequency. The relative frequency change for dimensional changes may be expressed in terms of linear dimensions or in terms of temperature changes. Most of the substrates which are generally used for microwave applications like Polytetra Fluroethylene (PTFE) based materials, Teflon/Fiberglass reinforced materials, and ceramic powder filled TFE (epsilon) materials exhibit a decrease in dielectric constant with an increase in temperature [6].

## 3. Design Procedure of Antenna

### 3.1. Antenna Design Specifications

The rectangular microstrip patch antenna has been designed by following procedure which assumes that the specified information includes;
- Resonant frequency ($f_r$): 3 GHz
- Substrate thickness ($h$): 1.6mm.
- Material used for patch and ground plane: Copper.
- Material used for dielectric substrate: GaAs, FR4, Quartz, Polyimide, Neltech, polyethylene, Rogers and Teflon.
- Substrate permittivity ($\varepsilon_r$): 12.9, 4.4, 3.78, 3.5, 2.6, 2.25, 2.17 and 2.1 respectively.

The design of the whole structure of microstrip antenna is explained below:
- Initially, select the desired resonant frequency of operation, height of substrate and dielectric constant of the substrate.
- Obtain width ($W$) of the patch.
- Obtain Length ($L$) of the patch after determining the Length Extension ($\Delta L$) and Effective dielectric constant ($\varepsilon_{eff}$) using following expressions; [7] & [8].

$$W = \frac{\lambda}{2}\left(\frac{\varepsilon_r}{2}\right)^{-1/2} \quad (1)$$

$$L = \frac{1}{2f_r\sqrt{\varepsilon_{eff}}\sqrt{\mu_0\varepsilon_0}} - 2\Delta L \quad (2)$$

Where

$$\Delta L = 0.412h\frac{(\varepsilon_{eff}+0.3)}{(\varepsilon_{eff}+0.258)}\frac{\frac{W}{h}+0.264}{\frac{W}{h}+0.8} \quad (3)$$

$$\varepsilon_{eff} = \frac{(\varepsilon_r+1)}{2}\frac{(\varepsilon_r-1)}{2}\frac{1}{(1+12\frac{h}{W})^{\frac{1}{2}}} \quad (4)$$

$$\frac{\Delta l_{th}}{l} = 7.2\times10^{-8}T^3 + 3.5\times10^{-8}T^2 + 0.013T - 0.26 \quad (5)$$

The relationship between temperature and $\varepsilon_r$ is given by

$$\varepsilon_r = 0.00072\,T + \varepsilon_r(\,at\,T = 27^oC) \quad (6)$$

As a result new length/width of the microstrip antenna will be calculated by

$$l = l_0 + \Delta l_{fringing} + \Delta l_{thermal} \quad (7)$$

## 4. Simulation Environment

The software used to model and simulate the microstrip patch antenna is High Frequency Structure Simulator (HFSS) software. HFSS is a high-performance full-wave electromagnetic (EM) field simulator for arbitrary 3D volumetric passive device modeling. Ansoft HFSS employs the Finite Element Method (FEM), adaptive meshing, and brilliant graphics to give unparalleled performance and insight to all 3D EM problems. Ansoft HFSS can be used to calculate parameters such as S parameters, resonant frequency, and fields.

The length and the width of the patch and the ground plane found to be: $L = 30.4$ mm, $W = 24.16$ mm, $L_g = 90$ mm, $W_g = 76$ mm.

## 5. Patch Antenna without Temperature Variation: Simulation Results

The rectangular microstrip patch antenna is designed for the mentioned antenna specifications and also simulated for the substrate materials. Without any Temperature variation the simulation results for eight substrate material are given in Tables 1 and 2.

*Table 1. List of materials used with their dielectric constant, resonant frequency and return loss*

| | **Without temperature variation** | | | |
|---|---|---|---|---|
| S.No. | Name of material | Dielectric constant ($\varepsilon_r$) | Resonant frequency (GHz) | Return loss (dB) |
| 1 | GaAs | 12.9 | 1.671119 | -13.4131 |
| 2 | FR4 | 4.4 | 2.823873 | -16.8336 |
| 3 | Quartz | 3.78 | 3.030467 | -14.2122 |
| 4 | Polyimide | 3.5 | 3.151085 | -11.6956 |
| 5 | Neltec | 2.6 | 3.616027 | -12.4734 |
| 6 | Polyethylene | 2.25 | 3.802671 | -13.7565 |
| 7 | Rogers | 2.17 | 3.850083 | -14.7122 |
| 8 | Teflon | 2.1 | 4.010851 | -14.0404 |

*Table 2. List of materials used with their dielectric constant, VSWR and Gain*

| S. No. | Name of material | Dielectric constant (εr) | VSWR | Gain (dB) |
|--------|------------------|--------------------------|------|-----------|
| 1 | GaAs | 12.9 | 1.542828 | 3.51 |
| 2 | FR4 | 4.4 | 1.336412 | 4.19 |
| 3 | Quartz | 3.78 | 1.483581 | 7.36 |
| 4 | Polyimide | 3.5 | 1.703245 | 5.99 |
| 5 | Neltec | 2.6 | 1.624204 | 6.27 |
| 6 | Polyethylene | 2.25 | 1.51635 | 5.31 |
| 7 | Rogers | 2.17 | 1.450435 | 5.76 |
| 8 | Teflon | 2.1 | 1.495633 | 5.40 |

# 6. Patch Antenna with Temperature Variation: Simulated Results

The results of four materials for variation of temperature from 27 0C to 117 0C are discussed below with their graphs between resonance frequencies vs. return loss. Here, the effect of temperature variation on dielectric constant, resonant frequency and return loss is discussed. The results for another four materials are same. The results tabulated in table 3 are obtained by variation of temperature:

Case (1): Variation of temperature on the antenna for material (M1): FR4 (εr =4.4)

In case (1), the dielectric material FR4 (εr = 4.4) has been considered for designing of the antenna, and simulated results are represented in Figure (1) and Table 3. Figure (1) shows that the first curve of resonant frequency is the actual resonant frequency of the antenna without temperature variation.

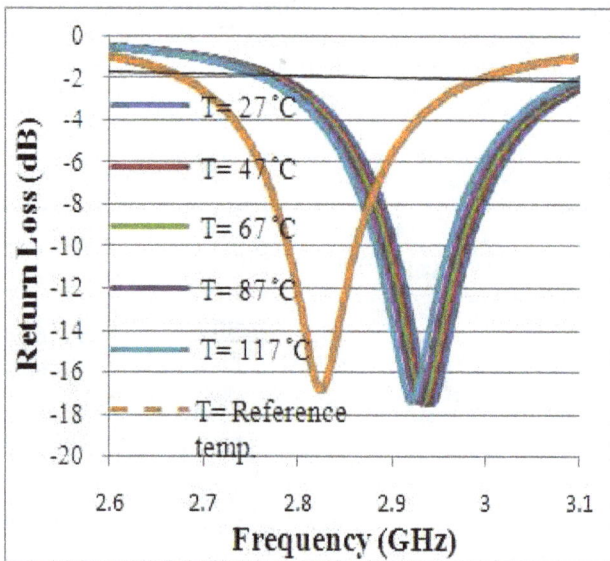

*Figure 2. Gain of FR4 at various temperatures*

*Table 3. Effect of temperature variation on FR4*

| S. No. | Temperature ($^0$C) | $\varepsilon_{rt}$ | $f_r$ (GHz) | $S_{11}$ | VSWR | Gain (dB) |
|--------|---------------------|--------------------|-------------|----------|------|-----------|
| 1. | 27 | 4.0506 | 2.9415 | -17.503 | 2.329 | 4.40 |
| 2. | 47 | 4.0650 | 2.9365 | -17.471 | 2.338 | 4.41 |
| 3. | 67 | 4.0794 | 2.9315 | -17.439 | 3.469 | 4.40 |
| 4. | 87 | 4.0938 | 2.9265 | -17.407 | 2.355 | 4.39 |
| 5. | 117 | 4.1154 | 2.9190 | -17.357 | 2.369 | 4.38 |

It is clear that on increasing the temperature the value of dielectric constant of substrate material increases and resonant frequency decreases. Decrease in resonance frequency led to the increase in return losses. The actual resonance frequency of FR4 material without any temperature variation is 2.823873 GHz, return loss of -16.8336 and VSWR of 1.336412. It is also clear that on increasing the temperature the resonant frequency decreases towards the value of actual resonant frequency. Figure (2) shows the gain of antenna for FR4 substrate from temperature 27 $^0$C to 117 $^0$C. The gain of antenna is not much affected by temperature variation.

Case 2: Variation of Temperature on the antenna for material (M$_2$): Quartz (ε$_r$ = 3.78)

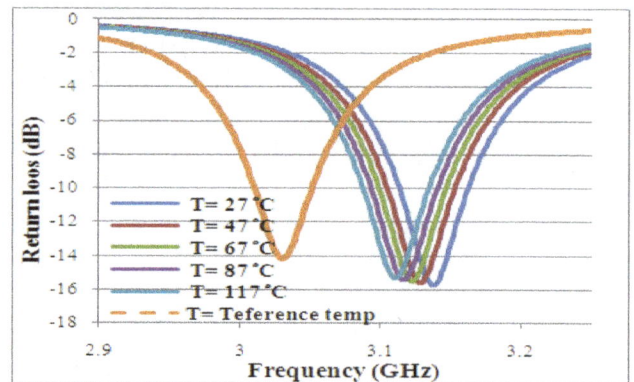

*Figure 1. Return loss at various temperature of the antenna for FR4 substrate*

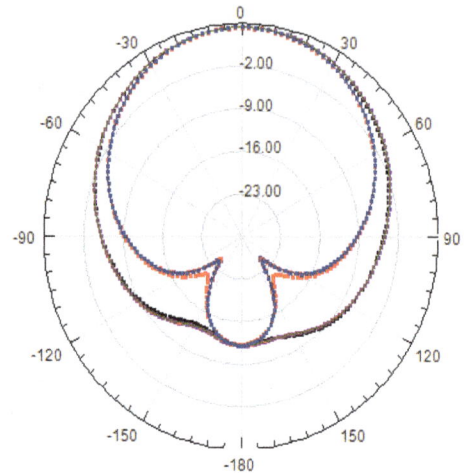

*Figure 3. Return-loss of various temperature of the antenna for Quartz substrate*

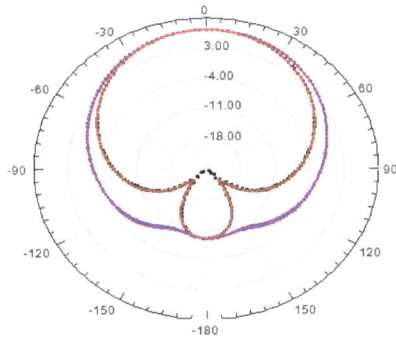

*Figure 4. Gain of Quartz at various temperatures*

**Table 4.** *Effect of temperature variation on Quartz*

| S. No. | Temperature ($^0$C) | $\varepsilon_{rt}$ | fr (GHz) | $S_{11}$ | VSWR | Gain (dB) |
|---|---|---|---|---|---|---|
| 1. | 27 | 3.5105 | 3.1373 | -15.750 | 1.389 | 7.30 |
| 2. | 47 | 3.5249 | 3.1298 | -15.634 | 1.396 | 7.31 |
| 3. | 67 | 3.5393 | 3.1261 | -15.577 | 1.402 | 7.31 |
| 4. | 87 | 3.5537 | 3.1183 | -15.461 | 1.407 | 7.32 |
| 5. | 117 | 3.5753 | 3.1106 | -15.345 | 1.412 | 7.32 |

In case 2, the dielectric material Quartz ($\varepsilon r = 3.78$) has been considered for designing of the antenna, and simulated results are represented in Figure (3) and Table 4. Figure (3) shows that the first curve of resonant frequency is the actual resonant frequency of the antenna without temperature variation. It is clear that on increasing the temperature the value of dielectric constant of substrate material increases and resonant frequency decreases. Decrease in resonance frequency led to the increase in return losses. The actual resonance frequency of Quartz material without any temperature variation is 3.030467 GHz, return loss of -14.2122 and VSWR is of 1.483581. It is also clear that on increasing the temperature the resonant frequency decreases towards the value of actual resonant frequency. Figure 4 shows the gain of antenna for Quartz substrate from temperature 27 0C to 117 0C. The gain of antenna is not much affected by temperature variation.

Case 3: Variation of Temperature on the antenna for material (M3): Polyimide ($\varepsilon r = 3.5$)

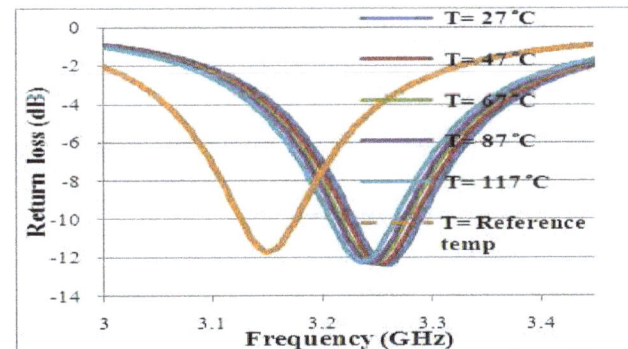

*Figure 5. Graph of Frequency (GHz) vs. Return Losses of Polyimide*

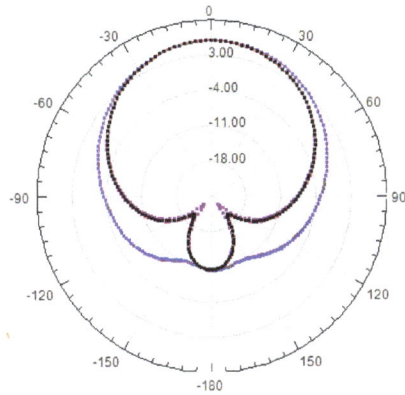

*Figure 6. Gain of Polyimide at various temperatures*

**Table 5.** *Effect of temperature variation on Polyimide*

| S. No. | Temperature ($^0$C) | $\varepsilon_{rt}$ | fr (GHz) | $S_{11}$ | VSWR | Gain (dB) |
|---|---|---|---|---|---|---|
| 1. | 27 | 3.2651 | 3.2587 | -12.390 | 1.632 | 5.94 |
| 2. | 47 | 3.2795 | 3.2537 | -12.357 | 1.635 | 5.95 |
| 3. | 67 | 3.2939 | 3.2462 | -12.312 | 1.639 | 5.96 |
| 4. | 87 | 3.3083 | 3.2412 | -12.265 | 1.641 | 5.97 |
| 5. | 117 | 3.3299 | 3.2312 | -12.202 | 1.645 | 5.97 |

In case 3, the dielectric material Polyimide ($\varepsilon r=3.5$) has been considered for designing of the antenna, and simulated results are represented in Figure (5) and Table 5. Figure (5) shows the first curve of actual resonance frequency of the antenna without temperature variation. The variation is found to be same as in previous case. The actual resonance frequency of Polyimide material without any temperature variation is 3.151085GHz, return loss= -11.6956 and VSWR= 1.703245. Figure (6) shows the gain of antenna for Polyimide substrate from temperature 27 0C to 117 0C. The gain of antenna is not much affected by temperature variation.

Case 4: Variation of Temperature on the antenna for material (M4): Teflon ($\varepsilon r = 2.1$)

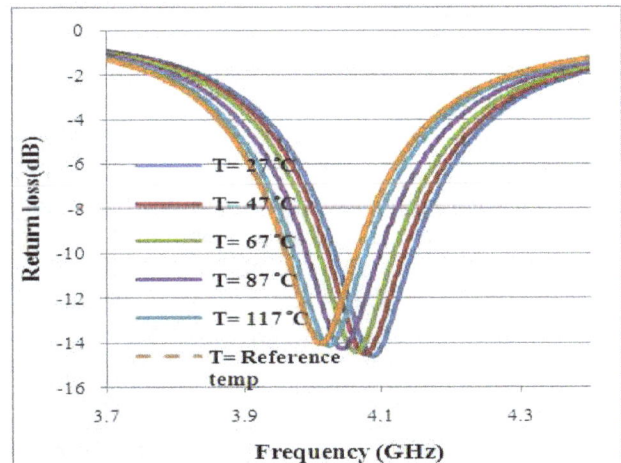

*Figure 7. Return loss of various temperature of the antenna for Teflon substrate*

In case 4, the dielectric material Teflon ($\varepsilon_r = 2.1$) has been considered for designing of the antenna, and simulated results are represented in figure (7) and Table 6 and as usual variations have been noticed.

The results for other four materials are also same.

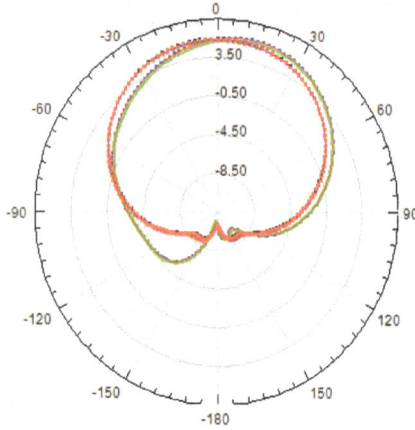

**Figure 8.** *Gain of Teflon at various temperatures*

**Table 6.** *Effect of temperature variation on Teflon material*

| S. No. | Temperature ($^0$C) | $\varepsilon_{rt}$ | fr (GHz) | $S_{11}$ | VSWR | Gain (dB) |
|--------|---------------------|--------------------|----------|----------|------|-----------|
| 1. | 27 | 2.0218 | 4.0834 | -14.748 | 1.458 | 5.19 |
| 2. | 47 | 2.0362 | 4.0734 | -14.488 | 1.463 | 5.20 |
| 3. | 67 | 2.0506 | 4.0534 | -14.340 | 1.470 | 5.24 |
| 4. | 87 | 2.0650 | 4.0434 | -14.270 | 1.479 | 5.31 |
| 5. | 117 | 2.0866 | 4.0233 | -14.128 | 1.489 | 5.37 |

# 7. Temperature Compensation

To compensate the decrease in resonance frequency due to variation of temperature, the height of substrate is increased. Here, the substrate height of four materials (FR4, Quartz, Polyimide and Teflon) has been varied. The formula used for Temperature Compensation with variation of height of substrates is as follows;

$$h = \frac{W}{12}\left[\left(\frac{1}{\varepsilon_{eff}}\frac{(\varepsilon_r + 1)}{2}\frac{(\varepsilon_r - 1)}{2}\right)^2 - 1\right] \quad (8)$$

As the height of substrates is increased from its original value (1.6 mm), the dielectric constant decreases for FR4 and Quartz materials, hence, the resonance frequency decreases. But the dielectric constant increases for Polyimide and Teflon substrate and resonance frequency increases. In this way at a particular height of substrate for each material the actual resonance frequency is obtained. The results are tabulated in tables 7-10 are obtained with variation of height of the substrate for four materials:

### 7.1. Temperature Compensation with variation of height of substrate of FR4

Since, the original height of the substrate is 1.6 mm. Here, the height of FR4 Substrate is increased from 1.6 mm to 4. 0

and then to 4.8 mm and found that at $h = 4.6$ mm, $\varepsilon_{rt} = 3.7976$, $fr = 2.823873$ GHz. This is the actual resonance frequency of FR4 substrate material without temperature variation. This confirms that temperature compensation can be achieved by increasing the height of substrate material.

**Table 7.** *Temperature compensation of FR4 material*

| S. No. | Height of Substrate(mm) | $\varepsilon_{rt}$ | fr (GHz) | $S_{11}$ |
|--------|-------------------------|--------------------|----------|----------|
| 1. | 4 | 3.8431 | 2.851419 | -23.8728 |
| 2. | 4.5 | 3.8048 | 2.831386 | -17.2563 |
| 3. | 4.6 | 3.7976 | 2.823873 | -16.5136 |
| 4. | 4.7 | 3.7906 | 2.821369 | -15.6219 |
| 5. | 4.8 | 3.7837 | 2.818865 | -14.8711 |

### 7.2. Temperature Compensation with Variation of Height of Substrate of Quartz

Here, the height of Quartz Substrate is increased from 1.6mm to 6.7mm and found that at $h = 6.6$ mm, $\varepsilon_{rt} = 3.2227$, $fr = 3.030467$GHz. This is the actual resonance frequency of Quartz substrate material without temperature variation.

**Table 8.** *Temperature compensation of Quartz material*

| S. No. | Height of Substrate(mm) | $\varepsilon_{rt}$ | fr (GHz) | $S_{11}$ |
|--------|-------------------------|--------------------|----------|----------|
| 1. | 5 | 3.2969 | 3.126461 | -18.4251 |
| 2. | 6 | 3.2481 | 3.070952 | -27.1579 |
| 3. | 6.5 | 3.2268 | 3.041736 | -18.4557 |
| 4. | 6.6 | 3.2227 | 3.030467 | -17.5575 |
| 5. | 6.7 | 3.2187 | 3.02379 | -16.6711 |

### 7.3. Temperature Compensation with Variation of Height of Substrate of Polyimide

Here, the height of Polyimide Substrate is increased from 1.6mm to 3.5mm and then to 4mm and found that at $h = 3.73$ mm, $\varepsilon_{rt} = 3.1509$, $f_r = 3.151085$ GHz. This is the actual resonance frequency of Polyimide substrate material without temperature variation.

**Table 9.** *Temperature compensation of polymide material*

| S.No. | Height of Substrate(mm) | $\varepsilon_{rt}$ | fr (GHz) | S11 |
|-------|-------------------------|--------------------|----------|-----|
| 1. | 3.5 | 3.1657 | 3.163606 | -38.0245 |
| 2. | 3.7 | 3.1528 | 3.153589 | -26.7789 |
| 3. | 3.72 | 3.1515 | 3.153589 | -26.3730 |
| 4. | 3.73 | 3.1509 | 3.151085 | -25.8475 |
| 5. | 4 | 3.1345 | 3.138564 | -20.3086 |

### 7.4. Temperature Compensation with Variation of Height of Substrate of Teflon

Here, the height of Teflon Substrate is increased from 1.6 mm to 2 mm and then to 4 mm and also found that at $h = 2.8$ mm, $\varepsilon_{rt} = 2.0401$, $fr = 4.010851$GHz. This is the actual resonance frequency of Teflon substrate material without temperature variation.

**Table 10.** *Temperature compensation of Teflon material*

| S.No. | Height of Substrate(mm) | $\varepsilon_{rt}$ | fr (GHz) | $S_{11}$ |
|-------|------------------------|------|----------|----------|
| 1. | 2 | 2.0694 | 3.904508 | -12.8160 |
| 2. | 2.5 | 2.0504 | 3.928548 | -13.1583 |
| 3. | 2.8 | 2.0401 | 4.010851 | -11.2272 |
| 4. | 3 | 2.0336 | 3.933222 | -19.1445 |
| 5. | 4 | 2.0055 | 3.826377 | -21.1603 |

The Tables 7, 8, 9, and 10 also shows that the decrease in resonance frequency due to variation of temperature can be compensated by increasing the height of the substrate. But the variation of height is different for each material.

# 8. Conclusions

In this paper, the rectangular microstrip patch antenna is designed and simulated with the temperature variation on different substrates materials suitable for wireless sensor network. The effect of temperature dependent of the substrate varies its dielectric constant and resonance frequency. Due to increase in temperature the dielectric constant of substrate material decreases from original dielectric constant of substrate because of this effect the resonance frequency increases from the value of operating frequency for all the substrate materials. For FR4 material, the actual resonance frequency without any temperature variation is 2.823873 GHz. After increasing the temperature up to 117 $^0$C the resonance frequency increases to 2.9190 GHz, Which is compensated by increasing the height of substrate material from 1.6 mm to 4.6 mm. For Quartz material, the actual resonance frequency without any temperature variation is 3.030467 GHz. After increasing the temperature up to 117 0C the resonance frequency increases to 3.1106 GHz, Which is compensated by increasing the height of substrate material from 1.6 mm to 6.6 mm. For Polyimide material, the actual resonance frequency without any temperature variation is 3.151085 GHz. After increasing the temperature up to 1170C the resonance frequency increases to 3.2312 GHz, Which is compensated by increasing the height of substrate material from 1.6 mm to 3.73 mm. For Teflon material, the actual resonance frequency without any temperature variation is 4.010851 GHz. After increasing the temperature up to 117 0C the resonance frequency increases to 4.0233 GHz, Which is compensated by increasing the height of substrate material from 1.6mm to 2.8mm. Therefore, it has been

concluded that the increase in resonant frequency due to temperature variation can be compensated by the variation of height of substrate material. It has been clear from the above analysis that if the temperature of environment changes (increases) the temperature compensation can be done by increasing the height of substrate material and microstrip patch antenna can be used in that environment successfully. Also the above substrate materials can be preferred for antenna designing in that environment where the temperature conditions changes considerably.

# References

[1]   M., Kumar, Sinha, M. K., Bandyopadhyay, L. K., & Kumar, S. (n.d.). "Design of a Wideband Reduced Size Microstrip Antenna". In. Retrieved from Union Radio-Scientifique Intemationale.

[2]   N. I .Voytovich, A. V. Ershov, V .A. Bukharin, N. N. Repin, "Temperature effect on cavity antenna parameters," General Assembly and Scientific Symposium, 2011 XXXth URSI , 13-20 Aug. 2011,pp-1-4.

[3]   P. Kabacik, and M. Bialkowski, "The temperature dependence of substrate parameters and their effect on microstrip antenna performance," IEEE Trans. on Antennas and Propag., Vol. 47, No. 6,Jun. 1999, pp. 1042-1049.

[4]   R. K. Yadav, J. Kishor and R. L. Yadava, "Effects of Temperature Variations on Performances of Microstrip Antenna", International Journal of Networks and Communication 2013, Vol.3, issue 1, pp.21,24.

[5]   M. A. Weiss, "Temperature compensation of microstrip antennas," in IEEE Antennas Propagation Soc. Int. Symp. Dig., Losa Angeles, CA, June 1981, vol. 1, pp. 337–349.

[6]   K. Carver, and J. Mink, "Microstrip antenna technology," IEEE Transaction on Antennas and Propagation, Vol. AP.29, No. 1, January 1981 ,pp.1-24.

[7]   Md. M. Ahamed, *etal.* "Rectangular Microstrip Patch Antenna at 2GHZ on Different Dielectric Constant for Pervasive Wireless Communication". International Journal of Electrical and Computer Engineering (IJECE).Vol.2, No.3, June 2012, pp. 417 - 424.

[8]   A. Elrashidi, K. Elleithy and H. Bajwa, "The performance of a cylindrical microstrip printed antenna for TM10 mode as a function of temperature for different substrates," International Journal of Next-Generation Networks (IJNGN), Vol.3, No.3, September 2011, pp.1-18.

[9]   S. Babu and G. Kumar, "Parametric study and temperature sensitivity of microstrip antennas using an improved linear transmission line Model, " IEEE Transactions on Antennas and Propagation, Vol. 47, No. 2, pp.221-226, February 1999.

# 7

# Location agent: a study using different wireless protocols for indoor localization

**Ana Régia de M. Neves[1], Humphrey C. Fonseca[2], Célia G. Ralha[2]**

[1]Dept. of Electrical Engineering, University of Brasília, Brazil
[2]Dept. of Computer Science, University of Brasília, Brazil

**Email address:**

regianeves@unb.br (A. R. de M. Neves), humphrey.fonseca@gmail.com (H. C. Fonseca), ghedini@cic.unb.br (C. G. Ralha)

**Abstract:** Context-aware systems have received greater interest in the computing community. In order to provide relevant services at context-aware applications, the first task is to locate the user, what can be done preferably dynamically and intelligently. However, indoor mobile users localization is not a trivial problem, since it involves checking various devices, transmitting signals simultaneously on the same radio frequency, with possibly the three existing wireless network protocols: Wi-Fi, Bluetooth and ZigBee. In this direction, this paper presents an agent-based architecture with the Location Agent module defined for context-aware applications that uses three artificial neural network algorithms trained for the different protocols: backpropagation, backpropagation with momentum and levenberg–marquardt. Considering the research experimental aspects, a study is presented to compare the neural network algorithms including performance, regression analysis, precision and accuracy. The results indicate that the backpropagation algorithm trained with Bluetooth provides better accuracy (the average error of 0.42 meters) and the backpropagation trained with Wi-Fi provides better precision (73%). We consider our approach promising since the Location Agent has a quality of service component associated with the neural network algorithms that can choose the best received signal strength to locate indoor users.

**Keywords:** Indoor User Location, Context-Aware Systems, Multiagent System, Quality Of Service, Artificial Neural Network

## 1. Introduction

Mobile devices allow user's mobility and give seamless access to computing resources while moving from one point to another. In this way, there is an increasing interest in context-aware systems that exploit the context to understand various current aspects of users situation to interact with the environment in a more intelligent way [1].

One of the most popular mobile services context aware applications are location based services (LBS) [2]. The LBS are value-added services that use the mobile location to provide relevant information or service to the user at an specific location. Such services can be required outdoor and indoor environments. In outdoor environments LBS are possible due to global positioning system (GPS) that enables accurate positioning. Nowadays, most of mobile devices are equipped with a GPS-receiver. But GPS is not suitable to track mobile users (MU) in indoor environments

with acceptable accuracy, since signals might be attenuated by roofs and walls. In this case, the usage of other sensor on mobile devices, such as wireless local area network (WLAN), Bluetooth and ZigBee can be exploited as alternative positioning sensors in indoor environments. At present, indoor positioning remains an open research problem and is our focus of study.

According to [3] existing indoor positioning techniques can be grouped in two main approaches: (i) their level of precision and installation of specialized additional infrastructure such as ultra wide band or ultrasonic which the precision is often high, but are expensive and unsuitable for large scale deployment; and (ii) exploiting already existing network infrastructure, for instance WLAN or inertial sensors for positioning, which the precision is limited, but the system are more economical and can be deployed with few additional expenses. Obviously, wireless network is not designed for the purpose of indoor user localization. However, measurements of the received signal strength (RSS), which is founded on the decay law of the received signal versus the distance, imply the location of any MU. On the other hand, suffer from signal attenuation and noise due to hardware characteristics, exacerbated by environmental factors such as walls, furniture and people in motion.

In this research work we focus on methods using existing infrastructures, such as: (i) Wireless fidelity (Wi-Fi) over IEEE 802.11, that includes IEEE 802.11a/b/g standards for WLAN and allows users to access the Internet at broadband speeds when connected to an access point (AP) or in ad hoc mode; (ii) Bluetooth over IEEE 802.15.1, based on a wireless radio system designed or short-range and cheap devices to replace cables for computer peripherals; and (iii) ZigBee over IEEE 802.15.4, defines specifications for low rate *wireless personal area network* (LR-WPAN) for supporting simple devices that consume minimal power and typically operate in the personal operating space (POS) of 10 m.

According to [4], there are two main groups for indoor location based on wireless network existing infrastructure: (i) signal propagation model and information about the geometry of the building to convert RSS to a distance measurement, with knowledge of the coordinates of the WLAN access points (APs), the method of trilateration can be used to estimate the location of the MU; and (ii) location fingerprinting technique that matches the obtained RSS values with a database containing previously captured RSS patterns in the area of interest. According to [5] location fingerprinting includes two phases: offline and online. In the offline phase, the area of interest is divided into grid points and values of the RSS from multiple APs. The RSS data is collected for a certain period of time and stored in a database, called radio map. During the online phase, the server compares by an algorithm the measured RSS fingerprint to fingerprints stored in the radio map to obtain mobiles location on the grid. The coordinates associated with the fingerprint that provides the smallest distance, for instance Euclidean distance, is returned as the estimated position.

Artificial neural networks (ANN) can also be used to establish a relationship between pattern of RSS samples and location [6]. In [7] an ANN is defined as a massively parallel model, with distributed processors made up of simple processing units called neurons. A variety of ANN models have been proposed and all of them must be trained. Basically, there are two types of training, supervised and unsupervised. Whereas the ANN supervised training knows the desired output, the unsupervised explores correlations between patterns in the data, and organizes into categories from these correlations.

A multilayer feed-forward ANN consists of multiple layers of units connected by directed links and uses supervised training. In this research work there are three supervised learning algorithms being used for training our networks: backpropagation (BP) [8], backpropagation with momentum (BPM) [9] and Levenberg-Marquardt (LM) [10].

The BP algorithm has been widely used as a supervised learning algorithm in feed-forward multilayer ANN based on the Gradient Descent method. That attempts to minimize the error of the network by moving down the gradient of the error curve as stated. However, the BP has a slow convergence. Consequently, many faster algorithms were proposed to speed up the convergence of the BP and can be grouped in two main categories [11]: (i) uses heuristic techniques developed from an analysis of the performance of the standard steepest descent algorithm, for instance BPM, to prevent instabilities caused by a too-high learning rate; and (ii) uses standard numerical optimization techniques, such as LM algorithm, which is an approximation to Newton's method, suitable for training small and medium-sized problems.

According to [12], the use of ANN improved performance and accuracy since they are capable of tackling noisy measurements and are widely used when the correlation between the input and output values of a system is unclear or subject to noise data.

Indoor location accuracy can be measured by the error between the estimated position and the actual position of the mobile device. T*his characteristic* can be improved by Quality of Service (QoS), since it chooses the best RSS to locating users and being one of the key evaluations for LBS. Generally, the QoS is measured in relation to accuracy, response time, availability and consistency [13].

The multi-agent system (MAS) approach is also interesting to apply for indoor localization using together with the ANN and QoS [14]. According to [15], a MAS is characterized by the existence of multiple agents that interact autonomously and work together to solve a problem or to achieve a common goal. In this way, a MAS can be used as an alternative to deal with the complexities of developing an indoor location system, which aggregates multiple wireless protocols. The agents have intrinsic attributes, such as: (i) the ability to perceive and act on the environment; (ii) the possibility of achieving individual goals; (iii) the ease of communicating with other agents; (iv) the ability to perform actions with some level of autonomy; and (v) the ability to provide services. Besides, the MAS architecture can incorporate reactive and deliberative agents that take decisions automatically at execution time.

Our previous work presented and evaluated an agent-based user location module -- the Location Agent Module (LAM), based on fingerprinting technique that uses ANN and QoS in the existing wireless network infrastructure to improve accuracy of indoor *user* location [14]. Furthermore, we compare results of three ANN supervised learning algorithms (BP, BPM and *LM)* to indoor localization trained with data of three wireless protocols: Wi-Fi, Bluetooth and ZigBee.

The rest of the paper is organized as follows: in Section 2, we discuss the state of the art related to indoor localization; the LAM architecture is presented in Section 3; Section 4 discuss experimental results, while conclusions and future works are presented in Section 5.

## 2. State of the Art

Wireless RSS fingerprinting has become the most promising indoor positioning technique because of its easier deployment and lower cost compare to other methods [5]. In this section some approaches to fingerprint-based indoor localization are going to be presented.

To overcome the limitations suffer from signal attenuation and multipath in Wi-Fi signatures, [16] proposes a new approach to fingerprint based indoor localization that leverages FM broadcast radio signals by achieving localization accuracies similar or better to the one achieved by Wi-Fi signals. Besides augmenting the wireless signature, the SI4735 FM receiver has been used and provides three additional signal quality indicators: signal-to-noise ratio, multipath and frequency offset. Authors show that localization accuracy can be further improved by more than 5%. Moreover, combined FM and Wi-Fi signals to generate wireless fingerprints, the localization accuracy increases as much as 83% compared to Wi-Fi RSSI used alone as a signature.

Considering the ANN domain, [17] presents a system to find the location of mobile sensor nodes in the harsh, uncertain, dynamic and noisy conditions using some beacon nodes. To achieve this an ANN was developed and validated through some experiments in real world prone to different sources of noise and signal attenuation. The ANN is trained using BMP algorithm and the results are compared with the trilateration technique.

The system presented in [18] combines two different Wi-Fi approaches to user localization: fingerprinting and trilateration using three known AP coordinates detected on the user's device to derive the position. The combination of the two approaches enhances the accuracy of the user position in an indoor environment allowing LBS to be deployed more effectively.

[19] present a software architecture designed for a hybrid location system supporting multiple technologies simultaneously. To demonstrate the application of the architecture and its platform, the paper introduces two case studies based on real deployments: (i) associates ZigBee and Ultra Wide Band mobile nodes, plus the accelerometer; and (ii) uses RSS measurements and a fingerprinting location algorithm.

Another approach is based on collaborative localization of mobile users; for instance, using Bluetooth protocol to improve accuracy and coverage indoors and improve power consumption by duty-cycling GPS outdoors from nearby neighbors [20].

Differing from the presented initiatives, this researchwork focus in the modularization of complexity and interaction of multiple wireless protocols in the indoor environment through the use of an agent-based model. The agent model is appropriate to represent complex interactions among different entities in the indoor environment and permits that decisions are taken automatically at execu-

tion time. In addition, the different wireless protocols can be contextualized into different agent types that interact in the environment. Another clear difference in our approach from the cited work [16-20] is the implicit way to deal with the QoS mechanisms, without additional hardware and extra costs.

## 3. Architecture Overview

In [21] previous work, we developed a prototype for indoor user localization using MAS approach. This prototype allowed to define the necessary characteristics for a more complex architecture to context-aware systems. Therefore, we improved our previous architecture with the definition of the LAM, together with the ANN and QoS [14].

Fig. 1 presents LAM composed of three modules and one knowledge base:

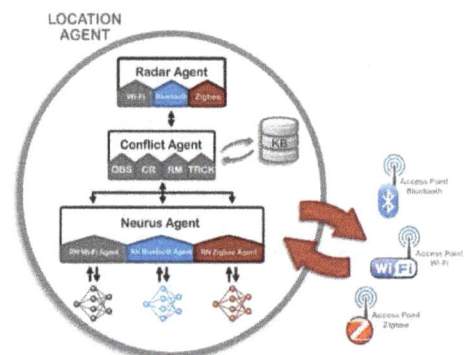

***Figure 1.*** *LAM agent-based architecture.*

- Radar Agent – it starts the process of indoor users loctlization being composed of three sub-modules: (i) Wi-Fi; (ii) Bluetooth; and, (iii) ZigBee. These sub-modules are responsible for monitoring the environment to gather RSS information and send to the Conflict Agent;
- Conflict Agent – request the RSS for the Radar Agent and is composed of four sub-modules: (i) Observation (OBS), responsible for the RSS request; (ii) Conflict Resolution (CR), responsible to decide which position use, giving access to Knowledge Managed-based; (iii) Knowledge Managed-based (RM), responsible for the inference using If-Then rule statements; and (iv) Tracking (TRCK), responsible for monitoring the user;
- Neurus Agents - receive the RSS of Wi-Fi, ZigBee and Bluetooth infrastructure and transfer them to its respective ANNs, in order to check its own QoS and return location for Conflict Agent.

A prototype was implemented to validate the LAM architecture presented in Fig. 1, which is composed of a set of layers to analyze the map of the RSS signals and the absolute coordinates as illustrate in Fig. 2. Each layer functionality is described in the sequence:

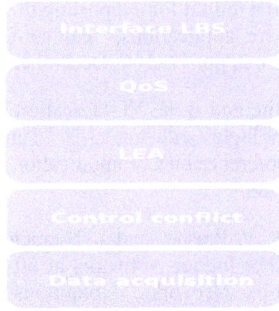

*Figure 2. Modular layered architecture.*

- LBS interface – an interface that enables communication between the prototype and other applications that make use of the indoor service localization;
- QoS – analyze data from the LEA layer using two levels of QoS predefined: (i) the maximum acceptable error, defined as 1.5 meters according to [11]; and (ii) the signal strength evaluation for assessing the proximity of an access point AP with the acceptable error level defined as 1.0 meters;
- LEA – each independent ANN are responsible to calculate indoor location and the result is obtained from absolute coordinates of users mobile device;
- Control conflict – prepare data obtained at the data acquisition layer and implement two phases: (i) calibration of the account for the hardware differences; and (ii) online to prepare the input values to be used in LEA layer;
- Data acquisition – collect the RSS signals of mobile users in a particular area in the indoor environment.

The LAM agent-based architecture was defined to be a flexible, adaptable and extensible one, since new agents can be added considering other wireless protocols, ANN algorithms and different QoS levels.

## 4. Experimental Results and Discussion

The results presented in [14] show that LAM architecture using different wireless protocols and QoS is suitable for indoor localization process. The use of QoS allows to choose the best signal and leads to more accurate location; also improve the service levels offered. Moreover, as cited in Section I, we compare three ANN algorithms, such as BP, BPM and *LM*. Our experimental tests evaluated the ANN performance, regression analysis, precision and accuracy.

The first experimental test for both ANNs was empirically defined with three types of layers (input, hidden and output) varying the number of neurons at the hidden layer from 100 to 10. The MLPs are trained using the BP, BPM and LM algorithms. The activation function used for the hidden and output layers is the hyperbolic tangent (tansig). The training parameters were defined with: (i) 10,000 epochs; (ii) target of mean squared error (MSE) equal to zero, since the process of training was intended to be permanent, MSE is the av-

erage squared difference between outputs and target; and (iii) the learning rate was set to 0.1 according to the literature [7].

Training automatically stops when generalization stops improving, as indicated by an increase in the MSE. Moreover, regression analysis was performed to measure the correlation between outputs and targets.

In order to calculate the accuracy ($\rho$), our work uses Euclidean distance that measures the distance between an online RSS value (X,Y) obtained (ob) and the offline training database RSS (X,Y) expected (expc) to each point of the grid, as presented in Equation 1.

$$\rho_i = \sqrt{\left(X_{obt} - X_{expec}\right)^2 + \left(Y_{obt} - Y_{expc}\right)^2} \quad (1)$$

Afterwards, we calculate the value of the average error for the point $\rho$, defined by $\rho_p$, as in Equation 2. Note that, n represents samples per point in the environment.

$$\rho_P = \frac{1}{n}\sum_{ip}^{n}\rho_i \quad (2)$$

The accuracy ($\rho$) is calculated as in Equation 3.

$$\rho_P = \frac{1}{n}\sum_{pp}^{n}\rho_p \quad (3)$$

In order to determine the precision ($\delta$), the standard deviation of the samples ($\rho$i) grouped by the expected point ($\delta$p) is calculated as in Equation 4.

*Table 1. ANN algorithms comparison.*

| ANN algorithms and Wireless protocols | Performance | Regression R | Precision (%) | Accuracy |
|---|---|---|---|---|
| Backpropagation Wi-Fi | 0,0055 | 0,9859 | 73 | 3,35 |
| Backpropagation Bluetooth | 0,0068 | 0,985 | 63 | 0,42 |
| Backpropagation ZigBee | 0,0064 | 0,9844 | 67 | 2,03 |
| Backpropagation with momentum Wi-Fi | 0,0059 | 0,9844 | 72 | 0,64 |
| Backpropagation with momentum Bluetooth | 0,085 | 0,9859 | 65 | 2,95 |
| Backpropagation with momentum ZigBee | 0,061 | 0,9595 | 71 | 1,83 |
| Levenberg-Marquardt Wi-Fi | 0,0062 | 0,998 | 69 | 0,89 |
| Levenberg-Marquardt Bluetooth | 0,0064 | 0,995 | 70 | 2,15 |
| Levenberg-Marquardt ZigBee | 0,0095 | 0,886 | 67 | 1,94 |

$$\sigma = \frac{1}{m} \sum_{pp}^{n} \left(1 - \frac{\sigma_p}{\rho_p}\right) \times 100 \qquad (4)$$

Table 1 summarizes the main differences among the three ANN algorithms considering performance, regression, precision and accuracy.

For all ANN, the test results to regression analysis presents a stronger level of correlation with a positive increasing value that indicates a good fit. In terms of precision, BP trained with Wi-Fi protocol presents a better result: 73%, followed by 72% to BPM also trained with Wi-Fi and 71% to BPM trained with ZigBee protocol. Considering the accuracy, the average error of 0.42 meters to BP trained with Bluetooh, 0.64 meters to BPM trained with Wi-Fi and LM 0.89 meters to LM trained with Wi-Fi.

The results presented in [14] are based on our best ANN architectures and additional level of QoS that improved accuracy in 17% to the Wi-Fi, 11% to the Bluetooth and 21% to the ZigBee protocols.

## 5. Conclusions

As cited in Section I, the goal of context-aware applications is to perceive the users mobile location and dynamically offer them personalized services. In order to achieve this goal, devices and agents must be integrated and cooperate what is possible through the use of an agent-based approach.

Our previous approach presented and evaluated an agent-based user location module -- the LAM, based on fingerprinting technique that uses ANN and QoS in the existing wireless network infrastructure [14]. These experiments were a consequence of our first attempt to develop a prototype for indoor localization using MAS approach [21]. Furthermore, we compared results of three supervised ANN learning algorithms (BP, BPM and LM) to indoor localization, trained with data of three wireless protocols (Wi-Fi, Bluetooth and ZigBee) based on fingerprinting location technique.

Analyzing the experimental results, we can say that even though BP algorithm has a slow convergence, it has better performance in indoor localization than the BPM and the LM algorithms. In addition, the use of QoS together with ANN can improve the accuracy of the results as presented in the previous work [14]. For future work, we intend to study the accuracy improvement possibilities by adding other levels of QoS and integrating them to the ZigBee, WiFi and Bluetooth protocols. Also using semantic resources, such as ontologies, to characterize different contexts we plan to improve the provision of relevant context-aware services to mobile users.

## References

[1]  A. K. Dey, G. D. Abowd and D. Salber, "A conceptual framework and a toolkit for supporting the rapid prototyping of context-aware applications," Human-Computer Interaction, vol. 16, no. 2, pp. 97–166, December 2001.

[2]  S. Wang, J. Min, and B. K. Yi, "Location based services for mobiles: technologies and standards," Tutorial for IEEE International Conference on Communication (ICC'08), Beijing, China, May 2008.

[3]  V. Q. Bien, R. V. Prasad, and I. Niemegeers, "Handoff in Radio over Fiber Indoor Networks at 60 GHz," Journal of Wireless Mobile Networks, Ubiquitous Computing, and Dependable Applications, vol. 1, no. 2/3, pp. 71–82, 2010.

[4]  Y. Gu, A. Lo, and I. Niemegeers, "A survey of indoor positioning systems for wireless personal networks," IEEE Communications Surveys & Tutorials, vol. 11, no. 1, pp. 13–32, First Quarter 2009.

[5]  V. Honkavirta, T. Perälä, S. Ali-Löytty, and R. Piché, "A comparative survey of WLAN location fingerprinting methods," Proceedings of the 6th Workshop on Positioning, Navigation and Communication (WPNC'09), pp. 243–251, March 2009.

[6]  C. S. Chen, "Artificial Neural Network for Location Estimation in Wireless Communication Systems," Sensors, vol. 12, no. 3, pp. 2798–2817, March 2012.

[7]  S. O. Haykin, "Neural Networks and Learning Machines," 3rd ed., Prentice Hall, 2008.

[8]  Y. Chauvin, and D. E. Rumelhart (ed.) "Backpropagation: Theory, Architectures, and Applications," 1st ed., Psychology Press, 1995.

[9]  V. Phansalkar, and P. S. Sastry, "Analysis of the back-propagation algorithm with momentum," IEEE Transactions on Neural Networks, vol. 5, no. 3, pp. 505–506, May 1994.

[10] M. T. Hagan, and M. B. Menhaj, "Training feed forward networks with the Marquardt algorithm," IEEE Transactions on Neural Networks, vol. 5, no. 6, pp. 989–993, November 1994.

[11] S. Lahmiri, "A comparative study of backpropagation algorithms in financial prediction," International Journal of Computer Science, Engineering and Applications (IJCSEA), vol. 1, no. 4, pp. 15–21, August 2011.

[12] M. Altini, D. Brunelli, E. Farella, and L. Benini, "Bluetooth indoor localization with multiple neural networks," Proceedings of the 5th IEEE International Synposium on Wireless Pervasive Computing (ISWPC'10), pp. 295–300, May 2010.

[13] Y. Chen, J. Qi, Z. Sun, and Q. Ning, "Mining user goals for indoor location-based services with low energy and high QoS," Computational Intelligence, vol. 26, no. 3, pp. 318–336, August 2010.

[14] H. C. Fonseca, A. R. d. M. Neves, and C. G. Ralha, "A user location case study using different wireless protocols," Proceedings of the 9th ACM international symposium on Mobility management and wireless access (MobiWac'11), pp. 143–146, October 31–November 4, 2011.

[15] M. Wooldridge, "Introduction to Multi-Agent Systems," 2nd ed., John Wiley & Sons, Ltd., 2009.

[16] Y. Chen, D. Lymberopoulos, J. Liu, and B. Priyantha, "FM-based indoor localization", Proceedings of the 10th International Conference on Mobile Systems, Applications, and Services (MobiSys12), pp. 169–182, June 2012.

[17] M. Gholami, N. Cai, R. W. Brennan, "An artificial neural network approach to the problem of wireless sensor network localization," Robotics and Computer-Integrated Manufacturing, vol. 29, no. 1, pp. 96–109, February 2013.

[18] S. Chan, and G. Sohn, "Indoor localization using Wi-Fi based fingerprinting and trilateration techniques for LBS applications", Proceedings of the 7th International Conference on 3D Geoinformation, pp. 1–5, May 2012.

[19] J. Rodas, V. Barral, and C. J. Escudero, "Architecture for multi-technology real-time location systems," Sensors, vol. 13, no. 2, pp. 2220–2253, February 2013.

[20] A. Barreira, P. Sommer, B. Kusy, and R. Jurdak, "Collaborative localization of mobile users with bluetooth: caching and synchronisation," Special Issue on the 3rd International Workshop on Networks of Cooperating Objects, CONET 2012, vol. 9, no. 3, pp. 29–31, July 2012.

[21] A. R. de M. Neves, L. T. Maia, C. G. Ralha, and R. P. Jacobi, "Prototype for indoor localization basedMon multiAgent System," Proccedings of the International Conference on Intelligent and Advanced Systems (ICIAS'10), pp. 1–4, IEEE Computer Society, June 2010.

# 8

# Routing protocol of wireless sensor network (ED-LEACH)

**Elnaz Shafigh Fard, Mohammad H. Nadimi**

Faculty of Computer Engineering, Najafabad branch, Islamic Azad University, Isfahan, Iran

**Email address:**

shafighfard@azaruniv.edu (E. S. Fard), nadimi@iaun.ac.ir (Md. H. Nadimi)

**Abstract:** This paper presents a new version of leach protocol called "ED-LEACH" which aims to conserve energy considering BS distance and central distance of cluster by the passage of time. Compared to leach protocol, ED-LEACH has more longevity. The researcher evaluates both LEACH and ED-LEACH in Matlab. Results proves that energy consumption decreases about 20% Moreover in protocol of wireless sensor the first node lasts 6 times longer than leach protocol.

**Keywords:** Clustering, Energy, Distance, LEACH Protocol, ED-LEACH Protocol

## 1. Introduction

Wireless Sensor Network (WSN) consists of hundreds and even thousands of tiny devices called sensor nodes distributed autonomously to monitor physical or environmental conditions such as temperature, sound, vibration, pressure and motion at different locations. Energy plays an important role in wireless sensor networks because nodes are battery operated. Consequently many protocols have been proposed in order to minimize the energy consumption of these nodes. [1][2]

Each node in a sensor network is typically equipped with one or more sensors, a radio transceiver or other wireless communications device, a small microcontroller, and an energy source, since in most WSN applications battery supplies the required energy [6] plays an important role in wireless sensor network. Preserving the consumed energy of each node is an important goal that must be considered while developing a routing protocol for wireless sensor networks.

Many routing protocols have been proposed in the literature such as LEACH, PAMAS.[4][6][12].

Leach is considered as the most popular routing protocol that use cluster based routing in order to minimize the energy consumption. IN this paper, we propose an improved version of on the Leach Protocol that further enhance the Power consumption, simulation results show that our protocol outperforms Leach protocol in term of energy consumption and overall throughput.(potential)

In section 2-3the researcher will discuss the Leach protocol in details. Section 4 will present the related literature, in section 5; investigator will introduce our proposed protocol ED-LEACH. In section 6 evaluation of protocol and presentation of the simulation results will be done. In section 7the conclusion will be given

## 2. Energy Routing Analysis of Routing Protocols

### 2.1. Directs Communication Protocol:

Each sensor sends its data directly to base station. If base station is far away, large amount of transmission power from each node will quickly drain to nodes+ reduction system longevity.

### 2.2. Minimum-Energy Routing Protocol:

Each node acts as a router for other nodes data in addition to sensing data.

- Some variations of this protocol only consider energy of transmitter + neglect energy dissipation of receivers.
- The intermediates are chosen in this waybecause transmit amplifier energy is minimized –as shown in formula.
- Every node sends a message to the closest node on its way to base station.
- When transmission energy is on the same order as receive energy (transmission destination is short), direct method of transmission is more energy efficient.

### 2.2.1. Cluster-Based Routing

The basic objective of any routing protocol is to make the network useful and efficient. A cluster is based onthe sensor nodes of protocol groups where each group of nodes has a CH or a gateway. Sensed data is sent to the CH instead of being sent it to the BS; CH performs some aggregation function on data it receives, then sends it to the BS where these data are needed.[2][3][4].

Two of the most well-known hierarchical protocols are LEACH, PAMAS and PEGASIS. All these show significant reduction in the overall network energy over other non-clustering protocol.

Hierarchical routing protocols are designed to reduce energy consumption by localizing communication within the cluster and data are aggregated to reduce transmissions to the BS.

# 3. Leach Protocol

Low Energy Adaptive Clustering Hierarchy (LEACH) is the first hierarchical cluster-based routing protocol for (WSN) which partitions the nodes into clusters, in each cluster a dedicated node with extra privileges called Cluster Head (CH) is responsible for creating and manipulating a TDMA (Time Division Multiple Access) schedule and sending aggregated data from nodes to the BS where these data are needed using CDMA (Code Division Multiple-Access). Remaining nodes are cluster members.

The operation of LEACH is divided into rounds. Each of these rounds consists of a set-up and a steady-state phase.

During the set-up phase, cluster-heads are determined and the clusters are organized. During the steady-state phase data transfers to the base station occur. This paper presents an improvement of LEACH's cluster-head selection algorithm; the formation of clusters is not the topic of this paper.

We use the same radio model as stated in [10] with Eelec=50nJ/bit as the energy being dissipated to run the transmitter or receiver circuitry and εamp=100pJ/bit/m² as the energy dissipation of the transmission amplifier. Transmission ($ETx$) and receiving costs ($ERx$) are calculated as follows:

$$E_{Rx}(k) = E_{elec}k \qquad E_{Tx}(k,d) = E_{elec}k + \varepsilon_{amp}kd^{\lambda}$$

Set-up Phase
(1) Advertisement Phase
(2) Cluster Set-up Phase
Steady Phase
(1) Schedule Creation
(2) Data Transmission

### 3.1. Setup Phase

Each node decides independent of other nodes if it will convert into a CH or not. This decision is taken into account when the node served as a CH for the last time (the node that hasn't been a CH for a long time is more likely to elect itself than nodes that have been a CH recently).

In the following advertisement phase, the CHs inform their neighborhood with an advertisement packet that they become CHs. Non-CH nodes pick the advertisement packet with the strongest received signal strength.

In the next cluster setup phase, the member nodes inform the CH that they become a member to that cluster with "join packet" containing their IDs using CSMA.

$$T(n) = \frac{P}{1 - P \times \left( r \bmod \frac{1}{P} \right)} \qquad \forall n \in G$$

$$T(n) = 0 \qquad\qquad\qquad \forall n \notin G$$

After the cluster-setup sub phase, the CH knows the number of member nodes and their IDs. Based on all messages received within the cluster, the CH creates a TDMA schedule, pick a CSMA code randomly, and broadcast the TDMA table to cluster members. After that steady-state phase begins.

### 3.2. Steady-State Phase

Data transmission begins; Nodes send their data during their allocated TDMA slot to the CH. This transmission uses a minimal amount of energy (chosen based on the received strength of the CH advertisement). The radio of each non-CH node can be turned off until the nodes are allocated to TDMA slot, thus minimizing energy dissipation in these nodes.

When all the data have been received, the CH aggregates these data and sends them to the BS.

LEACH is able to perform local aggregation of data in each cluster to reduce the amount of data that is transmitted to the base station.

Although LEACH protocol acts in a good manner, it suffers from many drawbacks namely;

• CH selection is random, that does not take into account energy consumption.

• It can't cover a large area.

• CHs are not evenly distributed; where CHs can be located at the edges of the cluster.

Since LEACH has many drawbacks, many researches have been conducted to make this protocol performs better.

### Longevity of a Micro Sensor Network

The definition of the longevity of a micro sensor network is determined by the kind of service it provides. Hence, three new approaches of defining longevity are proposed. In some cases it is necessary that all nodes survive as long as possible, since network quality decreases considerably as soon as one node dies. Scenarios for this case include intrusion or fire detection. In these scenarios it is important to know when the first node dies. The new metric First Node Dies (FND) de-notes an estimated value for this event for a specific network configuration. Furthermore, sensors can be placed in proximity to each other. Thus, adjacent sensors could record related or identical data. Hence, the loss of a single or few nodes does not automatically diminish the quality of service of the network.

In this case the new metric Half of the Nodes Alive (HNA) denotes an estimated value for the half-life period of a network. Finally, the metric Last Node Dies (LND) gives an estimated value for the overall lifetime of a network

# 4. Related Work

## 4.1. AHP Protocol

Three parameters; energy, movement, and distance of the central cluster have an impacts on cluster head selection .This algorithm undertakes doe central method.[7]

## 4.2. LEACH-C protocol

LEACH offers no guarantee about the placement and/or number of cluster heads. In [13], an enhancement over the LEACH protocol was proposed. The protocol, called LEACH-C, uses a centralized clustering algorithm and the same steady-state phase as LEACH. LEACH-C protocol can produce better performance by dispersing the cluster heads throughout the network. During the set-up phase of LEACH-C, each node sends information about its current location (possibly determined using GPS) and its energy level to the sink. In addition to determining good clusters, the sink needs to ensure that the energy load is evenly distributed among all the nodes. To do this, the sink computes the average node energy, and determines which nodes have energy below this average.

Once the cluster heads and associated clusters are found, the sink broadcasts a message that obtains the cluster head ID for each node. If a cluster head ID matches its own ID, the node is a cluster head; otherwise the node determines its TDMA slot for data transmission and goes to sleep until it is time to transmit data. The steady-state phase of LEACH-C is identical to that of the LEACH protocol.

## 4.3. E-LEACH Protocol

Energy-LEACH protocol improves the CH selection procedure. It makes residual energy of node as the main metric which decides whether the nodes turn into CH or not after the first round [9]. Same as LEACH protocol, E-LEACH is divided into rounds, in the first round, every node has the same probability to turn into CH, it means that nodes are randomly selected as CHs, in the next rounds, the residual energy of each node is different after one round communication and it is taken into account for the selection of the CHs. In other words, nodes with more energy will become a CHs rather than nodes with less energy.

## 4.4. TL-LEACH

In LEACH protocol, the CH collects and aggregates data from sensors in its own cluster and passes the information to the BS directly. CH might be located far away from the BS, so it uses most of its energy for transmitting because it is always on it will die faster than other nodes.

A new version of LEACH called Two-level Leach was proposed. In this protocol; CH collects data from other cluster members as original LEACH, but rather than

transfer data to the BS directly, it uses one of the CHs that lies between the CH and the BS as a relay station.[10]

## 4.5. M-LEACH Protocol

In LEACH, Each CH directly communicates with BS no the distance between CH and BS does not matter. It will consume lot of its energy if the distance is far. On the other hand, Multichip-LEACH protocol selects optimal path between the CH and the BS through other CHs and it uses these CHs as a relay station to transmit data over through them.

First, multi-hop communication is adopted among CHs. Then, according to the selected optimal path, these CHs transmit data to the corresponding CH which is nearest to BS. Finally, this CH sends data to BS.

M-LEACH protocol is almost the same as LEACH protocol, only makes communication mode from single hop to multi-hop between CHs and BS.[8]

## 4.6. HEED Protocol

That is a distributed protocol that is independent from how nodes are deployed and it depends on how much energy is remaining in every node.[11]

## 4.7. ICLA Protocol

A CLAB based is another protocol that counts of neighbors and energy are two parameters to select cluster head.

## 4.8. Distributed Clustering Algorithm

This algorithm, considers that nodes are fix and with especially weight. This algorithm combines some factors to select head cluster.

## 4.9. Minimal tree algorithm

In this algorithm, energy is not a factor for being head cluster.

## 4.10. Adaptive Clustering for Mobile Wireless

In this algorithm, the rank of connectivity and lower id of nodes are important factors to be cluster head.

## 4.11. Learning Automata-Based Clustering

By using learning automata, a way is introduced that remaining energy of nodes and count of neighbors are important factors to be cluster head.

# 5. Proposed Work

In this protocol, there are some assumptions that are included:

1. Time (round) has divided to equal parts.
2. All nodes work synchronously in related to sending data to CH.
3. All nodes are deployed in half cycle.
4. Radius of all cycles is equal.

5. Nodes are not mobile.

6. Energy of nodes is not the same and they can be different.

7. Nodes were deployed in half cycle monotonously.

Hypothesis of protocol: With considering the distance of nodes to the centre of cycle and distance to BS and energy of nodes and putting them in especially formula, energy consumption can be reduced compared to leach protocol.

### 5.1. ED-LEACH Protocol

According to last part, in LEACH protocol selecting cluster head is random, so in this way nodes which have a low energy can be selected at first and longevity of network will be shorter, but in ED-LEACH, to be a cluster head it has considered to distance between node and its neighbors, duty cycle of every section and energy of nodes will be explained in the next section.

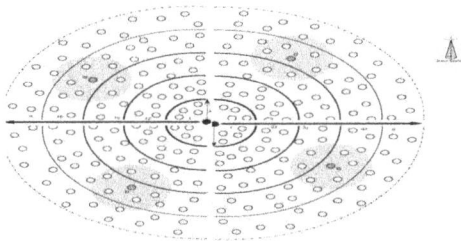

*Figure 1*

### 5.2. Algorithm Of Selecting Cluster Head

In this proposed protocol three parameters are very important to select cluster head energy, distance of BS or Sink, and distance of cycle centre are three fundamental parameters in selecting cluster head for the proposed protocol.

$f(t)$ is a very important factor to select cluster head it is calculated according to:

$$f(t) = C_{ave} * (1/sqrt(D_{ave})) * (2E_{ave}/t)$$

1-Init $D_{ave}$ is an average distance between $node_x$ that is the nearest one to base station and $node_y$ that is the furthest one to base station that it can be calculated at a central cycle.

$$D_{ave} = \sqrt{(X_{BS} - X_i)^2 + (Y_{BS} - Y_i)^2}$$

2-$C_{ave}$: shows average distance between the nearest and the furthest nodes to the centre of cycle.

$$C_{ave} = \sqrt{(X_f - X_n)^2 + (Y_f - Y_n)^2}$$

There are some expressions in relation to formula:

2- $D_{ave}$: is average value of the nearest distance and the furthest distance node to base station

3-$E_{ave}$: is an average energy of all nodes that will be decreased by node energy, it will be reduced.

$f(i)$ is a parameter that every node after calculating of its distance to base station and its distance to centre of

cycle and its energy is calculated according to the expression below:

$$f(i) = C_i * (1/sqrt(D_i)) * (E_i)$$

Starts to be compared to $f(t)$, and if it's $f(i)$ becomes more than $f(t)$, it can be selected as a head cluster, and in every t, some nodes can be selected as a head cluster. In this expression:

$D_i$: is a parameter that shows distance $node_i$ to BS.

$C_i$: is a parameter that shows distance $node_i$ to centre of cycle.

$E_i$: is a parameter that shows energy of $node_i$.

For the next round, $f(t)$ will be calculated again. Additionally, every node starts to calculate $f(i)$ and tries to compare with each other, if $f(i)$ is less than $(t)$, it loses its chance to be a cluster head.

### 5.3. Comparison of Routing Algorithms

Finally, in order to understand this proposed work with previous works, table1 shows some properties of algorithms and compares them with each other ,that priority of CH selection and how data are sent to BS and type of protocol have been shown in table1.

*Table 1. comparisons of proposed work with previous works*

|  | CH Selection | Data Transmission | Mobility |
|---|---|---|---|
| LEACH | Random | All CH sends directly to BS | Stationary |
| C-LEACH | Energy | CH sends directly to BS | Stationary |
| AHP | energy, movement, distance of central of cluster, | CH sends directly to BS | Mobility |
| TL-LEACH | Energy, distance | In Two level | station |
| E-LEACH | Two round(randomly, Energy) | CH sends directly to BS | Stationary |
| Proposed work | Energy, distance of BS, distance of central cluster | CH sends directly to BS | Stationary |

## 6. Experimental Study

### 6.1. Simulation and Result:

MATLAB is used as a simulation platform.

### 6.2. Simulation parameters:

*Table 2. Summery of the parameters used in the simulation experiments.*

| parameters | Values |
|---|---|
| Rounds | 1000 times |
| Topology size | 60*60 m^2 |
| Number of nodes | 300 |
| Node distribution | Randomly |
| Base station position | 65*70 (x,y) |
| E(Amp) | 0.0013*0.000000000001 |
| E(tx) | 50*0.000000001 |
| Initial node power | Randomly 0.01<power<0.9 joule |
| k | 400b |

### 6.3. First Evaluation

For evaluation of our proposed work, we compare our algorithm with leach protocol. Input data in first experimentation is our expression and our output variable is count of live nodes that shows lifetime of sensor network.

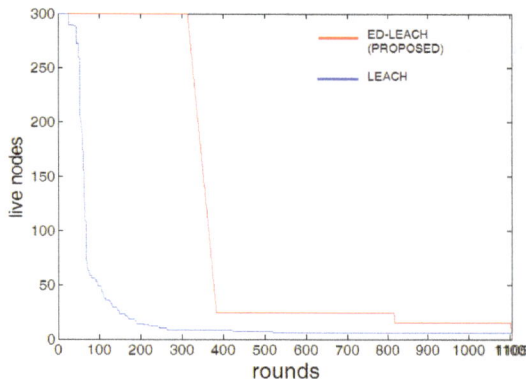

*Figure 2*

As can be seen in this picture, our proposed work has three times more lifetime than leach protocol.

### 6.4. Second Evaluation

In under figure below, consumed energy until 500 rounds, is shown.

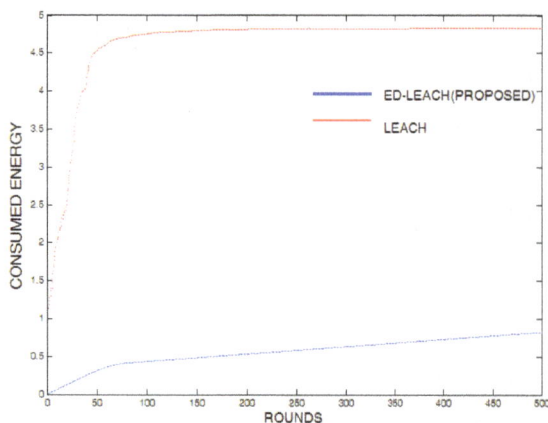

*Figure 3*

The figure shows that, consumed energy in leach is more than ED-leach (proposed work).Almost consumed energy in ED-LEACH is 1/5 times less than LEACH protocol.

## 7. Conclusion

In this paper ‹ to make longevity duration for (WSN) a very important factor of quality service, ED-LEACH is reduced as a way of clustering that considers about distance to BS, and distance to Centre of cycle and energy of nodes. Results of experimental energy consumption has been decreased about 20% and in this proposed protocol the first node death times 6 times longer in comparison with leach protocol.

## 8. Future Work

A lot of good solution ‹ introduced but in the future being a cluster head can be with using fussy logical ways.

## References

[1]   B. Williams and T. Camp."Comparison of Broadcasting Techniques for Mobile Ad hoc Networks", 2002.

[2]   Xiang-Yang Li and Ivan Stojmenovic. "Broadcasting and topology control in wireless ad hoc networks", July 8, 2004.

[3]   W. Heinzelman, A. Chandrakasan and H. Balakarishnan, "Energy-Efficient Communication Protocols for Wireless Microsensor Networks," *Proceedings of the Hawaaian International Conference on Systems Science*, January 2000.

[4]   SeapahnMegerian and MiodragPotkonjak, "Wireless sensor networks," Book Chapter in Wiley Encyclopedia of Telecommunications, Editor: John G. Proakis, 2002.

[5]   Thiemo Voigt, Adam Dunkels, Juan Alonso, Hartmut Ritter and JochenSchiller."Solar-aware clustering in Wireless Sensor Networks", 2004.

[6]   Heinzelman W., Chandrakasan A., and Balakrishnan H.: "Energy-Efficient Communication Protocol for Wireless Microsensor Networks".2000.

[7]   E.Y.K wang,C.C.Youn.Analytic Hierarchy Process Approach for Identifying Relative Importance of Factors to 235 Improve Passenger Security Checks at Airports[C].Air Transport Management,2006,12(3):135-142.

[8]   Dissertation, Hang Zhou, Zhe Jiang and Mo Xiaoyan, "Study and Design on Cluster Routing Protocols of Wireless Sensor Networks", 2006.

[9]   Fan Xiangning, Song Yulin. "Improvement on LEACH Protocol of Wireless Sensor Network", 2007.

[10]  V. Loscri, G. Morabito, and S. Marano, A Two-Level Hierarchy forLow-Energy Adaptive Clustering Hierarchy, DEIS Department, University of Calabria

[11]  O. Younis and S. Fahmy, "HEED: A Hybrid Energy-Efficient Distributed Clustering Approach for Ad Hoc Sensor Networks," IEEE Transactions on Mobile Computing, vol. 3, no. 4, Oct-Dec 2004.

[12]  S. Singh and C. S. Raghavendra,"PAMAS – Power Aware Multi-Access protocol with Signaling for Ad Hoc Networks,"ACM SIGCOMM, Computer Communication Review, July, 1998.

[13]  W. B. Heinemann et al., "An Application-Specific Protocol Architecture for Wireless Micro sensor network.

[14]  Farajzadeh, N. and meybodi, M .R,"learning Automata – based clustering algorithm for sensor networks", proceedings of 12 the annual csi computer conference of Iran, shahidbeheshti university, Tehran.2007

[15]  Esnaashari, M and Meybodi, M.R.,"A Cellular learning Automata based Clustering Algorithm for Wireless sensor Networks", sensor letters, 2008.

# 9

# Technical issues for IP-based telephony in Nigeria

A. A. Ojugo.[1], R. Abere.[1], B. C Orhionkpaiyo.[1], E. R. Yoro[2], A. O. Eboka[3]

[1]Department of Mathematics/Computer, Federal University of Petroleum Resources Effurun, Delta State
[2]Department of Computer, Delta State Polytechnic Ogwashi-Uku, Delta State
[3]Dept. Of Computer Sci., Federal College of Edu, (Technical), Asaba, Delta State

**Email address:**
ojugo_arnold@yahoo.com(A.A. Ojugo), arnoldojugo@yahoo.com(A. A. Ojugo), areuben2001@yahoo.com(R. Abere),
orhionkpaiyo@yahoo.com(B. C Orhionkpaiyo), rumerisky@yahoo.com(E. R. Yoro), andre_y2k@yahoo.com(A. O. Eboka)

**Abstract:** Data transmission in IP-networks, with all its challenges and associated prospects, is today a major milestone in the convergence of information and communication technology (ICT). IP-telephony simply refers to the data transfer that encapsulates data (in forms of audio, data, voice and video) media or traffic so that they are transported over data network via IP network technologies. IP telecommunications defines a wider range of applications, technologies and convergence networking that refers to a multi-service network that allows the integration of data, voice and video solutions onto a converged network infrastructure via the use of hardware, software and open-source apps/protocols (such as H.323, Megaco/H.248 signaling and transport protocols) used to setup, control and manage both data, voice and video sessions. The study selected area is the Federal University of Petroleum Resources Effurun in Delta State, in which the VoIP was implemented and issues in its implementation were discussed. These include: delay, packet loss and jitters. Jitters and Packet loss can be curbed via increase in bandwidth allocation; latency can be minimized but not totally eradicated by constantly upgrading the network equipment in terms of speed. The many benefits of IP communications cannot be over-emphasized as it includes advanced dial tone, rich media streaming, unified messaging, IP contact center etc while providing the advantages such as resilience, economy, flexibility, mobility and productivity for users. Even with its many pending challenges, IP communications continues to proffer advancements in mobile computing, apps to support mobile users and telecommuting apps that provide users the much needed merits with soft-phones, soft agents and VPN services. It is recommended that an organization intending to harness its full potentials should join forums and user-groups that will constantly update them such as Multi-Switch Forum (MSF) – in a bid to help them improve the efficiency and effectiveness of their IP implementation.

**Keywords:** Communications, Data, Media, Traffic, Voice, Data Services

## 1. Introduction

IP-based telephony is simply data transfer or transport of data (in forms of voice, file, audio and video) over telecommunication media using the Internet Protocol (IP). Communication is the key to fast development with shared ideas, giving birth to new ones to provide a platform for exponential innovation in ICT. The development in IP has drastically provided a new, different alternative set for provision of long distance calls and data transfer with 3-basic factors: (a) IP is a packet network that only uses resources on need, (b) methods exist that allows transmission of voice in a highly compressed format while retaining quality that the consumer demands cum allows the integration of speech, and (c) technology is available that allows integration of telephone networks with the speech compression systems and in turn, with IP. Thus, bringing closer the intent and concept of IP universality namely low cost, speech compression, quality and ease of implementation for IP long distance calls [10, 14, 16].

The birth of 3G-telecommunications has brought about use of interactive services, originally made possible via PSTN (public switched telephone network) – so that transfer is now *expansive* (over long distances) without video interactions. The advent of IP reinvents communications via new technologies, allowing users to exchange data in real time – via *IP telephony* and alternatives, which are fast gaining wide acceptance as open-source solutions to users as well as opening new functionalities for markets that previously required expensive PBXs – noting that IP telephony is more, software-based rather than hardware implemented [1, 4].

## 1.1. PSTN Overview

PSTN are dumb devices driven by an intelligent network – with signals sent between phone switches to indicate terminal status involved in calls and to connect or terminate calls.

PSTN/Traditional PBX still dominate the market, helping to provide allied and various ICT solutions to small/medium enterprises. PSTN employs devices such as plain old telephones (POTs), phone lines, fibre-optics, cables, microwave and satellites, cellular and undersea telephone cables, all aimed at interconnecting switch-centers, allowing all the users across the world to interact [8].

Its signals are carried over separate data network known as Common Channel Signalling (CCS) that uses Signalling System 7 protocol and Signalling Control Points (SCP) as its *database* to provide necessary data needed for advanced call processing. It also has a Centrex (Service Provider's end) that provides business class telephony feats, as built on a carrier's central office switch, allowing users to have either an analog or digital phone [3, 10]. Such telephones are connected via an copper pair back to a central office switch via:

1. Centrex lines (copper pairs access nodes linked to a central switch with PSTN Connectivity
2. Analog and/or digital handsets or terminals
3. Centrex feats such as caller ID, call centre feats, etc and an optional voice mail

## 1.2. Converged Network

The basic foundations of a converged network is in its capabilities and tools that allows users flexibility, secure and cost effectively data transfer services (in form of or a combination of data, voice and video) packets across the same transmission links via the same switching, routing and gateway platforms. Thus, the converged-network is more fault tolerable, flexible, secure, resilient, scalable and showcases easy manageability of data services and network devices [4, 16] – as its applications use open source underlying intelligent network services to ensure data transfer service quality, availability, reliability and security [5]. With electronic data systems invading every frontier, access becomes increasingly bothersome to have users being tethered with wires. Thus, the need for wireless devices with merits such as: (a) low-cost deployment, (b) broadcast the same data to many locations simultaneously, (c) deployment ease in hostile environment, and (d) mobile communication. Its demerits: (a) lesser data rates, (b) lesser reusable frequencies, and (c) more susceptible to interference [5].

PSTN today, is being replaced with IP, packet networks (with PSTN still used by non-IP devices connected to such network). Thus, PSTN conveniently support various data services via SCP and by Internet devices wishing to connect to such PSTN network. Conversely, IP telephony is a call process that uses signal technology built on Open IP-standards, providing users with end-to-end communication or data transfer (file, data, voice and video formats) – to aid data transfer services for public carrier networks and Internet users in general with an inter-operable networks [7].

[2] Note IP telephony involves a large family of communication standard to deliver voice and video services via open packet network and uses H.323 protocol to setup, control and manage sessions. The many benefits of IP-network as easily deployed services over legacy PSTN includes:

- IP Telephony allows call processing services to be located anywhere on network and use packet networks rather than TDM for services.
- Allows service delivery over converged network so that dual cabling and network equipments for connections to PBX or IP-PBX are not required.
- Carry traffic across different areas and various vendors spanning various countries - interfacing a variety of Internet and Telephone technolgies more flexibly, with greater benefits at reduced cost of implementation and operation.

Thus, the converged network simply helps to extend the capabilities of such an intelligent IP network over or into a PSTN network (when built to use underlying network) protocols and are based upon a server or network appliance [9, 15].

## 1.3. Applications of IP Telephony

[21] Note IP-telephony as applied in various forms:
1. Private users who employ IP telephony in VoIP for end-to-end voice calls over the Internet as they constantly trade quality, features and reliability for low cost. The density of users in this group is low even though many users (globally) take advantage of the technology, when compare with the PSTN call volumes.
2. Business users on private networks provided by telecomm/datacom providers – with high quality, feature rich services and reliability that comes at high cost. In contrast, when compared to users of PSTN call volumes – they are small.
3. IP trunking solution are used by long haul voice providers to connect islands of PSTN together to private IP networks, and users access these services via traditional black phones but the data is carried over an IP-network.

These applications continues to have its place in the future of IP networks and in general, telecommunications; But, we are yet to deal with issues of how the wider PSTN can be mitigated to an end-to-end voice infrastructure. Simply put, how do we provide a voice over IP solution that scales to PSTN call volumes and offer PSTN call quality, equivalent services and support such new innovative services.

## 1.4. Benefits of IP-Network

The benefits of converged intelligent network are derived

from its fundamental capabilities to provide these advantages:

1. Economy – Traditional PSTN uses expensive legacy systems as connecting elements, apps and technologies like DS1/DS0 line cards, trunk cards and digital signalling technologies. Conversely, IP networks allow users to build data services via Ethernet economics and Moore's law for rapid advances in computing performance. A merit of IP-network over traditional PSTN is cost in connecting of enterprise PBX system to PSTN via ports, cards and circuits. Thus, cost is lesser to provide connections to other sites and to other apps [17].

2. Flexibility – PSTN element/app(s) are mostly proprietary, monolithic and restrictive. Whereas IP network connections are made from virtual reach with resources distributed on demand to anywhere needed, and economies are gained via centralized gateway and server resources. Use of many types of media and applications to be brought together to facilitate communication. It supports broadband voice, front/back office integration and apps, outsourcing operation, mobility requirements, centralised management, telecommuting, moves/adds/changes, extension mobility, desktop integration and automation, enterprise directories and takes advantage of a plethora of emerging web services such as instant messaging, presence and mobility [19].

3. Security – It aims at achieving a higher degree of security for secure data transfer that are vastly superior over legacy voice technologies; while deploying and integrating wireless LAN apps, video surveillance, IP video on demand, streaming and rich media conferencing applications [17, 18].

4. Resilience – With business continuity and disaster recovery high, resiliency of connectivity and abilities provided by IP network keeps an organization connected make it an ideal candidate for survivable services. Its redundancy is built into intelligent layer2 and layer3 networking technologies and apps. Clustering and hot standby technologies, fault tolerant storage technologies like RAID, dual power supply and UPS systems are now common in the industry. IP offer superior failover, self-healing and redundant abilities that are easy to deploy, open standards based, and can support an organization's communication services. It offers reliability, availability and superior alternatives over legacy PSTN that are far more expensive, and are unable to provide same overall system resiliency needed for as broad a range of services and apps as in IP-network [18-19].

5. Productivity – Its focus shift from cost savings to enabling users become more productive with apps to

help accomplish higher quality communications more quickly and easily, has yielded a network of phone apps that employ any existing web or enterprise database on an IP network. Thus, end-users can take advantage of emerging web innovations, enterprise directories, e-mails, voice mail, fax, and general tools for programming communication rules. Voice recognition and soft-phone support via user desktops can be added to an IP-communications environment. While some of these possibilities exist in legacy PSTN, they are more expensive, less scalable, and more difficult to deploy [6].

# 2. Objectives

The study aims to address IP telephony issues implemented at Federal University of Petroleum Resources Effurun, Delta State, Nigeria; while ensuring that its many benefits are fully harnessed.

# 3. Methods and Materials

A critical, structured analysis of the existing hybrid PSTN/PBX system indicates that these must be addressed before effective implementation namely:

1. Difficulty to accomodate differences in disparate technologies and equipment
2. Difficulty to traverse geographic boundaries
3. Manage many sites centrally
4. Change the way resources are used on a network
5. Traverse regulatory boundaries
6. Deliver such new communication services using different media types
7. Provide level of integration, ease of use, access and management found in IP telephony systems.

### 3.1. Network Planning

Planning an IP-based network on FUPRE's existing hybrid PSTN/PBX – took into account that services must be accessed by 5000 users (both staff and students). The Nigerian University Commission (NUC) note that to aid effective learning, each department is encouraged to have a teacher-student ratio of 1:25; FUPRE has a ratio of 1:45 – except in General Studies and its related courses where number exceed. Currently, with eleven departments, four levels in Science and five levels in Technology respectively – the institution has a total of 2300 students, and 1210 staff (academic and non-academic) – a total of 5000-subscribers simultaneously.

### 3.2. Network Components

The Federal University of Petroleum Resources has 2-campuses of 3km apart. The VoIP design implemented will cater for 5000-user simultaneously and effectively provide the needed IP solution. This can be deployed in many different segments but mostly on backbones and enterprise networks. We adopt [3] next generation VOIP end-to-end architecture with additional constraints, implemented on FUPRE as in fig 1. There are three (3) service providers (SP) with varied functions as: (a) Service Provider 1 – will offer local access as a LEC to support IP phones and IP PBX systems using SIP and POTS phones via either an access gateway (Next-Gen DLC) or a subscriber gateway using the H-248 or MGCP protocol, (b) Service Provide 2 – will act as an interchange to support SIP and SIP –T or BICC signaling through its network, and (c) Service Provider 3 – will offer local access as LEC that only supports POTS phones using an Access Gateway. SIP signaling is supported but is terminated by the SIP server rather than using a SIP phone or other CPE devices. The functional specification of the network components, depends on the particular network architecture and LAN cabling – as some of these components may be combined into a single solution.

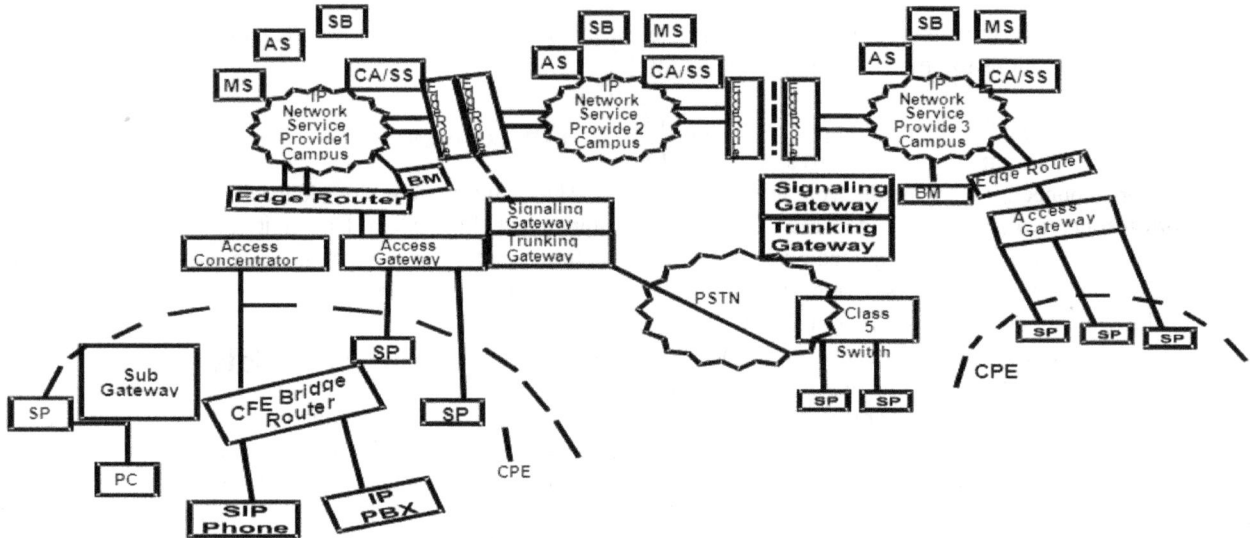

*Fig 1.* IP-Network Solution in FUPRE

For example, a combined signaling and trucking gateway. The *vision* of the project still remains – implementation of IP telephony, while the *scope* remains FUPRE and the function of each network component is described as thus:

1. Call Agent and SIP: Call agents (may include service logic to) provide an IP network with call logic, call control functions, maintains state of every call on network and offer supplementary services (like Call ID, waiting, forwarding etc) and also interact with servers to supply services that are not directly hosted on call agents. They participate in signaling, device control flows and provide details of each call to support building and reconciliation. Call agents are also called soft-switches, media gateway controller and call controllers – all of which conveys the emphasis on its ability to maintains call state as the common function.

2. SIP server provides equivalent function as the call agent with its primary role being to route and forward SIP requests, enforce policies (like care admission control) and maintain call details records; while SIP client conversely, provides similar function but originates and terminates SIP signaling rather than forwarding it to a SIP phone or other CPE devices.

3. Service Broker (SB) provides service distribution; coordination and control between all the application servers, media server, call agents and services that exist on alternate technologies (that is, create a parlay gateway and SCPS). It also allows consistent repeatable approach for controlling applications in conjunction with their service data and media resources to enable services and allow service to be reused with other services, to create value added services.

4. Application Server (AS) provides the service logic and execution for one or more applications or services that are not directly hosted on the call agent. It may provide voice mail or conference calling facilities typically, the call agent will route to the appropriate application server when a service is involved that which call agent itself cannot support.

5. Media Server use control protocol such as H.248/Megaco/MGGP, as supervised by a CA or AS to provide functions such as: (a) Play announcement, (b) provide 3-way call support, (c) Codec transcoding and voice detection, (d) Tone detect/generation, (e) Interactive Voice Response processing and (f) jaw processing.

6. Signaling/Trunking Gateway: Signaling gateway acts as a gateway between the call agent signaling and SS7 protocol-based PSTN, to provide the required signaling translation and a cross platform for and between different packets carrier domains.

7. Trunking gateway provides the gateway between the carrier IP network and Time-Division Multiplexing PSTN. Thus, provides transcoding from the packet based VOIP onto a TDM network (a reason for our choice of design – merging both into one single solution). Trunking gateway is controlled by a Call Agent via a device control protocol (H.248/MGCP).

8. Access Gateway/Concentrator: Access gateway provides support for POTS; while access concentrator helps to terminate service provided from WAN links used. Thus, in a DSL (Digital Subscriber Link) network, its function is to combine the capabilities of DSLAM (Digital Subscriber Line for Autonomous Machines) with direct POTS (plain old telephone systems).

9. Edge Router helps to route IP traffic onto the carrier backbone network and provides many other functions when combined with Access Concentrator into one single solution.

10. Bandwidth Manager is responsible for providing the required QOS (Quality Of Service) from the network, setting up and teasing down of bandwidth within the network as well as controlling the access of individual calls to this bandwidth while also helping to install the appropriate policy in edge routers to police media/data flow on a per call basis.

11. Gateways/Router/Bridge helps terminate user sessions from the WAN link (such as DSL, T1, fixed wireless, cable) and provides both voice ports and data connectivity. Bridges/routers and gateways perform same function with a simple difference that bridges/routers does not provide any native voice support, though voice services such as SIP phones can be bridged/routed via the device(s).

12. IP Phone/ PBX: IP phones, softphones and PBX systems provide voice services. They interact with the Call Agent/SIP Server in signaling protocol SIP, H.323 or a device control protocol Megaco (i.e. H.248).

In summary, these components are required to build the converged, and can be grouped into four as:

a. Intelligent Network infrastructure (like switches, routers, gateways etc), services, software and protocols – mostly dominated at SP's end (in hosted IP solution) to form the physical infrastructure and deliver intelligent network services such as security, QOS and resiliency.

b. Application provides new capabilities via integrated (audio-data-voice video) apps with improved call feats such as Media conferencing. Unified messaging and IP contacts. All of which work better in a secure IP network due to the trunkless/portless nature of IP telephony to use such new innovations and apps that continues to yield greater productivity.

c. End Point Client Devices are the access point from where users take advantage of IP network apps such as IP-phones, soft-phones (desktop systems with speakers), PDA (Personal Digital Assistants), mobile phones etc.

d. Call Processing – is the software that drives it all so as to run effectively network appliance servers on third party servers.

# 4. Implementation Issues

IP network growth has brought about radical transformation and improvement in the market, technology and open source software, allowing them to gain wide acceptance. The incremental design of implementation adopted, the technical issues experienced that threatens service quality are [4, 5, 14]:

### 4.1. Service Set

A crucial decision facing IP network deployment is the service set and design to be supported, which can be either minimal set, full scale PSTN equivalence or advance services for operators and carriers wishing to replace their current infrastructure with a new converged network for all subscribers. An important feat in the design is the choice of user terminal to be supported by the services to be offered to include POTS/black phones, IP phones, PBX and key systems, PC soft-phones (with web apps) etc.

### 4.2. Security

PSTN became resistant to security attacks with the advent of SS7 out-of-band signaling. IP nets are more susceptible to attacks to address 3-issues:

1. Invasion of Privacy – Callers in PSTN expect calls are private with no third party eavesdrop. PSTN achieves this via a physical security mechanism (wire from a user's home is only connected to local exchange or digital loop carrier and cannot easily accessed). Whereas, IP network uses different encryption measure to cater for such security issues via its cable/wireless media. E.g A5 cipher used in GSM or CDMA.

2. Denial of Services attack prevents a legitimate user access to the network feats and services. Though rare and extremely difficult in PSTN; But, are common in IP networks. Example includes sending false signaling message so that a call agent is fooled and bombarded with pings from a soft phone or other packets so frequently that it has no spare processing power to process legitimate request. A consequence of this is that sometimes – the soft-phone(s) can no longer pull data from the IP network. Also, hacking a subscriber gateway to send ftp or other data traffic as high priority voice traffic.

3. Theft of Service – is aimed at SP where the attacker simply wants to use a service without paying for it. Its most common form in PSTN is called subscriber fraud – where a user sets up an account with a

service provider (SP) using false billing data such as stolen credit card. other forms are more technical that utilizes black boxes or similar to fool the network into providing free service. With VoIP, bandwidth is still a limited resource even with low packet loss and jitters required for good voice quality.

Thus, converged network needs to be protected from users who misuse high-priority bandwidth. E.g. Two SIP user agents can setup a direct call between them, to access high-priority bandwidth and bypass SIP server(s) to not get billed.

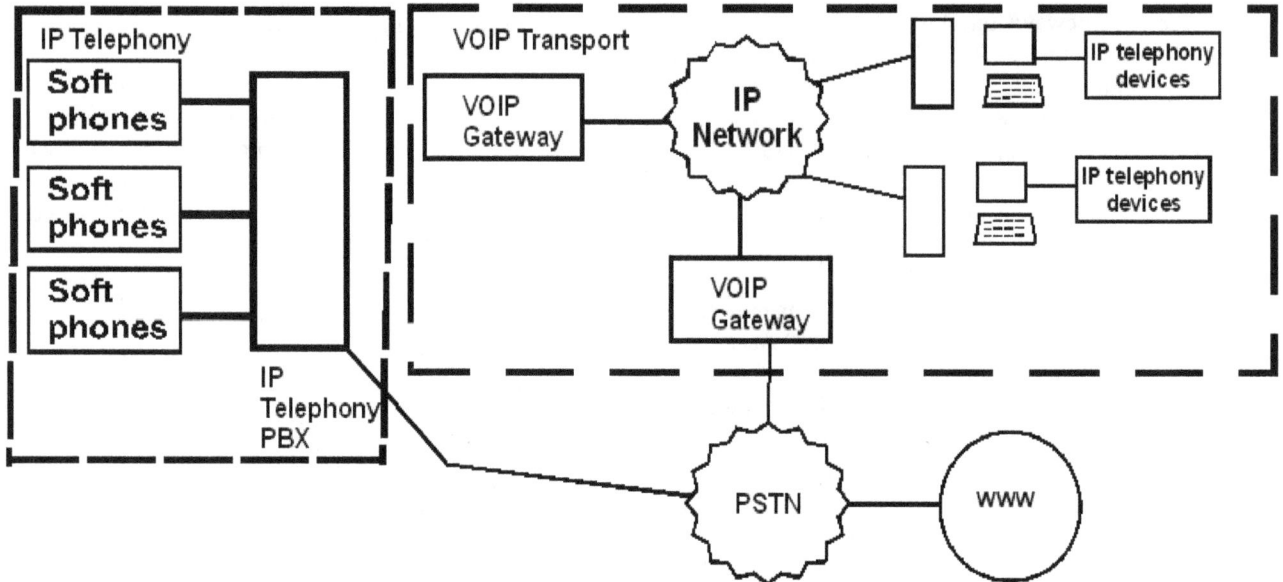

*Fig 2. Pragmatic schema of the Hybrid IP/PSTN/PBX Implementation at FUPRE*

### 4.3. Jitters

Jitter is difference between the expected time and actual arrive time of packet. Real time voice communications are sensitive to delays and variations in packet arrival times. Codecs require a steady, dependable stream of packets to provide reasonable playback quality. Packets arriving late too early, too late or out of sequence result in jerky and/or jumbled playback. This phenomenon is termed jitters. Since there is no network a perfectly steady stream of packets in real-world apps, soft-phones use buffers to smooth out the problem.

With a constant packet transmission rate of 20ms (as in design), new packets is expected to arrive at destination every 20ms. However, jitters are caused by queuing variations of ongoing changes in traffic loads as well as when one or more packets takes a different equal cost link not physically (electrically) the same length as link used in other voice packets.

To curb this, we employed media with play-out buffer to buffer packet stream so that the reconstructed voice wave is not affected by jitters (or rather minimized). A jitter buffer is simply a First-in, First-Out memory cache that collects the packets as they arrive, forwards them as evenly spaced to the codec and in proper sequence for accurate playback. This enables it to successfully mask mild delays and jitters. Although, sever jitters can also overwhelm the buffer and result in packet loss. Increasing the buffer size can help to a point.

### 4.4. Call Management Problem

The IP call-manager is many times overwhelmed by request or its network connections is impaired, call setup delays can reach the point where users abandon calls before they are able to connect to other party. If a soft-phone is misconfigured or its server connection is impaired, calls remain open in the call queue long after the parties have disconnected.

### 4.5. Packet Loss / Bandwidth

This results from network congestion, router/switch buffer overflow, discarded packets, undesirable file transfer, retransmission ability from protocols etc. Real-time application using RTP protocol for TCP is more tolerant to packet loss than UDP applications. The bandwidth allocated to the net will help a curb it. Network managers can calculate the bandwidth required to support voice traffic, deciding how much to allocate to each service in a converged voice/data network requires careful consideration of an organization's priorities, available bandwidth and its cost. We noted with this implementation, it was noted that packet loss can result from jitter buffers overwhelmed, media failure and poor wireless signal quality such that regardless of its source – the soft-phones, soft agents and gateways attempts to conceal signal degradation by duplicating the packets to fill-in missing data. Packet loss has always been characterized as burst phenomena – as the network tends to either sporadically drop single packets called gaps or large number of contiguous packets in a burst. Thus, such factors are considered when making bandwidth calculation as thus:

a. Impact of bandwidth priority
b. Trade-off in compression and voice quality, and
c. Projected peak usage by users

Thus, the selected bandwidth needed support 1,000 full-duplex G.711 encoded voice channels of 20ms packet creation and 200bytes packet size grouped as (160B payload + 40B IP header) is thus:

$$\text{Samples per Secs} = 1000ms \ / \ \text{packet creation rate}$$
$$= 1000 \ / \ 20 = 50bps$$

$$\text{Bps} = \text{samples/secs} * \text{packet size} * \text{no calls} * 8bps$$
$$= 50 * 200 * 1000 * 8 \ \text{bits/secs} = 80Mbps$$

Thus, the raw measure of the IP traffic but does not take into account the overhead used by the transport media (links between the routers) and the data link protocol. To determine the link speed needed to support this number of calls, network managers must add this raw IP value to that of the overhead. As implemented, bandwidth requirements vary depending on the rate at which the calls are generated and the signaling protocol used. If a large number of calls are initiated in a relatively short time, the peak bandwidth needs for the signaling can be high. Note the maximum amount of bandwidth required by IP signal protocol is roughly 3% of all bearer traffic. Thus, 1,000 calls initiated in one second is approximately 2.4Mbps (i.e 3% of 80Mbps).

### 4.6. Network Interconnection

PSTN is not a single network but a collection of networks operated by various service providers. At each boundary, a network interface is required for interconnection. Interconnection agreements are put in place to cover interconnection points, signaling, timing, billing and tariffs, bearer transport, regulatory requirement etc. In addition, these require approval from the relevant regulators. Scalability constraints and established business model implies that in the nearest future, IP network will become expansive as PSTN (collection of networks and network interconnections will be drawn up).

### 4.7. Delays/Latency

*Latency* is time taken to reach its destination by the packet through the network. Large delay may not degrade sound quality in phone calls, they can disrupt the rhythm of such conversation – making it difficult to interact. Several factors are known to contribute to such delay in such a multiservice network to include: (a) *packet* creation, (b) *propagation*, (c) Serialization, (d) *queuing*, and (e) packet *forwarding*.

In designing the multiservice network, total delay a packet exhibits is sum of all latency contributors. An accepted end-to-end latency should be less than 150ms for toll quantity calls. To mitigate latency, as implemented at FUPRE, we have that:

1. *Packet Creation* Delay is time destination takes to create packets used in the voice services, and exists at both source and destination units of a voice connection. At the source, this delay varies based on amount of time it takes to fill the packet with data. Thus, voice packets tend to be smaller to help minimize amount of delay in the creation process. At destination, media gateway must remove and re-process the packet. Thus, *chosen* protocol must enforce that all criteria equal, all/any gateway cannot exceed 30ms.

2. *Serialization Delay* is time taken to serialize digital data onto physical link that interconnects the equipments and it is inversely proportional to the link speed. Thus, the faster the media, the lesser the time it takes to serialize digital data onto physical link and the lower the delay – all of which is dependent on technology used and its access method. Thus, we adopted TCP cable, to keep number of links small and use a high bandwidth interface, to reduce overall delay.

3. *Propagation* is time a signal takes to travel the conductor's length. Electrical or photonic signal speed via a conductor is always slower than the speed of light – due to propagation delay in net. Computing the delay on 3000km (2-way) is:

$$PD = \frac{Circuit \ km}{(29300 * 0.6)} = \frac{(3000 * 2)}{(299300 * 0.6)} = 33.4$$

4. *Queuing* is packet buffering time as it awaits transfer. It varies on traffic load. Managers can configure length of time packet waits in a network – as this delay is dependent on the amount of traffic network element trying to pass via a given link – and thus, it increases with load. It is curbed with increased bandwidth and resources for voice traffic, as queues not serviced fast enough eventually grows into greater delay.

5. Packet Forward is time it takes router, switch, firewall or other network device to buffer a packet and make a forwarding or drop decision – via accessing which interface to forward the packet to, and whether to drop/forward packet against an Access Control List or security policy. This delay varies and is based on function and architecture of the networked device. So, if a packet is re-buffered as part of processing, the greater the delay incurred.

### 4.8. Cost and Reliability

Certain tradeoffs were considered as adequately explained to management that the cost of this solution cannot equal the cost of implementation of a full-interact access. Thus, the cost of re-access fee and the cost of voice calls services (as used by all staff/students within the institution) will seem expensive. After due consideration, FUPRE's management opted for the IP solution with its many benefits to cut down cost of implementation.

PSTN is reliable and handles millions of calls simultaneously – achieved via redundant and load sharing equipment and networks via call agents, access gateways media servers. In addition to fault tolerant feat, equipments, quality – other feats are: (i) no single point of failure, (ii) Hot-swap capability (iii) redundant hardware, (iv) redundant connection and (v) software and/or firmware that are upgraded without loss of service.

### 4.9. Quality of Service

A key requirement to the widespread deployment of VOIP is in its ability to offer toll free quality service equivalence of the existing PSTN. With many users (staff/students) connected, the perceived voice quality became very sensitive in three (3) key performance areas as the network started experiencing: data transfer *delays, jitters* and packet *loss*. However, IP by its nature provides a best effort service and does not however, provide guarantee about which key criteria. To *cub* this, we implemented ATM (Asynchromous Transfer Mode) –as a means to guarantee prioritization of voice media streams over best-effort data, and to ensure that the VoIP service is not compromised by unforeseen traffic patterns.

### 4.10. Bandwidth Utilization

Using an incremental, implementation approach as earlier noted, gave the researcher a way to improve the quality of service caused by jitters, delays and packet loss – as the researcher (via the ICT unit) decided to reduce system overhead cost by implementing the compression of RTP-UDP (Real-Time Protocol/Universal datagram protocol) and IP headers. A typical sample voice is less than 100bytes but its combined header equals about 40bytes-so that for lower bandwidth WAN links such as DSL and cable, the header overhead is significant and reduces the number of voice channels or data bandwidth available. Since a major merit of VOIP is the ability to use lower bit coders so that bandwidth is sowed-this compression mechanism is put in place with a point-to-point link (the goal is to maintain the state for each compressed RTP flow).

### 4.11. OSS Support

PSTN has very extensive Operations Support System providing feats such as: (i) flow through provisioning (ii) fault isolation (iii) loop testing (iv) alarms, (v) performance monitoring and (vi) policy definition and enforcement. Thus, with these in mind – the IP network must be designed to offer the same level of OSS support with PSTN integration via protocols and gateway accesses.

### 4.12. Signaling Protocol

Numerous protocols have been deployed applicable to VOIP. These includes: (a) for *device* control, we have H248,

MGCP, NCS amongst others, (b) for access service signaling protocol, examples are SIP, H-323 etc, and lastly (c) for network service, examples include SIP, SIP-T, BICC, CMSS etc. The choice of which protocol to use is very much dependent both on the service set being offered alongside the equipment types available to which services. Thus, we *implemented* the H.248 (Megaco) for *device* control, H.323 for *access* service signaling and SIP-T for network service. These were carefully chosen – so that in the event that in the nearest future with upgrades, users include SIP phones, the SIP-T protocol becomes handy.

## 5. IP-network Supports

Managing an IP-based network implies measurement or metrics of some parameters to ensure efficient network running. Thus, the network is and must be subjected to user assessment of quality and monitoring via IP-monitoring tools and analyzers that tends to compute how much various impairments factors like codec compression, jitters, delays and packet loss, will affect the typical user's perception of call and/or service quality. Such support to curb such issues of section 5 is Network Monitoring Tools, that accesses IP quality management from an ICT administrator's view rather than from a telecom engineer – as it aims to track, store and analyze long-term trends, network performance, and maintains a database of Call Detail Records from which the administrator can generate reports for management or service providers. The tools should also be capable to automatically notify when some selected parameter or statistics are developing a problem. An example as used is Network Instrument "Observer". Major components or statistics monitored by the observer includes [11]:

1. Tracks network performance and alerts if MOS (mean option score) or R-factor score falls to 3.5, jitters and delay crosses MPLS mesh of 20ms and 80ms respectively.
2. Troubleshoots Connection Problems
3. Traffic and call summary
4. Evaluates Jitters
5. Reconstructs and reviews Calls
6. Compares VoIP to network performance
7. Monitors VoWLAN
8. Manages Voice, Video, Audio and Data Quality
9. Monitors Quality of Service (QoS)
10. Measures Bursts and Gaps
11. Tracks and Decodes VoIP and Video

## 6. Discussion of Results

There are lots of benefits accrued to this hybrid implementation as proposed, and the nature of this work compelled the researchers to join the Multiservice Switching Forum (MSF), which is committed to aggressive technical solutions for a full PSTN replacement network with next generation IP infrastructure as well as QoS cum security in a way that scales to the many billions of busy hour calls that a typical PSTN must handle.

MSF's approach to problem solving: (i) proposes a coherent and pragmatic network vision (ii) identifies existing protocols that delivers its vision, enhance them for the purpose while eradicating barriers to easy interoperability, and (iii) develop program of easily interoperability testing, with carrier grade equipment supported by detailed to relevant test plans.

Frequent reviews are released (though a major reason for implementing our incremental approach design/model – so as to accommodate the problems as they surfaced in bits). Other issues currently undertaken by MSF are:

1. Scaleable IP-QoS (voice/multimedia) over IP – defines a solution to provide scalable QoS to the many billions of BHCA that PSTN supports in order to create an interaction between SIP services and QoS mechanism that will address the finality required in IP edge router.
2. Security – end-to-end IP implies vulnerability to a variety of attacks as mentioned. This can be addressed using cryptographic schemes.
3. Service/Feat Interactions – MSF service layer will aim to identify and prove new innovative network services and verifying that the services can be supported over such scalable next generation VOIP,
4. Call-routing mechanisms are to be considered a great power when implementing pure IP, mixed IP and/or legacy PSTN network,
5. Management related issues will hinge on a design that will motivate the overall architecture and act as a start point for protocol profiling and interoperability testing.

## 7. Summary / Conclusion

IP is ubiquitous and cost-effective. Thus, my moving towards an IP-based network, a carrier can:

1. Managing Call Quality:
2. Deploy new voice/data services, removing need to manage separate voice and data networks.
3. Reap the benefits of a standard, highly flexible network, giving competitive market to equipment vendors, and encompassing a wide range of equipment for different market niches.
4. Utilize cheaper IP-based backbone equipments to carry voice data.

A number of IP solutions exists with limitations that are dependent on solution type, bandwidth etc. These can be attributed to implementation around early versions of standards that provide restricted interoperability with and between other vendors.

## 8. Recommendations

[12-13] note these recommendations:

1. To improve service quality, network administrators and managers must: (a) Understand and measure call quality component, (b) conduct site surveys, (c) deploy analysis tools strategically for maximum visibility, (d) implement quality of service prioritization, (e) implement VLAN to isolate/monitor VoIP issues, (f) monitor rollouts to ensure positive user experience, (g) compare jitters to overall network bandwidth utilization to understand network response time, (h) set up analyzer to automatically and proactively monitor VoIP activities, (i) automate problem resolution and (j) set up baseline network traffic.

2. MSF (multiservice switching forum) is committed to providing next generation network that provides both full multi-vendor interoperability as well as the support for a full featured, secure PSTN. Thus, institutions seeking such solutions are encouraged to register with MSF so that their personnel can take advantage of such forum in proffering solutions to issues in the implementation of their solution type.

## References

[1] Abraham, M., Jajodia, S and Podell, H., "Data security", IEEE Transaction on Comp., 1995, 13(2), CA: Los Alamitos.

[2] Brennen, R and Dipak, G., "Secure IP telephony via multi-layered protection", 2009, Technical University of Denmark report, Centre for ICT, Denmark: Lyngby.

[3] Drew, P and Gallon, C., "Next generation VoIP network architecture: MSF whitepaper report", 2003, www.msforum.org, accessed April 2013.

[4] Eung-Ha, K., Cho, K.S and Ryu, W., "Amendment to MSF whitepaper on personalized converged services: network-to-network", [online]: www.msforum.org/techninfo/reports.shtml, last accessed March 2013.

[5] Garg, V., Smolik, K and Wilkes, J., "CDMA application in wireless communications", 1997, Prentice Hall publications, New Jersey: Upper Saddle River.

[6] Hafner, K and Lyon, M., "Where wizards stay up late", 1996, Simon and Schuster, New York.

[7] Helgert, H., "Integrated services digital net: Architecture, protocols and standards", 1991, Readings: MA: Addison-Wesley.

[8] International Telecommunications Union Supplement Handbook., "Rural telecommunications", 2010, pp12-48, Geneva.

[9] Martins, J.A.H., "Telecommunications and the computer", 1990, Prentice Hall publications, New Jersey: Upper Saddle river

[10] Matthews, V., Shakunle, J and Adetiba, E., "Hybrid cellular mobile network for rural telecommunications in Nigeria", 2007, Int. J. Research in Physical Sci.,

ISSN: 1597-0823, 4(1), pp 34-42.

[11] Network Instrument., "A White paper guide to troubleshooting IP-based networks, 2007, [online]: www.networkinstruments.com.

[12] Ojugo, A.A, "Introducing VoIP in Nigeria", 2010, J. Res. Phy. Sci., ISSN-1597-8028, 6(1), pp 43-51.

[13] Ojugo, A.A., Yoro, E.R., Eboka, A.O., Yerokun, M.O and Iyawa, I.J.B., "Implementation issues of VoIP for rural telephony in Nigeria", Journal of Emerging Trends in Comp. and Information System, ISSN: 2079-8407, 2012, 4(2), pp 172-179, [online]:www.cisjournals.org

[14] Osuagwu, O.E., Anyanwu, E and Amaeshi, L., "Deployment of computer-assisted radar technology: a case study for improved military security and surveillance in Nigeria", 2005, J. Computer Sci. App., 9(1), pp.71 – 81.

[15] Rosa, J., "Rural telecommunications via satellite", J. of Telecommunications, 2005, 3(2), pp 75 – 81.

[16] Schwartz, R., "Wireless communications in developing nations: cellular and satellite net systems", 1997,

Artech house. MA: Boston.

[17] Stallings, W and Van Slyke, R., "Business data communications", 6th Ed., 2010, Prentice Hall Int. Ed., ISBN – 1-13-761230-3, pp316-340.

[18] Williams, M., Ewan, S and Reza, T., "Convergence, IP telephony and regulations: issues and opportunities for network development in India", 2005, Technical Univeristy of Denmark Center for ICT, Denmark: Lyngby.

[19] Conte, R., "Rural telephony: new appraoch via mobile satellite communication", Proceedings of Pacific telemmunication conf., 1994, Hawai: Honolulu.

[20] Salamsi, A and Gilhousen, K.S., "On system design aspects of CDMA applied to digital cellular and personal communication network", 1991, IEEE Proc. Vehicular Tech. Conference, pp 57-63.

[21] Yoro, E.R., "Next Generation VoIP deployment: a case study of Delta State Polytechnic Ogwashi-Uku ", 2012, an unpublished Masters thesis, Department of Computer Science, Benson Idahosa University: Benin City, Nigeria.

# 10

# Performance analysis of MSE-OFDM system

**Poonam Singh[1], Saswat Chakrabarti[2]**

[1]Electronics & Comm. Engg. Dept. National Institute of Technology Rourkela, India
[2]GSSST, Indian Institute of Technology, Kharagpur, India

**Email address:**

psingh@nitrkl.ac.in(P. Singh), saswat@ece.iitkgp.ernet.in(S. Chakrabarti)

**Abstract:** In this paper we analyze the effects of interchannel interference (ICI) and intersymbol interference (ISI) on the performance of a Multi-Symbol Encapsulated Orthogonal Frequency Division Multiplexing (MSE-OFDM) system. MSE-OFDM is a bandwidth efficient OFDM scheme, where a number of OFDM symbols are grouped together as a frame and protected by one single cyclic prefix. This reduces the extent of redundancy caused by the CP and increases the bandwidth efficiency of the system. We have derived expressions for probability of error for MSE-OFDM in presence of ICI and ISI, which result from the time variation and delay spread of mobile channels. Both analysis and simulation results are presented for the MSE-OFDM system and are found to be almost identical.

**Keywords:** OFDM, MSE-OFDM, Bandwidth Efficiency, Synchronization Errors, PAPR

## 1. Introduction

Orthogonal Frequency Division Multiplexing (OFDM) has been applied widely in wireless communication systems due to its high data rate transmission capability with high bandwidth efficiency and its robustness to multipath delay. Another advantage of OFDM is its simple receiver structure using a frequency domain equalizer with only one complex multiplication per subcarrier. This is achieved by introducing a cyclic prefix, which is a cyclic extension of the output sequence, to eliminate intersymbol interference among the symbols [1]. Therefore, an appreciable amount of redundancy is introduced in an OFDM system which reduces the bandwidth efficiency. To avoid this redundancy, an OFDM system called Multi-Symbol Encapsulated Orthogonal Frequency Division Multiplexing (MSE-OFDM) has been proposed in literature [2-6]. OFDM is also sensitive to frequency and timing synchronization errors. The peak-to-average power ratio (PAPR) is also high, which limits the efficiency of the power amplifier. The effects of these phenomena can be reduced by using MSE-OFDM.

Two different implementations of the MSE-OFDM scheme have been proposed [2] [5]. The first implementation, termed the CP-reduced system, is designed to improve the bandwidth efficiency for static channels by reducing the number of CP insertions. The

bandwidth efficiency is improved as the MSE-OFDM frame size increases. The other implementation, termed FFT-size reduced system, is designed to keep the MSE-OFDM frame duration same as that of a conventional OFDM symbol i.e. the symbol duration reduces for the MSE-OFDM system, while the bandwidth efficiency remains the same. This reduces the number of subcarriers and the FFT size of MSE-OFDM system. So the PAPR and robustness to frequency offset improves.

An accurate channel estimation is required for the receiver design. Channel estimation for OFDM is usually performed in the frequency domain by either inserting pilot tones into all subcarriers of the OFDM symbols with a specific period or by inserting some pilot tones into each OFDM symbol. Several pilot-aided channel estimation schemes have been investigated for OFDM applications. Similar channel estimation methods can be used for MSE-OFDM systems.

Similar to conventional OFDM systems, MSE-OFDM are also very sensitive to frequency offsets due to the loss of orthogonality among the subcarriers. Various techniques have been proposed for frequency offset estimation and correction of OFDM systems [7-8]. Timing acquisition detects the presence of a new frame in the received data stream and once the frame is detected, it provides a coarse estimate of the timing error to find the correct position of the received DFT window. A popular timing acquisition algorithm was proposed by Schmidl and Cox [9]. A joint estimation of the frequency offset and the channel response were investigated [10] to improve the system performance and minimize the estimation errors. Similar channel estimation and frequency and timing synchronization schemes can be implemented for MSE-OFDM systems. There may be residual frequency and timing offsets even after synchronization techniques are used. The intercarrier interference (ICI) due to residual frequency offset will affect the accuracy of the OFDM channel estimation. We have analyzed and simulated the effects of these residual offsets on MSE-OFDM systems.

In this chapter, we analyze the performance of MSE-OFDM systems in AWGN and frequency selective Multipath Fading Channels. Channel variations severely degrade the performance of MSE-OFDM by introducing both a complicated multiplicative distortion and an additive intercarrier interference. The effects of carrier frequency offset and timing offset on MSE-OFDM are studied. We have done extensive simulations for these cases and verified the results using analysis. Section 2 gives a general description of the MSE-OFDM system. In Section 3 we have derived expressions for probability of error in presence of synchronization errors in multipath fading channels. Simulation results are given in Section 4 and some conclusions are given in Section 5.

## 2. The MSE-OFDM System Description

MSE-OFDM is a bandwidth efficient scheme, where a number of OFDM symbols are grouped together and a single cyclic prefix is used [5]. In CP-reduced MSE-OFDM, a number of OFDM symbols are grouped together in a frame and a single cyclic prefix is used in one frame. This reduces the redundancy caused by cyclic prefix in each symbol and therefore improves the bandwidth efficiency. In FFT-reduced MSE-OFDM system, each OFDM symbol is divided into smaller IFFT blocks, which are protected by one single cyclic prefix. The number of subcarriers is reduced, which reduces the effects of frequency offset and also reduces the PAPR. The bandwidth efficiency is same as that of OFDM.

The block diagram for the transmitter of the MSE-OFDM system is shown in Fig.1, where N and M denote the size of IFFT modulator and the total number of OFDM symbols in one MSE-OFDM frame respectively.

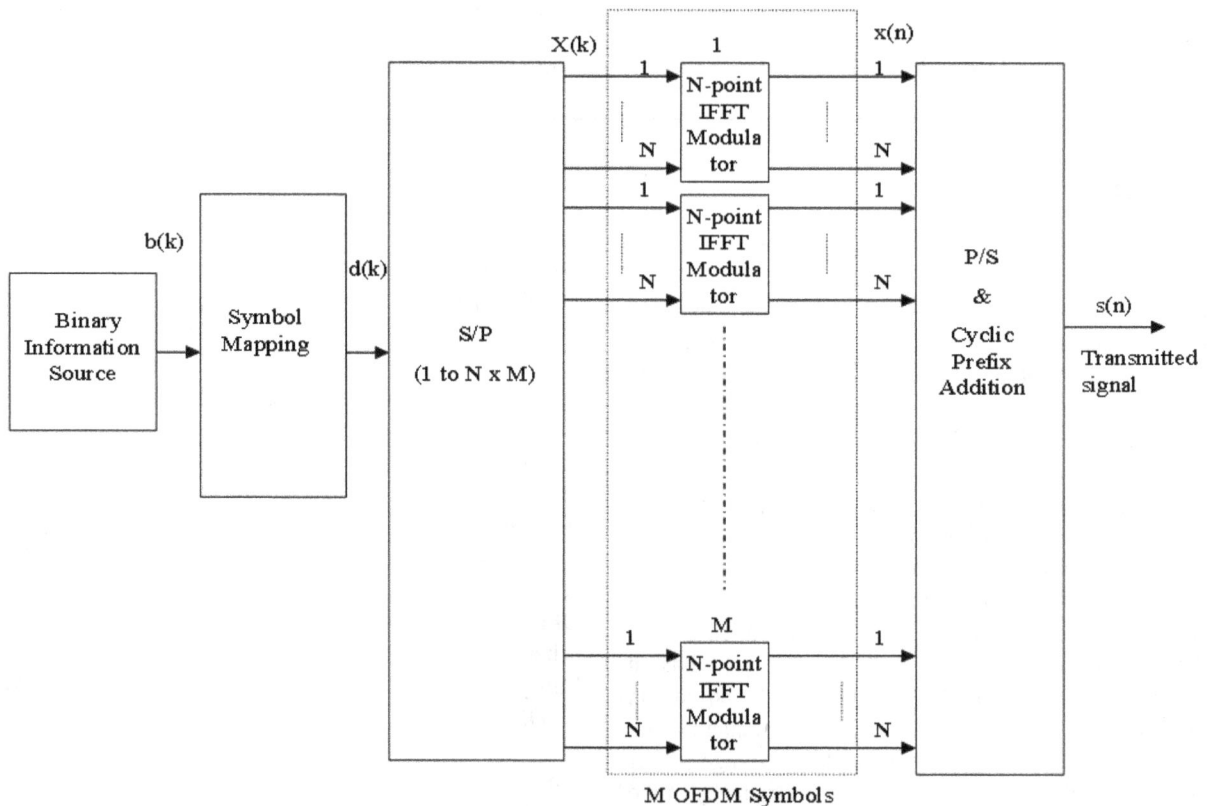

*Fig 1. Transmitter of MSE-OFDM, N: Number of data samples used to generate one OFDM symbol, M: Number of OFDM symbols in one MSE-OFDM frame, $N_{cp}$: number of samples used as cyclic prefix.*

To generate one frame of MSE-OFDM signal, M OFDM symbols are generated and then a cyclic extension of Ncp samples of last OFDM symbol in the same frame is inserted as the cyclic prefix. Here, one cyclic prefix is used for one frame which consists of M OFDM symbols. The OFDM signal consists of N complex exponentials or subcarriers which have been modulated with the complex input data X(k). Each OFDM symbol is generated by taking the N point complex modulation sequence through IDFT.

The subcarriers of the transmitted signal pass through frequency non-selective Rayleigh fading channel, and are also subjected to additive white Gaussian noise (AWGN) n(t), with a double-sided power spectral density No/2. The statistics of all the Rayleigh fading channels are assumed to be identical. Furthermore, we assume that the channel fading is slow and remains unchanged over two consecutive symbols. The complex impulse response of the Rayleigh fading channel for the m-th subcarrier can be written as

$$h_m(t) = \beta_m e^{j\theta_m} \delta(t)$$

where θm is a random phase introduced by the channel, it is modeled as uniformly distributed over the interval of [0, 2π] and is assumed to be i.i.d. for each subcarrier and each user, βm is a Rayleigh random variable with power E[β2m] = σ2 where E[.] is the expectation operator.

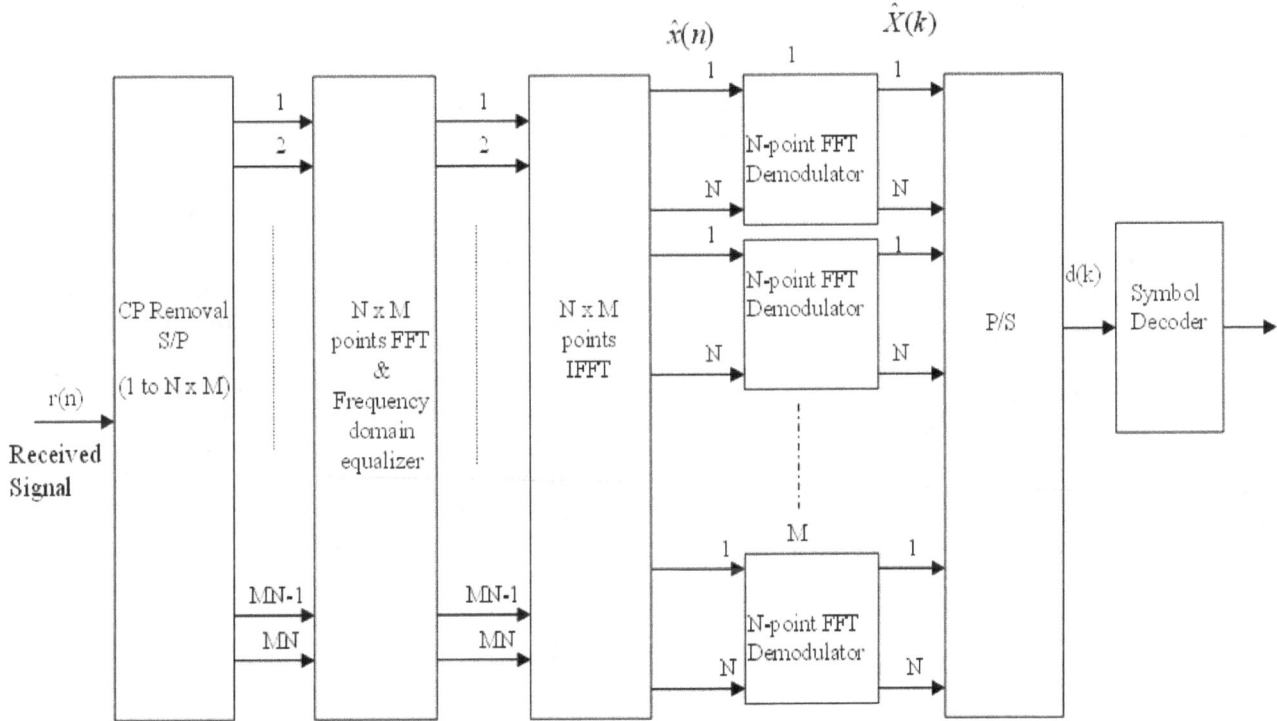

**Fig 2.** *MSE-OFDM Receiver*

### 2.1. Receiver

The MSE-OFDM receiver is shown in Fig.2. In OFDM receiver, after removing the cyclic prefix and taking inverse OFDM the received sequence is equalized to get the transmitted data. After channel estimation, the output of each subcarrier is equalized. Finally, the regenerated symbol sequences are parallel to serial converted to recover the transmitted binary data. With the help of cyclic prefix, simple frequency domain equalizer can be realized for the OFDM system. However, a new frequency domain equalizer has been employed due to the unique frame structure of the MSE-OFDM signal as shown in MSE-OFDM receiver in Fig.2 [2] [5]. Channel estimation is used to compensate for the amplitude and phase distortions associated with the received signal. To estimate the multiplicative channel response, pilot symbols are inserted among the transmitted data symbols. The receiver estimates the channel state information based on the received, known pilot symbols. MN point FFT is

performed to convert the whole frame to frequency domain. After channel estimation, one tap equalizer is used to compensate the channel distortions. For demodulation of each OFDM symbol in the same frame, the equalized frequency domain signal is converted back to time domain for IFFT demodulation. The equalized signal $\tilde{r}_l^{FEQ}$ is then split into M OFDM symbols for demodulation using N-point FFT.

To implement this frequency domain equalizer, a very large MN size Fast Fourier Transformation is needed, which increases the computational complexity of the system. However, the complexity can be reduced by using a time domain ISI cancellation process as proposed in [11]. In this case, the frequency domain equalization can be done on each individual OFDM symbols using only N point FFT.

## 3. Performance Analysis of MSE-OFDM

The performance analysis of OFDM signal has been carried out in several papers with and without synchronization errors [12] [13]. We have extended this work for MSE-OFDM schemes. The analysis of MSE-OFDM signal is similar to OFDM if there is no interference among the OFDM symbols in a frame. However, an ISI will occur among the OFDM symbols within the same MSE-OFDM frame prior to equalization. The accuracy of channel estimation is affected if the ISI is not suppressed. To reduce this ISI, a new frequency domain equalizer has been used due to the unique frame structure of the MSE-OFDM signal as explained in Section 1. In the receiver of MSE-OFDM systems, NM point FFT is performed to convert the whole frame to frequency domain. After channel estimation, one tap equalizer is used to compensate the channel distortions. For demodulation of each OFDM symbol in the same frame, the equalized frequency domain signal is converted back to time domain for IFFT demodulation. The equalized signal $\tilde{r}_l^{FEQ}$ is then split into M OFDM symbols for demodulation using N-point FFT.

In practical situations, Doppler shifts and oscillator instabilities result in a carrier frequency offset between the received carrier and the local sinusoids used for signal demodulation. A carrier frequency offset produces a shift of the received signal in the frequency domain and may result in a loss of mutual orthogonality among subcarriers. This causes ICI, which may cause severe performance degradation and so must be properly compensated.

In multicarrier systems, the DFT window should include samples from only one single block in order to avoid inter symbol interference (ISI). A timing offset in this DFT window may cause ISI among the symbols. If the length of the cyclic prefix (CP) is selected to be greater than the channel impulse response (CIR) duration, no ISI is present in the DFT output and it only results in a cyclic shift of the received OFDMA block. Thus the timing error $\Delta\theta$ appears as a linear phase across subcarriers and it can be compensated for by the channel equalizer, which cannot distinguish between phase shifts introduced by the channel and those caused by timing misalignments. We have analyzed the effects of carrier frequency offset and timing offset on the performance of MSE-OFDM.

Mathematical models have been used to find the analytical solutions for the problems. First, the analysis has been done for the ideal case, where we assume that there is only AWGN noise in the channel and there is no multipath fading or interference of any kind. Then the analysis is carried out for a multipath fading channel assuming no interferences. Then the effects of carrier frequency offset and timing offset in a multipath fading channel has been considered for the analysis. All the schemes have been analyzed and also been simulated and interpreted extensively.

The transmitter and receiver models for analysis of the MSE-OFDM system are shown in Fig.3 and Fig.4. Here, one cyclic prefix is used for one frame which consists of M OFDM symbols. The cyclic prefix is the cyclic extension of Ncp samples of the last OFDM symbol in the same frame. The IFFT size of modulator is N and the length of cyclic prefix is Ncp samples.

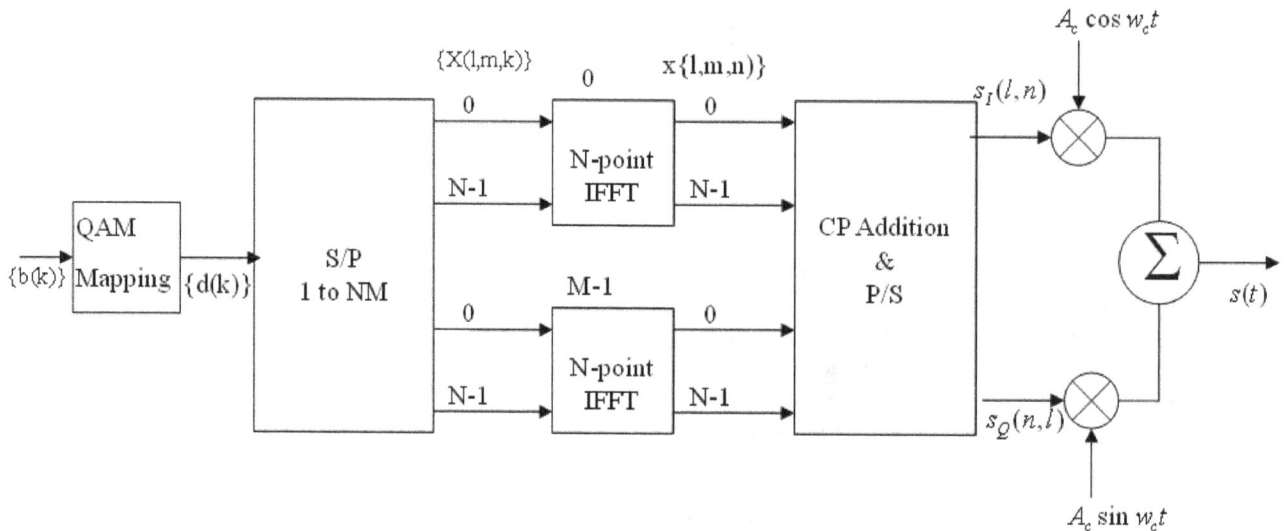

*Fig 3. Transmitter of MSE-OFDM indicating major signal processing blocks*

The serial binary data sequence {b(k)}, $0 \le k < aNM$, is the input data $b(k) = \pm 1$.

This is QAM modulated to get the sample sequence {d(k)}, $0 \le k < NM$,

$$d(k) = d_I(k) + jd_Q(k) , \ 0 \le k < NM \qquad (1)$$

where $d_I(k)$ and $d_Q(k)$ can have a value of either $\pm 1$.

After serial to parallel conversion, a complex sequence {X(m,k)} of N samples of d(k), $0 \le k < N$, form one set of input to the m-th IFFT block which generates the m-th OFDM sample set {x(m,n)}, $0 \le n < N$. Each OFDM sample is given by

$$x(m,n) = \sum_{k=0}^{N-1} X(m,k) \exp(j2\pi nk/N) , \ \begin{matrix} 0 \le n < N \\ 0 \le m < M \end{matrix} \qquad (2)$$

In MSE-OFDM, M OFDM symbols, each having N samples, are taken in one frame and $N_{cp}$ samples from the last OFDM sample are used as cyclic prefix. Thus one frame consists of NM + $N_{cp}$ samples.

The n-th sample of MSE-OFDM signal can now be written as

$$s(n) = \sum_{k=0}^{N-1} X(M-1,k) \exp(j2\pi k(N - N_{cp} + n)/N),\ 0 \le n < N_{cp}$$
$$+ \sum_{m=0}^{M-1} \sum_{k=0}^{N-1} X(m,k) \exp(j2\pi k(n - N_{cp} - mN)/N),\ N_{cp} \le n < NM + N_{cp} \quad (3)$$

where the first term represents the cyclic prefix and the second term represents the actual data to be transmitted.

As shown in Fig.4, the received signal r(t) is demodulated to baseband frequency using carriers recovered from the received signal, $\cos[w_c t + \theta(t)]$ and $\sin[w_c t + \theta(t)]$. The double frequency term is removed by using a filter and filtered signal is integrated and sampled at an interval of $T_s + \tau$ to get the in-phase and quadrature

components $r_I(l,n)$ and $r_Q(l,n)$. The received signal sequence {r(l,n)}, $0 \le n < NM + N_{cp}$ , is the transmitted signal sequence {s(l,n)} corrupted by channel noise and the additive white Gaussian noise w(l,n). After removing the cyclic prefix, {$\tilde{r}(l,n)$} is split into M OFDM symbols and demodulated using N-point FFT. The recovered complex samples $\hat{X}(l,m,k)$ are decoded to get a pair of numbers, ($\hat{d}_I, \hat{d}_Q$) which represent the maximum likelihood estimates of the quadrature coordinates of the transmitted symbols ($d_I, d_Q$). Thus, the symbols are detected from $\hat{X}(l,m,k)$ based on maximum likelihood (ML) detector and are parallel to serial converted to get the sequence { $\hat{d}(k)$ }. The binary bit sequence { $\hat{b}(k)$ } is recovered from { $\hat{d}(k)$ } using a binary mapping. The probability of symbol error is calculated as $\Pr(\hat{d}(k) \ne d(k))$ .

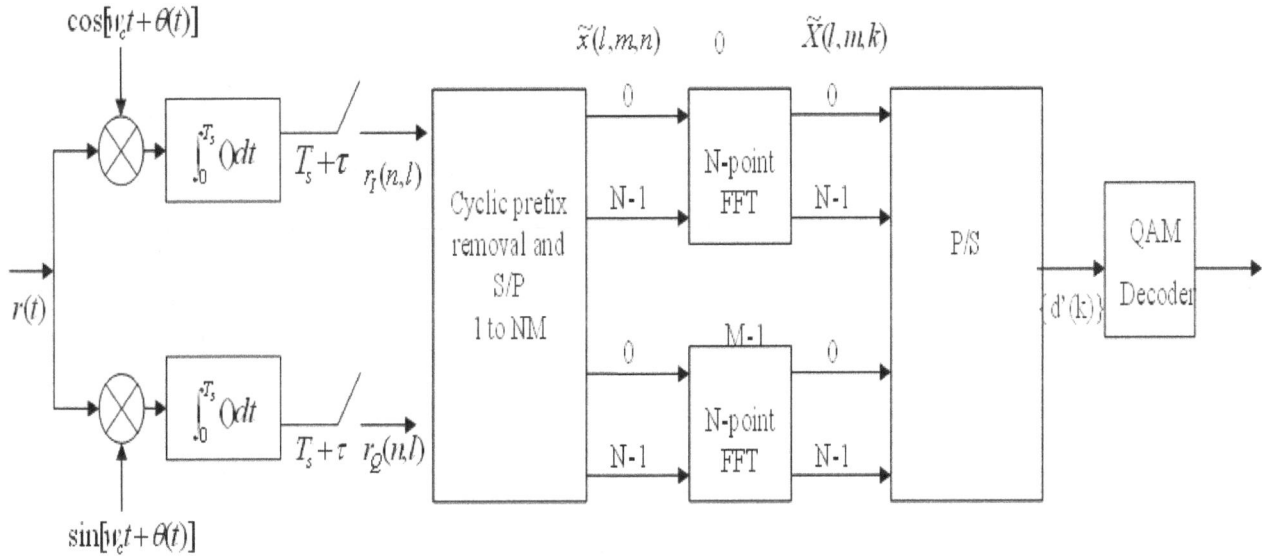

*Fig 4. Receiver of MSE-OFDM highlighting the signal processing aspects*

We observe that in AWGN channels the OFDM symbols can be separated from a MSE-OFDM frame without any interference. Therefore, the average probability of bit error for the MSE-OFDM system in AWGN channel using MQAM modulation will be same as that for an OFDM system.

In Rayleigh fading channels also, if the channel is slow fading and can be considered to be static over one frame duration, the average probability of bit error for the MSE-OFDM system using MQAM modulation will be same as that for an OFDM system. In presence of other interferences in addition to noise, the SNR is replaced by SINR, which is signal-to-interference noise ratio.

We reproduce the results derived for OFDM systems in literature [14-15] for comparison and to derive expressions for MSE-OFDM. The probability of symbol error of OFDM system using MQAM modulation in AWGN is given by [14],

$$P_e = 1 - \left[1 - \frac{2(\sqrt{M} - 1)}{\sqrt{M}} Q\left(\sqrt{\frac{3\gamma_s}{M-1}}\right)\right]^2 \quad (4)$$

where $\gamma_s$ stands for the SNR per symbol under AWGN.

The probability of bit error using QPSK Modulation in AWGN channel is given as [14],

$$P_e = Q\left(\sqrt{2\gamma_s}\right) = Q\left(\sqrt{\frac{2E_b}{N_o}}\right) \quad (5)$$

where $\gamma_s$ stands for the SNR per symbol under AWGN, Eb bhbh is the bit energy, No is the noise variance.

The average symbol error probability of an OFDM system employing MQAM constellation under Rayleigh fading channel can be computed by integrating the probability of symbol error under AWGN channel over the PDF of Rayleigh distribution.

The average probability of error under Rayleigh fading can be computed as [15]

$$P_e = \int_0^\infty P_e(\gamma) f_\gamma(\gamma) d\gamma \qquad (6)$$

where $f_\gamma(\gamma)$ is the PDF of the fading distribution.

The Rayleigh fading distribution is given by [15]

$$f_\gamma(\gamma) = \frac{\gamma}{\sigma^2} \exp\left(-\frac{\gamma^2}{2\sigma^2}\right), \ 0 \le \gamma < \infty = 0 \ \gamma < 0 \quad (7)$$

The probability of bit error for the MSE-OFDM system in AWGN channel is found to be same as that of an OFDM system in AWGN and fading channels. Thus we observe that the MSE-OFDM system gives better spectral efficiency without degradation in the performance.

The probability of bit error for the OFDM signal with frequency offset is given as [7]

$$P_e = \int_{-\infty}^\infty Q\left(\sqrt{\frac{\gamma_k^2 E_b}{(\sigma_{I_k}^2 + N_o)}}\right) f_\gamma(\gamma) d\gamma \qquad (8)$$

where $\gamma k$ is the fading parameter in the k-th subcarrier, Eb bhbh is the bit energy, No is the noise variance, $f_\gamma(\gamma)$ is the joint pdf of $\gamma_1, \gamma_2 .....\gamma_N$ and $\sigma_{I_k}^2$ is the variance of ICI given by [7]

$$\sigma_{I_k}^2 = X^2 \sum_{\substack{i=0 \\ i \ne k}}^{N-1} \gamma_i^2 \left\{\frac{\sin \pi \Delta k}{N \sin(\pi \Delta k / N)} e^{j\pi \Delta k(N-1)/N}\right\}^2 \qquad (9)$$

where $\Delta k$ is the normalized frequency offset defined as ratio of actual frequency offset to subcarrier spacing.

# 4. Joint Effects of Carrier Frequency and Timing Offsets

OFDM systems are very sensitive to frequency and timing errors. We consider the effects of carrier frequency and timing offsets on an MSE-OFDM signal in a multipath fading channel. The probability of bit error for the MSE-OFDM signal taking both frequency and timing offset into consideration is derived in this Section. The following assumptions are taken for this analysis:

i    Carrier frequency offset θ (t), uniformly distributed between 0 and $2\pi$ radians.
ii   Timing offset τ, uniformly distributed between 0 and $T_b$ (one bit duration).
iii  Input bits are random, independent and equiprobable,
iv   The channel is a ITU-R vehicular-A fading channel.

The received MSE-OFDM signal after demodulation and removing the double frequency term can be written as

$$r(t) = h(t)s(t-\tau) + I(t) + n(t) \qquad (10)$$

where s(t) is the transmitted signal, τ is the time delay, h(t) is channel impulse response, I(t) is inter-carrier interference due to carrier frequency offset and timing offset and n(t) is Gaussian noise with zero mean and power spectral density No/2.

After removing the cyclic prefix, the whole frame is equalized and split into M OFDM symbols $\hat{x}(m,n)$, each sample of the recovered OFDM symbol can be written as where H(k) is the channel transfer function at the k-th subcarrier frequency, $\Delta n$ is the relative timing offset (ratio of the timing offset to the sampling interval), and $\Delta k$ is the relative frequency offset (ratio between frequency offset to the subcarrier spacing) and w(m,n) is the Gaussian noise in n-th sample of m-th OFDM symbol.

$$\hat{x}(m,n) = \sum_{k=0}^{N-1} X(m,k)H(k)\exp(j2\pi(n+mN+\Delta n)(k+\Delta k))/N) + w(m,n), \begin{array}{l} 0 \le n < N \\ 0 \le m < M \end{array} \qquad (11)$$

These received samples are demodulated using N-point FFT to get samples $\hat{X}(m,k)$

$$\hat{X}(m,k) = \frac{1}{N}\sum_{n=0}^{N-1} \hat{x}(m,n)\exp(-j2\pi nk / N), \ 0 \le k < N \qquad (12)$$

substituting $\hat{x}(m,n)$ from eq.(11) in (12), we get

$$\hat{X}(m,k) = \frac{1}{N}\sum_{n=0}^{N-1}\left[\sum_{i=0}^{N-1} X(m,i)H(i)\exp(j2\pi(n+mN+\Delta n)(i+\Delta k) / N) + w(m,n)\right]\exp(-j2\pi nk / N)$$

$$= \frac{1}{N}\sum_{n=0}^{N-1}\sum_{i=0}^{N-1} X(m,i)H(i)\exp((j2\pi(n+mN+\Delta n)(i+\Delta k) - nk) / N) + \frac{1}{N}\sum_{n=0}^{N-1} w(m,n)\exp(-j2\pi nk / N)$$

$$= \frac{1}{N}\sum_{n=0}^{N-1}\sum_{i=0}^{N-1} X(m,i)H(i)\exp((j2\pi(n(i-k+\Delta k) + (mN+\Delta n)(i+\Delta k)) / N) + W(m,k)$$

$$= \frac{1}{N}\sum_{n=0}^{N-1} X(m,k)H(k)\exp(j2\pi(n\Delta k + (mN+\Delta n)(k+\Delta k)) / N)$$

$$+\frac{1}{N}\sum_{n=0}^{N-1}\sum_{\substack{i=0\\i\neq k}}^{N-1}X(m,i)H(i)\exp(j2\pi(n(i-k+\Delta k)+(mN+\Delta n)(i+\Delta k)/N)\ +W(m,k)\quad\begin{array}{l}0\leq m<M\\0\leq k<N\end{array}\quad(13)$$

using the properties of geometric series, we get

$$\hat{X}(m,k)=X(m,k)H(k)\frac{\sin\pi\Delta k}{N\sin(\pi\Delta k/N)}\exp(j\pi\Delta k(N-1)/N)\exp(j2\pi(k+\Delta k)\Delta n/N)\exp(j\pi(M-1)\Delta k)$$

$$+\sum_{\substack{i=0\\i\neq k}}^{N-1}X(m,i)H(i)\frac{\sin\pi(k+\Delta k)}{N\sin(\pi(i-k+\Delta k)/N)}\exp(j\pi(i-k+\Delta k)(N-1)/N).\exp(j2\pi(i+\Delta k)((M-1)+\Delta n)/N)+W(m,k)\quad 0\leq k<N\quad(14)$$

The fisrt term can be expanded using Taylor series as

$$X(m,k)H(k)\frac{\sin\pi\Delta k}{N\sin(\pi\Delta k/N)}\left[1+\frac{[\pi\Delta k(N-1)+2\pi(k+\Delta k)\Delta n+\pi(M-1)\Delta k]}{N}\right]\qquad(15)$$

Substituting (15), eq.(14) can be expressed as

$$\hat{X}(m,k)=X(m,k)H(k)S(0)+ISI+\sum_{i=0,i\neq k}^{N-1}X(m,i)H(i)S(i-k)+W(m,k)\quad 0\leq k<N\qquad(16)$$

where $X(m,k)$ denotes the transmitted symbol for the k-th subcarrier, whose amplitude and phase are modified by the frequency offset and timing offset, S(k) is due to frequency offset given by (16), $W(m,k)$ is the complex Gaussian noise sample and N is the number of subcarriers. Since N is always much greater than $\pi\Delta k$, $N\sin(\pi\Delta k/N)$ can be replaced by $\pi\Delta k$ and the first term reduces to

$$X(m,k)H(k)\frac{\sin\pi\Delta k}{\pi\Delta k}\exp(j\pi\Delta k(N-1)/N)\exp(j2\pi(k+\Delta k)\Delta n/N)\exp(j\pi(M-1)\Delta k)\qquad 0\leq k<N\quad(17)$$

The second term in (16) is due to ISI caused by the timing offset and is given by

$$ISI=X(m,k)H(k)\left[\frac{\pi\Delta k(N-1)+2\pi(k+\Delta k)\Delta n+\pi(M-1)\Delta k}{N}\right]\qquad 0\leq k<N\quad(18)$$

The third term in (16) is due to ICI caused by the frequency offset given by

$$I_{ICI}=\sum_{\substack{i=0\\i\neq k}}^{N-1}X(m,i)H(i)\frac{\sin\pi(k+\Delta k)}{N\sin(\pi(i-k+\Delta k)/N)}\exp(j\pi(i-k+\Delta k)(N-1)/N).\exp(j2\pi(i+\Delta k)((M-1)+\Delta n))/N)$$

$$=\sum_{i=0,i\neq k}^{N-1}X(m,i)H(k)S(i-k)\qquad 0\leq k<N\qquad(19)$$

The sequence S(k) depends on carrier frequency offset and timing offset and is given by

$$S(k)=\frac{\sin\pi(k+\Delta k)}{N\sin\frac{\pi}{N}(k+\Delta k)}\exp\left[j\pi\left(\frac{N-1}{N}\right)(k+\Delta k)((M-1)+\Delta n)\right]\quad 0\leq k<N\quad(20)$$

where $\Delta k$ is the normalized frequency offset and $\Delta n$ is the relative timing offset.

Since the QAM symbols X(m,i) are random variables, the interferences ICI and ISI are also random variables. For large values of N, the power spectral density of ICI and ISI can be approximated by Gaussian process using central limit theorem.

$$\sigma_{ISI}^{2}=X^{2}H^{2}\left|\frac{\pi\Delta k(N-1)+2\pi(k+\Delta k)\Delta n+\pi(M-1)\Delta k}{N}\right|^{2}\quad(21)$$

$$\sigma_{ICI}^{2}=$$

$$X^{2}\sum_{\substack{i=0\\i\neq k}}^{N-1}H_{k}^{2}\left\{\frac{\sin\pi(k+\Delta k)}{N\sin(\pi(i-k+\Delta k)/N)}\exp[j\pi\Delta k(N-1)/N]\exp(j\pi(M-1)\Delta k)\right\}^{2}\quad(22)$$

The probability of bit error for the MSE-OFDM signal taking both frequency and timing offset into consideration is given by

$$P_{e}=\int_{0}^{\infty}Q\left(\sqrt{\frac{\gamma_{k}^{2}E_{b}}{(\sigma_{ICI}^{2}+\sigma_{ISI}^{2}+N_{o})}}\right)f(\gamma)d\gamma\qquad(23)$$

where $\gamma_{k}$ is the fading parameter in the k-th subcarrier given by $\gamma_{k}^{2}=E[H(k)^{2}]$, $E_{b}$ is the bit energy, $N_{o}$ is the noise variance, $f_{\gamma}(\gamma)$ is the joint pdf of $\gamma_{1},\gamma_{2}.....\gamma_{N}$, $\sigma_{ICI}^{2}$ is the variance of ICI and $\sigma_{ISI}^{2}$ is the variance of ISI given by (21) and (22).

It is found that the signal amplitude and variance of ICI is multiplied by an extra exponential term in CP-reduced MSE-OFDM signal as compared to OFDM signal. This introduces a phase error due to multiple encapsulated symbols, which can be compensated for by the channel equalizer. The timing offset produces a phase shift in the received signal. The phase shift is more in MSE-OFDM as compared to OFDM systems. The phase error due to multiple encapsulated symbols can be compensated for by the channel equalizer and it does not cause much degradation in the performance.

For FFT-reduced MSE-OFDM system, the IFFT size is reduced keeping the bandwidth constant, so the subcarrier spacing is increased. This reduces the relative frequency offset $\Delta k$, which is defined as ratio of actual frequency offset to frequency spacing. This reduces the ICI variance and therefore we get better performance in FFT-reduced MSE-OFDM systems as compared to OFDM or CP-reduced MSE-OFDM in presence of carrier frequency offset. We have taken relative frequency offset values which are uniformly distributed with zero mean.

# 5. Simulation Results

**Table 1.** Simulation parameters for MSE-OFDM

| Parameters | Values |
|---|---|
| Number of subcarriers N | 128 |
| Number of OFDM symbols per frame M | 4 |
| Cyclic prefix $N_{cp}$ | 16 |
| Modulation | QPSK |
| Number of pilot carriers P | 16 |
| Relative frequency offset | Mean = 0.05, variance = 0.001 |
| Relative timing offset | Mean = 0.05, variance = 0.001 |

Simulations have been carried out to evaluate the bit error rate (BER) of MSE-OFDM system under different channel conditions. Programming languages C++ and MATLAB have been used for simulation. Stochastic models have been used to generate random numbers for input data. The simulation results are verified using extensive mathematical analysis for each case and the results are found to be close. The small discrepancy in the results is due to the assumptions taken in the analysis. Monte-Carlo simulation is used to get reliable simulation results. The analysis and simulation have been done for QPSK modulation. However, higher QAM modulation can be used to increase the spectral efficiency. The system parameters used are: N = 128, B = 4, Ncp = 16 and modulation scheme is QPSK. The channel is ITU-R defined vehicular A channel. The frequency and timing

offsets are assumed to be uniformly distributed random variables with zero mean and variances as mentioned in the simulation results. The relative frequency offset is a uniform random variable with zero mean and variance 0.05. The simulation parameters are given in Table.1.

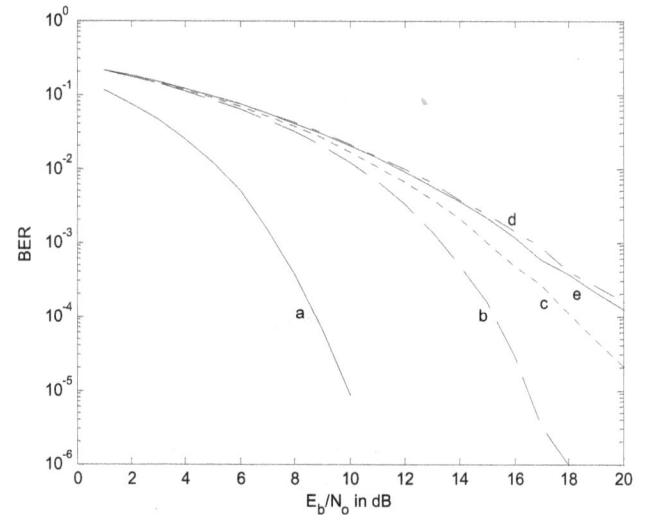

**Fig 5.** BER for MSE-OFDM system with frequency offset in fading channel (a) BER in AWGN channel (b) BER in fading channel without frequency offset (c) simulation result for BER of FFT-reduced MSE-OFDM signal with a relative frequency offset of 0.05 (d) simulation result for BER of CP-reduced MSE-OFDM signal with a relative frequency offset of 0.05 (e) analytical result for CP-reduced system with frequency offset.

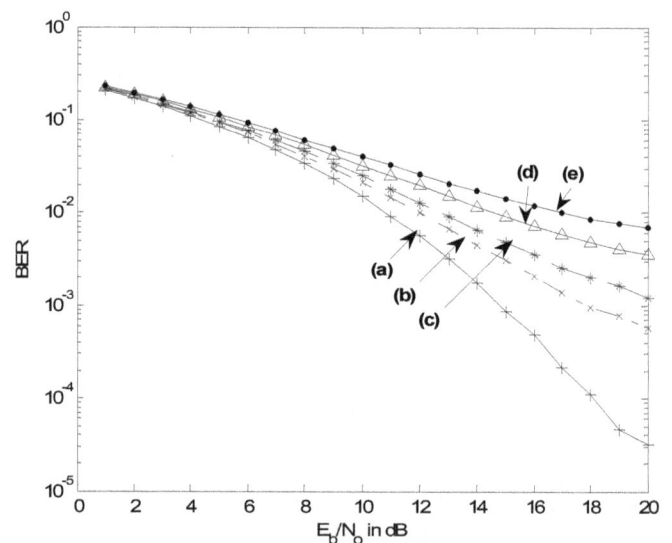

**Fig 6.** BER for CP-reduced MSE-OFDM system in fading channel for different values of frequency offset (a) without frequency offset (b) with a frequency offset of 0.01 (c) frequency offset of 0.05 (d) frequency offset of 0.08 and (e) frequency offset of 0.1.

The analytical and simulation results for the probability of bit error for both CP-reduced and FFT-reduced MSE-OFDM systems with frequency offset are shown in Fig.5. The BER in AWGN channel is shown in (a) and the BER in multipath fading channel is shown in (b), which are same for both the cases. The simulation result for BER of a

FFT-reduced MSE-OFDM with a relative frequency offset of 0.05 is shown in (c), the simulation result for BER of a CP-reduced MSE-OFDM with same relative frequency offset of 0.05 is shown in (d) and the analytical result is shown in (e). There is a degradation in BER from 10-5 to 10-4 at Eb/No of 16 dB due to frequency offset in CP-reduced MSE-OFDM. This degradation is less in case of a FFT-reduced scheme since the number of subcarriers is reduced.

The probability of bit error for the CP-reduced MSE-OFDM system with different values of frequency offset in fading is shown in Fig.6. The BER in fading channel without any frequency offset is shown in (a), the BER is found to be 10-4 at an Eb /No of 18 dB. The BER with a uniformly random relative frequency offset with mean 0.01 and variance 0.001 is shown in (b), which is found to be 10-3 at an Eb /No of 18 dB. In (c), the BER with a relative frequency offset of 0.05 is shown and is found to be 2 x 10-3 at the same Eb/No of 18 dB. With a relative frequency offset of 0.08, the BER becomes 5 x 10-3 at the same Eb/No as shown in (d) and with a relative frequency offset of 0.1, it is 10-2 as shown in (e) for the same value of Eb/No. Thus, we can see that a small change in frequency offset can degrade the system performance severely.

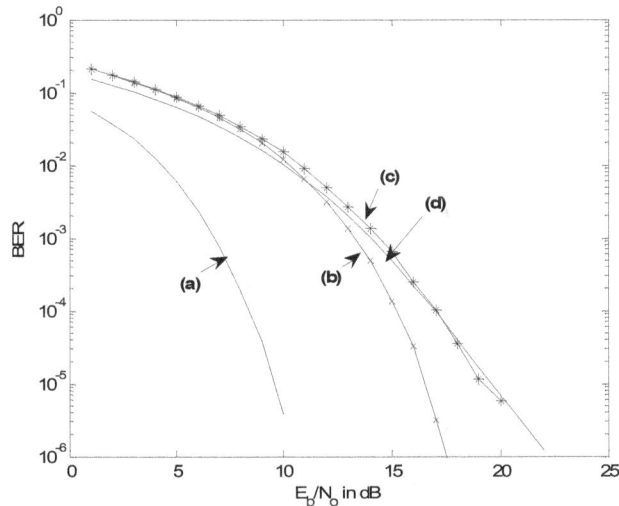

*Fig 8.* *BER for MSE-OFDM system in fading channel for different values of timing offset (a) without timing offset (b) with a timing offset of 0.01 (c) timing offset of 0.05 (d) timing offset of 0.08 and (e) timing offset of 0.1.*

The probability of bit error for the MSE-OFDM system with different values of timing offset in fading is shown in Fig.8. The BER of MSE-OFDM system in fading channel without any timing offset is shown in (a), the BER is found to be 10-5 at an Eb /No of 17 dB. The BER with a uniformly random relative timing offset with mean 0.01 and variance 0.001 is shown in (b), the BER is found to be 10-4 at an Eb /No of 17 dB. In (c), the BER with a relative timing offset of 0.05 is shown and is found to be 2 x 10-4 at the same Eb/No of 17 dB. With a relative timing offset of 0.08, the BER becomes 5 x 10-4 at the same Eb/No as shown in (d) and with a relative timing offset of 0.1, it is 10-3 as shown in (e) for the same value of Eb/No. It is found that a small change in timing offset can degrade the system performance severely.

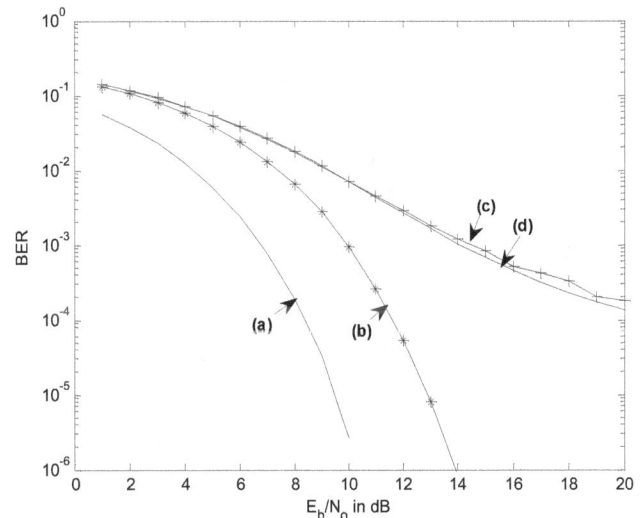

*Fig 7.* *BER for MSE-OFDM system with timing offset in fading channel (a) BER in AWGN channel (b) BER in fading channel without timing offset (c) simulation result for BER in fading channel with a relative timing offset of 0.05 (d) analytical result with timing offset .*

The analytical and the simulation results for the probability of bit error for the MSE-OFDM system with timing offset are shown in Fig.7. The BER in AWGN channel is shown in (a), the BER in fading channel is shown in (b), the simulation result for BER with a relative timing offset of 0.05 is shown in (c) and the analytical result for BER with a relative timing offset of 0.05 is shown in (d). There is a degradation in BER from 10-5 to 10-4 at Eb/No of 17 dB due to timing offset.

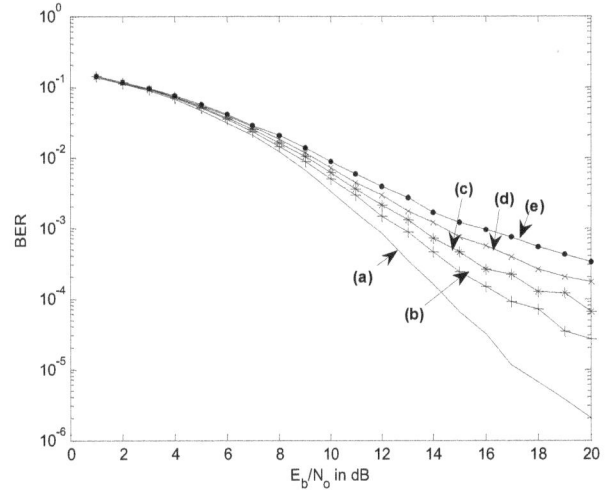

*Fig 9.* *BER for CP-reduced MSE-OFDM system with frequency offset and timing offset in multipath fading channel (a) BER in AWGN channel (b) BER in fading channel without frequency or timing offset (c) simulation result for BER with frequency and timing offset of 0.05 each and (d) analytical result for BER with frequency and timing offset.*

The analytical and the simulation results for the probability of bit error for the CP-reduced MSE-OFDM system with both frequency and timing offset are shown in Fig.9. The BER in AWGN channel is shown in (a), the BER in fading channel is shown in (b), the simulation result for BER with relative frequency offset of 0.05 and relative timing offset of 0.05 is shown in (c) and the analytical result for BER with frequency and timing offset of 0.05 is shown in (d). There is a degradation in BER from 10-6 to 10-3 at Eb/No of 14 dB due to frequency and timing offsets.

# 6. Conclusions

In this paper, we have analyzed the BER performance of MSE-OFDM systems. MSE-OFDM is very sensitive to synchronization errors. We have analyzed the effects of synchronization errors i.e. carrier frequency offset and timing offset on MSE-OFDM systems and derived expressions for probability of error in presence of these errors, which cause ICI and ISI.

We have done extensive simulations and verified the results analytically. Both analysis and simulation results are presented and are found to be very close. As may be expected, it is observed that the BER performance degrades in presence of synchronization errors. However, this degradation is less in case of FFT-reduced MSE-OFDM scheme as compared to conventional OFDM due to reduction in number of subcarriers. The performance improvement by 2-3 dB has been observed by using this scheme.

# References

[1]    S. B. Weinstein and P.M. Ebert, "Data transmission by frequency-division multiplexing using the discrete Fourier transform," IEEE transactions on Communication Technology, vol. 19, pp.628-634, Oct. 1971.

[2]    X. Wang, Y. Wu and J. Y. Chouinard, "On the Comparison between Conventional OFDM and MSE-OFDM Systems, "IEEE Global Telecomm. Conference, GLOBECOM'03, pp.35-39, Dec., 2003.

[3]    J. Y. Chouinard, X. Wang and Y. Wu, "MSE-OFDM: A New OFDM Transmission Technique with Improved System Performance", International Conference on Accoustics, Speech and Signal Processing, ICASSP 2005, vol.3, 18-23 March,2005.

[4]    X. Wang, Y. Wu and J. Y. Chouinard, "System Design and Implementation of Multiple-Symbol Encapsulated OFDM, "IEEE Vehicular Techonology Conference, 2005 VTC, vol.2, 30 May-1 June, 2005.

[5]    X. Wang, Y. Wu, J. Y. Chouinard and H. C. Yu, "On the Design and Performance of Multisymbol Encapsulated OFDM Systems," IEEE Transactions on Vehicular Technology, vol. 55, No. 3, May 2006.

[6]    E. Sun, K.Yi, B. Tian and X. Wang "A Method for PAPR Reduction in MSE-OFDM Systems,"International Conference on Advanced Information Networking and Applications (AINA'06), vol.2, April, 2006.

[7]    M. Morelli, C. C. J. Kuo and M. Pun, "Synchronization Techniques for Orthogonal Frequency Division Multiple Access (OFDMA): A Tutorial Review," Proceedings of the IEEE, vol. 95, No. 7, July 2007.

[8]    P.H. Moose, "A technique for orthogonal frequency division multiplexing frequency offset correction," IEEE transactions on communication, vol. 42, no.10 pp. 2908-2914, Oct. 2004.

[9]    Timothy M. Schmidl and Donald C. Cox, "Robust Frequency and Timing Synchronization for OFDM," IEEE Transactions on Communications, Vol. 45, No. 12, December 1997.

[10]   B. Chen, "Maximum likelihood estimation of OFDM carrier frequency offset," IEEE Signal Processing Letters, vol. 9, no.4 pp. 123-126, Apr. 2002.

[11]   X. Wang, Y. Wu, H.C.Wu and G. Gagnon, "An MSE-OFDM System with Reduced Implementation Complexity Using Pseudo Random Prefix, IEEE Global Telecomm. Conference, GLOBECOM'07, pp.2836-2840, Dec., 2007.

[12]   M. Chang and Y. T. Su, "Performance Analysis of Equalized OFDM Systems in Rayleigh Fading", IEEE Transactions on Wireless Communications, vol. 1, No.4, Oct. 2002.

[13]   Mishal Al-Gharabally and P. Das, "Performance Analysis of OFDM in Frequency Selective Time-Variant Channels with Application to IEEE 802.16 Broadband Wireless Access", ISSSTA2004, Sydney, Australia, 30 Aug.-2 Sep. 2004.

[14]   A Goldsmith, Wireless Communications, Singapore, Cambridge University Press, 2005.

[15]   T. S. Rappaport, "Wireless Communications: Principles and Practice," Prentice-Hall, 1996.

# 11

# Survey of routing protocols in wireless sensor networks

**Hussein Mohammed Salman**[1, 2]

[1]College of Material Engineering, Babylon University, Babil, Iraq
[2]Babylon University, Babil, Iraq

**Email address:**
hus12m@yahoo.com

**Abstract:** WSN is one of the most commonly communication tools used in many areas at the life, in both civilians and militaries. These networks composite from a large number of very small devices called sensor nodes. The sensor nodes communicate together by many wirelessly strategies. These communication strategies administrated by routing protocols. There are different types of routing protocol. This paper present and classify these protocols into many categories depending on set of metrics like their infrastructure, their functionalities, the level of privacy and security, or the application which used for it. This paper studied the availability and the reliability of each class of these routing protocols, and the energy consumption of each protocols. Depending on these criteria and other criteria, any future works may use this study to improving these protocols and used them in another types of networks.

**Keywords:** WSN, Routing Protocols, Sensors, Wireless Communications

## 1. Introduction

The original motivation for WSN research stemmed from the vision of Smart Dust in the late 1990s. This vision entailed an integrated computing, communication, and sensing platform consisting of many tiny devices, enabling applications such as dense environmental monitoring and smart home/office.[1]

A typical WSN encountered in the research literature consist of a large number of small, cheap, and resource-constrained sensor as well as a few base stations or sinks. In most WSN settings sensors collect data from the environment and forward it hop by hop to the sink. A sink is a powerful entity that may serve as a gateway to another network, a data processing or storage center, or an access point for human interface. WSN deployment can be ad hoc. The WSN might be often deployment on a large scale throughout a geographic region in hostile environments.

While many sensors connect to controllers and processing stations directly (e.g., using local area networks), an increasing number of sensors communicate the collected data wirelessly to a centralized processing station. This is important since many network applications require hundreds or thousands of sensor nodes, often deployed in remote and inaccessible areas [2]. A wireless sensor has not only a sensing component, but also on-board processing, communication, and storage capabilities. With these enhancements, a sensor node is often not only responsible for data collection, but also for in-network analysis, correlation, and fusion of its own sensor data and data from other sensor nodes. When many sensors cooperatively monitor large physical environments, they form a wireless sensor network (WSN). Sensor nodes communicate not only with each other but also with a base station (BS) using their wireless radios, allowing them to disseminate their sensor data to remote processing, visualization, analysis, and storage systems.

Wireless networks is an emerging new technology that will allow users to access information and services electronically, regardless of their geographic position.[3]

The sensor nodes have significantly lower communication and computation capabilities than do the full-featured computers participating in ad hoc networks. The problem of energy resources is especially difficult [4]. Due to their deployment model, the energy source of the sensor node is considered nonrenewable (although some sensor nodes might be able to scavenge resources from their environment). Routing protocols deployed in sensor networks need to consider the problem of efficient use of power resources.

Sensor networks are composed of resource constrained sensor nodes and more resourced base stations. All nodes in a network communicate with each other via wireless links, where the communication cost is much higher than the computational cost. Moreover, the energy needed to transmit a message is about twice as great as the energy needed to receive the same message. Consequently, the route of each message destined to the base station is really crucial in terms network lifetime: e.g., using short routes to the base station that contains nodes with depleted batteries may yield decreased network lifetime. On the other hand, using a long route composed of many sensor nodes can significantly increase the network delay.[5]

In the section 2; the paper discuss the components of wireless sensor networks, the section 3 review and classify the routing protocols into many categories, and there are some subsections. The conclusions will be at the end of this paper in section 4.

## 2. The Components of WSN

The main components of a general WSN are the sensor nodes, the sink (Base Station) and the events being monitored. Where the communication among the nodes is low-power wireless link while the communication between the base stations low latency and higher bandwidth link, as shown in the fig.1 [6,7].

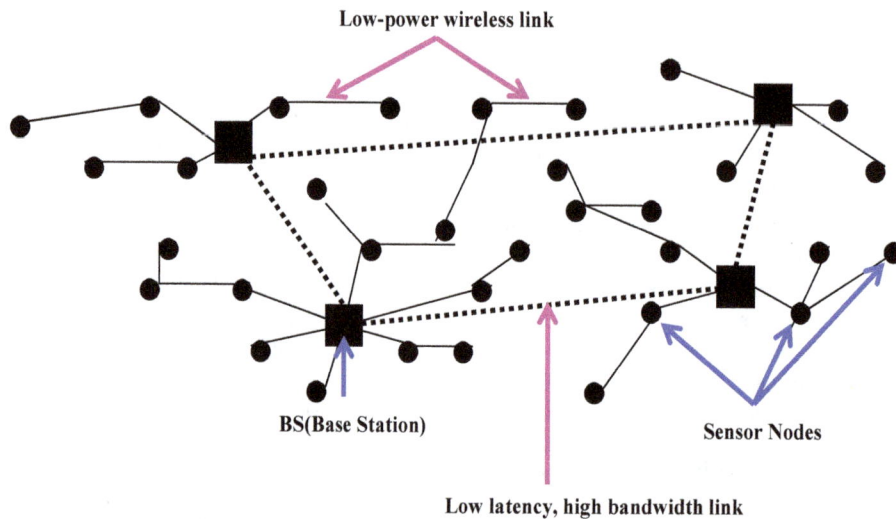

**Fig. 1:** A Representative Sensor Network Architecture

### 2.1. Base Station (Sink) (BS)

The sink (some time cluster head) is an interface between the external (management center) world and computational world (sensor network). It is normally a resourceful node having unconstrained computational capabilities and energy supply. There can be single or multiple base stations in a network. Practically, the use of multiple base stations decreases network delay and performs better using robust data gathering [7].

### 2.2. The Sensor Nodes

As shown in the Fig. 2, a sensor node is composed of four basic components: sensing unit, processing unit, transceiver unit and a power unit [2].

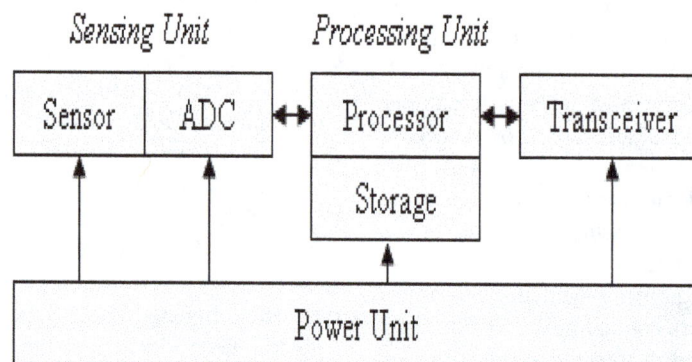

**Fig. 2:** Components of a Sensor

The sensing units are usually composed of two sub-units: Sensors and analogy-to-digital converters (ADCs). The analogy signals perceived by the sensor which are based in the observed phenomenon are converted to digital signals by the ADC, and then they are nourished to the unit of processing. The process unit, that is generally associated to a little storage device, manages / handles the procedures which make the sensor node collaborates with the others nodes in order to carry out the assigned sensor task.

# 3. WSN Routing Protocols

Routing is a process of determining a path between the source node and the sink(destination) node upon request of data transmission. In WSNs the network layer is mostly used to implement the routing of the incoming data. It is known that generally in multi-hop networks the source node cannot reach the sink directly. So, intermediate sensor nodes have to relay their packets. The implementation of routing tables gives the solution. These contain the lists of node option for any given packet destination. Routing table is the task of the routing algorithm along with the help of the routing protocol for their construction and maintenance [2].

WSN Routing Protocols can be classified into five ways, according to the way of establishing the routing paths, according to the network structure, according to the protocol operation, according to the initiator of communications, and according to how a protocol selects a next-hop on the route of the forwarded message, as shown in fig. 3.

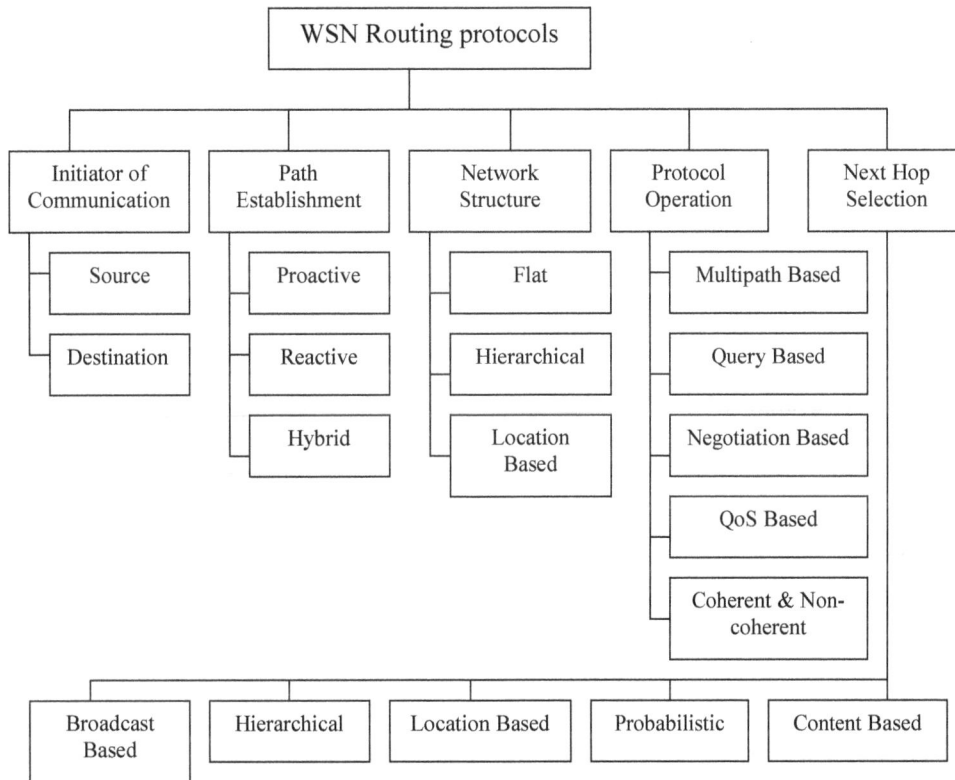

***Fig. 3:*** *WSN Routing Protocols*

## *3.1. Path establishment Based Routing Protocols*

Routing paths can be established in one of three ways, namely proactive, reactive or hybrid. Proactive protocols compute all the routes before they are really needed and then store these routes in a routing table in each node. Reactive protocols compute routes only when they are needed. Hybrid protocols use a combination of these two ideas [ 6].

• Proactive Protocols

Proactive routing protocols are maintain consistent and accurate routing tables of all network nodes using periodic dissemination of routing information. In this category of routing all routes are computed before their needs. Most of these routing protocols can be used both in flat and hierarchal structured networks. The advantages of flat proactive routing is its ability to compute optimal path which requires overhead for this computation which is not

acceptable in many environments. While to meet the routing demands for larger ad hoc networks, hierarchal proactive routing is the better solution [ 8].

• Reactive Protocols

Reactive routing strategies do not maintain the global information of all the nodes in a network rather the route establishment between source and destination is based on its dynamic search according to demand. In order to discover route from source to destination a route discovery query and the reverse path is used for the query replies. Hence, in reactive routing strategies, route selection is on demand using route querying before route establishment. These strategies are different by two ways: by re-establishing and re-computing the path in case of failure occurrence and by reducing communication overhead caused by flooding on networks [ 8 ].

• Hybrid Protocols

This strategy is applied to large networks. Hybrid routing strategies contain both proactive and reactive routing strategies. It uses clustering technique which makes the network stable and scalable. The network cloud is divided into many clusters and these clusters are maintained dynamically if a node is added or leave a particular cluster. This strategy uses proactive technique when routing is needed within clusters and reactive technique when routing is needed across the clusters. Hybrid routing exhibit network overhead required maintaining clusters [ 8 ].

### 3.2. Network Based Routing Protocols

Protocols are divided according to the structure of network which is very crucial for the required operation. The protocols included into this category are further divided into three subcategories according to their functionalities. These protocols are [ 6 ]

- Flat-Based Routing

When huge amount of sensor nodes are required, flat-based routing is needed where every node plays same role. Since the number of sensor nodes is very large therefore it is not possible to assign a particular identification (Id) to each and every node. This leads to data-centric routing approach in which Base station sends query to a group of particular nodes in a region and waits for response. Examples of Flat-based routing protocols are[ 5, 8, 9 ]:

- o Energy Aware Routing (EAR).
- o Directed Diffusion (DD).
- o Sequential Assignment Routing (SAR).
- o Minimum Cost Forwarding Algorithm (MCFA).
- o Sensor Protocols for Information via Negotiation (SPIN).
- o Active Query forwarding In sensor network (ACQUIRE).

- Hierarchical-Based Routing

When network scalability and efficient communication is needed, hierarchical-based routing is the best match. It is also called cluster based routing. Hierarchical-based routing is energy efficient method in which high energy nodes are randomly selected for processing and sending data while low energy nodes are used for sensing and send information to the cluster heads. This property of hierarchical-based routing contributes greatly to the network scalability, lifetime and minimum energy. Examples of hierarchical-based routing protocols are; [ 5, 8, 9 ]

- o Hierarchical Power-Active Routing (HPAR).
- o Threshold sensitive energy efficient sensor network protocol (TEEN).
- o Power efficient gathering in sensor information systems.
- o Minimum energy communication network (MECN).

- Location-Based Routing

In this kind of network architecture, sensor nodes are scattered randomly in an area of interest and mostly known by the geographic position where they are deployed. They are located mostly by means of GPS. The distance between nodes is estimated by the signal strength received from those nodes and coordinates are calculated by exchanging information between neighboring nodes. Location-based routing networks are; [ 5, 8, 9 ]

- o Sequential assignment routing (SAR).
- o Ad-hoc positioning system (APS).
- o Geographic adaptive fidelity (GAP).
- o Greedy other adaptive face routing (GOAFR).
- o Geographic and energy aware routing (GEAR).
- o Geographic distance routing (GEDIR).

### 3.3. Operation Based Routing Protocols

WSNs applications are categorized according to their functionalities. Hence routing protocols are classified according to their operations to meet these functionalities. The rationale behind their classification is to achieve optimal performance and to save the scarce resources of the network.

- Multipath Routing Protocols

As its name implies, protocols included in this class provides multiple path selection for a message to reach destination thus decreasing delay and increasing network performance. Network reliability is achieved due to increased overhead. Since network paths are kept alive by sending periodic messages and hence consume greater energy. Multipath routing protocols are [ 8 ] :

- o Multi path and Multi SPEED (MMSPEED).
- o Sensor Protocols for Information via Negotiation (SPIN).

- Query Based Routing Protocols

This class of protocols works on sending and receiving queries for data. The destination node sends query of interest from a node through network and node with this interest matches the query and send back to the node which initiated the query. The query normally uses high level languages. Query based routing protocols are [ 8 ] :

- o Sensor Protocols for Information via Negotiation (SPIN).
- o Directed Diffusion (DD).
- o COUGAR.

- Negotiation Based Routing Protocols

This class of protocols uses high level data descriptors to eliminate redundant data transmission through negotiation. These protocols make intelligent decisions either for communication or other actions based on facts such that how much resources are available. Negotiation based routing protocols are [8]:

- o Sensor Protocols for Information via Negotiation (SPAN).
- o Sequential assignment routing (SAR).
- o Directed Diffusion (DD).

- QoS Based Routing Protocols

In this type of routing, network needs to have a balance approach for the QoS of applications. In this case the application can delay sensitive so to achieve this QoS metric network have to look also for its energy consumption which is another metric when communicating to the base station. So to achieve QoS, the cost function for the desired QoS also needs to be considered. Examples of such routing are: [ 8 ]

- o Sequential assignment routing (SAR).
- o SPEED.
- o Multi path and Multi SPEED (MMSPEED).
- • Coherent and non-coherent processing:

Data processing is a major component in the operation of wireless sensor networks. Hence, routing techniques employ different data processing techniques. There are two ways of data processing based routing [ 6 ].

- Non-coherent data processing: In this, nodes will locally process the raw data before being sent to other nodes for further processing. The nodes that perform further processing are called the aggregators.

- Coherent data processing: In coherent routing, the data is forwarded to aggregators after minimum processing. The minimum processing typically includes tasks like time stamping, duplicate suppression, etc. When all nodes are sources and send their data to the central aggregator node, a large amount of energy will be consumed and hence this process has a high cost. One way to lower the energy cost is to limit the number of sources that can send data to the central aggregator node.

### 3.4. Initiator of Communication Based Routing Protocol

In this type of routing protocol, it depends on the communication between a network components, where they usually in sleep mode temporary. When any part of a network, the sink (destination, base station) node or the source node, needs service from other part, it will initiate the routing with other part to send or/and receive the control or data packets[ 6 ].

- • Source Initiator Routing Protocol.
- • Destination Initiator Routing Protocol.

### 3.5. Next-Hop Selection Based Routing Protocols

- • Content-based routing protocols

These protocols determine the next-hop on the route purely based on the query content. This type of routing protocols fits the most to the architecture of sensor networks, since the base station do not query specific nodes rather it requests only for data regardless of its origin[ 5, 9 ].

- o Directed Diffusion.
- o GBR.
- o Energy Aware Routing.
- • Probabilistic routing protocols

These protocols assume that all sensor nodes are homogeneous and randomly deployed. Using this routing protocol, sensor nodes randomly select the next-hop neighbor for each message to be forwarded. The probability of selecting a certain neighbor is inversely proportional to its cost[ 5 ].

- o Energy Aware Routing Protocol.
- • Location-based routing protocols

These protocols select the next-hop towards the destination based on the known position of the neighbors and the destination. The position of the destination may denote the centroid of a region or the exact position of a specific node. Location-based routing protocols can avoid the communication overhead caused by flooding, but the calculation of the positions of neighbors may result extra overhead. The local minimum problem is common for all decentralized location-based routing protocols: it might happen that all neighbors of an intermediate node are farther from the destination than the node itself. In order to circumvent this problem, every protocol uses different routing techniques[5].

- o GEAR (Geographical and Energy Aware Routing).
- • Hierarchical-based routing protocols

In case of hierarchical protocols, all nodes forward a message for a node (also called aggregator) that is in a higher hierarchy level than the sender. Each node aggregates the incoming data by which they reduce the communication overload and conserve more energy. Therefore, these protocols increase the network lifetime and they are also well-scalable. The set of nodes which forward to the same aggregator is called cluster, while the aggregator is also referred as cluster head. Cluster heads are more resourced nodes, where resource is generally means that their residual energy level is higher than the average. The reason is that they are traversed by high track and they perform more computation (aggregation) than other nodes in the cluster. Hierarchical routing is mainly two-layer routing where one layer is used to select cluster heads and the other layer is used for routing. [ 5, 9 ]

- o LEACH (Low Energy Adaptive Clustering Hierarchy) protocol.
- • Broadcast-based routing protocols

The operation of these protocols is very straightforward. Each node in the network decides individually whether to forward a message or not. If a node decides to forward, it simply re-broadcasts the message. If it declines to forward, the message will be dropped [ 5 ].

- o MCFA (Minimal Cost Forwarding Algorithm).

## 4. The Conclusions

In this paper, the researcher study the routing protocols in wireless sensor networks and classify them into many categories depending on many metrics, and concludes that there are many differents between these protocols and there are many application for some classes whereas other classes apply in special determine applications, because of the nature these protocols.

# References

[1]   Di Ma, and Gene Tsudik, 2010, "Security and Privacy in Emerging Wireless Networks", IEEE-Wireless Communication, p:12-21.

[2]   Dargie W. and Poellabauer C., 2010," WIRELESS SENSOR NETWORKS THEORY AND PRACTICE", John Wiley & Sons, 1$^{st}$ edition, USA.

[3]   Misra P., 2000, "Routing Protocols for Ad Hoc Mobile Wireless Networks", http://www.cis.ohio-state.edu/~misra

[4]   Boukerche A.,2009," ALGORITHMS AND PROTOCOLS FOR WIRELESS SENSOR NETWORKS", John Wiley & Sons, Canada.

[5]   Acs G. and Butty'an L., 2007, "A Taxonomy of Routing Protocols for Wireless Sensor Networks", Budapest University of Technology and Economics, Hungary.

[6]   Sharma G., 2009, "Routing in Wireless Sensor Networks", Master Thesis, Computer Science and Engineering Dept., Thapar Univ., Patiala.

[7]   Akyildiz I. F. ; Su W.; Sankarasubramaniam Y. and Cayirci C., 2002,"A Survey on Sensor Networks", IEEE communication magazine.

[8]   Ullah M. and Ahmad W., 2009, "Evaluation of Routing Protocol in Wireless Sensor Networks", master thesis, Department of School of Computing, Blekinge institute of technology, Sweden.

[9]   Luis J. *et al*, 2009, "Routing Protocols in Wireless Sensor Networks", sensor,Spain, www.mdpi.com/journal/sensors.

# 12

# Investigation of mutual coupling effects in conventional and fractal capacitive coupled suspended RMSAs

**Miti Bharatkumar Sukhadia[*], Veeresh Gangappa Kasabegoudar**

Post Graduate Department, Mahatma Basaveshwar Education Society's, College of Engineering, Ambajogai, India

**Email address:**
miti.sukhadia@yahoo.com(M. B. Sukhadia), veereshgk2002@rediffmail.com(V. G. Kasabegoudar)

**Abstract:** An array of coplanar capacitive coupled suspended microstrip antenna has been investigated to study the mutual coupling effect between its elements. Also, the work is extended to study the mutual coupling effect in Koch shaped coplanar capacitive coupled suspended fractal microstrip antenna. From the investigations it is observed that less coupling effect occurs between the elements in fractal array. This study reveals the use of fractal geometries in place of conventional (regular) shapes because of their numerous advantages like self similarity property, ability to excite multiple resonant modes etc. Antenna arrays are simulated using IE3D software from Zeland which is based on Method of Moments (MoM).

**Keywords:** Microstrip Antennas, Fractal Antennas, Miniaturization, Mutual Coupling

## 1. Introduction

Due to a dynamic advancement in wireless communication industry, modern antenna components are required to have compact size, small weight, low profile, and low cost of the production [1-3]. Based on the simplicity of design and circuit implementation as well as their ability to integrate with RF devices easily, conventional microstrip patch antennas are found to be extremely useful in many wireless communication applications [4,5].

However, microstrip antennas exhibit the drawback of poor bandwidth and gain in their conventional form. There are several researches reported in literature to improve the bandwidth [6-17]. It is well known that to increase the gain of microstrip antenna, multiple elements are used. However, the spacing between the elements plays important role in calculating the area required and mutual coupling effect between the elements.

It is necessary that efficient miniaturization of RF/microwave antenna circuit is expected from novel mobile devices with adequate radiating properties; the solution of this limitation can be reached by means of fractal geometry. Fractal geometry occupy less space and produces same radiation patterns and input impedance as conventional geometry.

However, such radiators suffer from narrow bandwidth and insufficient radiation pattern control [6]. In order to overcome this limitation, a new antenna array is designed which allow us to achieve a modified radiation patterns as well as broadband or multiband characteristics that would finally increased the efficiency[7-10].

On the other hand, mutual coupling adversely affects the feeding voltages of the array elements, typically raising the level of the back radiation and filling in the nulls of the antenna patterns [11].

This problem can be managed by using smaller array elements, which would maintain inter-element spacing while increasing the physical gap between elements. In this paper, mutual coupling effect between the array elements is presented for conventional and fractal elements. The microstrip patch, fractal patch and array are designed and optimized. Section 2 presents the basic geometries and microstrip fractal patch with their working. Simulation results and antenna array with mutual coupling have been presented in Section 3 followed by conclusions in Section 4.

## 2. Antenna Geometry

In this section geometry description of basic element is given. The basic geometries considered are conventional coplanar capacitive coupled probe fed suspended microstrip antenna and its fractal shape.

## 2.1. Conventional Patch

The microstrip geometry in its conventional form has limited bandwidth (~2-3%), which is inadequate for handling the high data traffic. To handle high data rates many broad banding techniques have been proposed by several researchers [10-17]. Among these a single layer coplanar capacitively coupled probe fed offers an impedance bandwidth of about 50%. The geometry of this antenna is shown in Figure 1. The optimized dimensions of the geometry are presented in Table 1 [13].

**Table 1.** *Optimized dimensions for capacitively coupled probe fed suspended patch antenna [13].*

| Parameter | Value |
|---|---|
| Length of the radiator patch ($L$) | 15.5mm |
| Width of the radiator patch ($W$) | 16.4mm |
| Length of the feed strip ($s$) | 3.7mm |
| Width of the feed strip ($t$) | 1.2mm |
| Separation of feed strip from the patch ($d$) | 0.5mm |
| Air gap between substrates ($g$) | 6.0mm |
| Relative dielectric constant ($\varepsilon r$) | 3.0 |

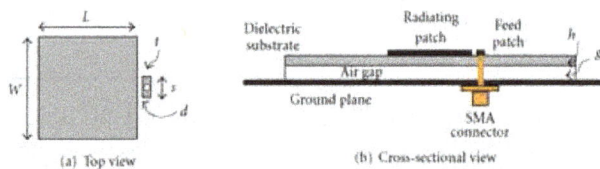

**Figure 1.** *Coplanar capacitively coupled probe fed suspended microstrip antenna (a) Top view (b) Cross-sectional view [12].*

As reported in [13], this geometry offers an impedance bandwidth of about 50% with good gain and patterns throughout the band of operation.

## 2.2. Fractal Antenna Geometry

To reduce the size of planner antenna, number of methods have been developed and reported [14-19]. It may be noted that the defects in the patch metallization lead to noticeable radiator miniaturization [14]. However, one can observe that such a technique decreases antenna resonant frequency, so antenna optimization is necessary. An alternative method of antenna size reduction is reported in [15] based on introducing a lumped element between the ground plane and signal line.

However, it is important to emphasis that such a solution adds complexity to the circuit. An integrating approach to antenna miniaturization involves the utilization of meta materials [16, 17]. However, the use of artificial materials leads to certain fabrication problems and the design process becomes complex. One of the solutions to above limitations can be reached by means of fractal curves. It has been reported in [18, 19] that the application of space filling fractal curves enables us to achieve considerable miniaturization and also wideband and multi band properties.

Figure 2 shows examples of fractal geometries proposed by Koch (1904) and Sierpinski (1916).

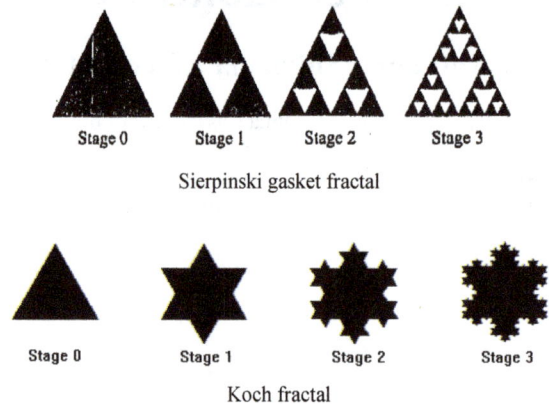

Sierpinski gasket fractal

Koch fractal

**Figure 2.** *Typical fractal geometries [7].*

Fractal has interesting properties like self-similar and space filling by iterating certain simple geometry, due to this they are use in many applications.

In this work we use the geometry presented in Figure 1 for the design of fractal antenna. This conventional antenna is treated as the base for the comparison in terms of size reduction.

Next, the structure is modified by the addition of multiple V-grooves along the length and width in three steps, which corresponds to the three iterations of the fractal generation. The addition of the groove is based upon the "Koch curve". The middle part is replaced by two straight lines meeting at 60 angles (a bent) and they fit into the original gap in an equilateral triangular fashion as shown in the Figure 3.

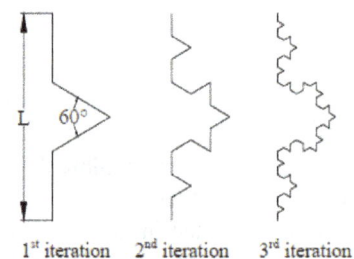

**Figure 3.** *Fractal generation based on Koch curve [7].*

Thus the dimension of each newly generated straight line is now one third of the original straight line and when each side of the square is stretched out, length increases by one third of the original length. The iterative process of dividing a straight line into three equal segments and replacing the middle by a bent curve is continued. In the true fractal, this process is repeated for infinite number of times. In the present work, three iterations are considered and the fractal patch obtained from this technique is shown in the Figure 4.

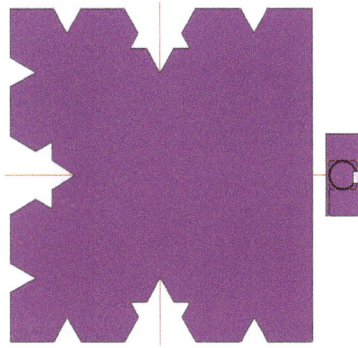

**Figure 4.** *Coplanar capacitively coupled fractal patch.*

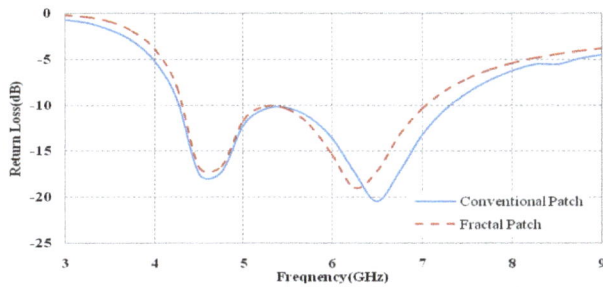

**Figure 5.** *Return loss comparisons of fractal and conventional geometries.*

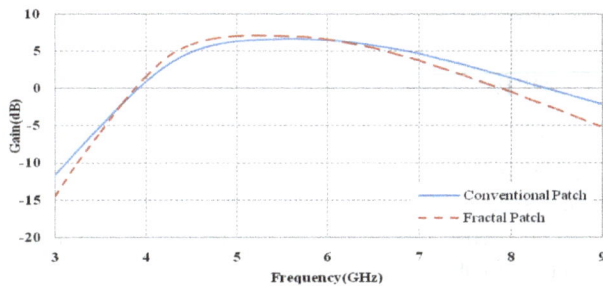

**Figure 6.** *Gain vs. frequency plot of conventional and rectangular plots.*

It is well known that microstrip antennas are good candidates for size reduction, as large electrical length can be fitted into the small physical volume. However, to make the antenna resonate at a particular frequency the useful range of size reduction lies only up to third iterations. Maximum size reduction results after first iteration only. Subsequent iterations result in decrease in percentage of size reduction such as only 5% or less than this is obtained after third iteration [7]. Therefore, one can limit the iterations up to third iterations only. Going for the higher iteration adds mostly the small edges, which is practically not useful in increasing the electrical length.

# 3. Simulation Results and Discussions

In this section two elements conventional and fractal elements arrays have been investigated to observe the coupling between the elements. All simulations have been carried out from Zeland's IE3D which is MoM based electromagnetic software. Distance between the array elements has been varied and optimized.

## 3.1. Design of Two Elements Array and Effect of Mutual Coupling

It is well known that arranging antenna elements in a predefined manner (array) can boost the performance of the radiating system like increasing the directive gain and reducing the beam width. However, the spacing between the elements plays a crucial role. Large spacing offers the enhanced gain at the expense of increased overall size and side lobes. The side lobes start appearing and become significant when spacing between the elements increases over one wavelength. Also, in applications where size is the prime constraint, increasing the spacing between the elements may not be feasible. On the other hand if the spacing is kept low results in the mutual coupling between the elements and thus disturbs the radiation characteristics and also reduces the gain of the antenna. Hence, in this work an effort has been made to find the optimum spacing between the elements of the proposed geometry and also the same is demonstrated for the two elements array with fractal geometry. Following paragraph explains the effect of mutual coupling in conventional as well as fractal arrays.

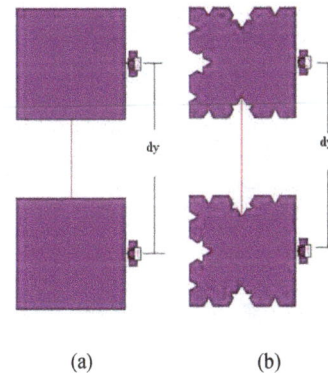

(a)                    (b)

**Figure 7.** *Two elements microstrip array placed 0.9λ$_g$ apart (a) Conventional array (b) Fractal array.*

Two elements conventional as well as fractal array are shown in Figure 7. The separation between the centres of element is varied from 0.9λ$_g$ to 2.25λ$_g$. Where guided wavelength (λ$_g$) is calculated at the resonant frequency of 5.75 GHz. The substrate used for design and simulation is RT Duriod make laminate with relative dielectric constant of 3.0 and thickness of 1.56mm. For comparison, we have chosen an array of rectangular patches with same number of elements. We have assumed that all patch elements are fed uniformly (their exciting current have the same amplitude and phase). Amount of coupling between the elements of conventional and fractal arrays is presented in Figures 8 (a) and (b). It may be observed that amount of coupling is less in fractal array elements than elements in the conventional array for all cases studied. Also, similar radiation (Figure 9) and gain (Figure 10) characteristics are observed in both fractal array and conventional array. From the parametric study it is clear that if mutual coupling is main constraint, fractal arrays may be preferred.

S$_{12}$ plots for conventional patch elements.

S$_{12}$ plots for fractal patch elements.

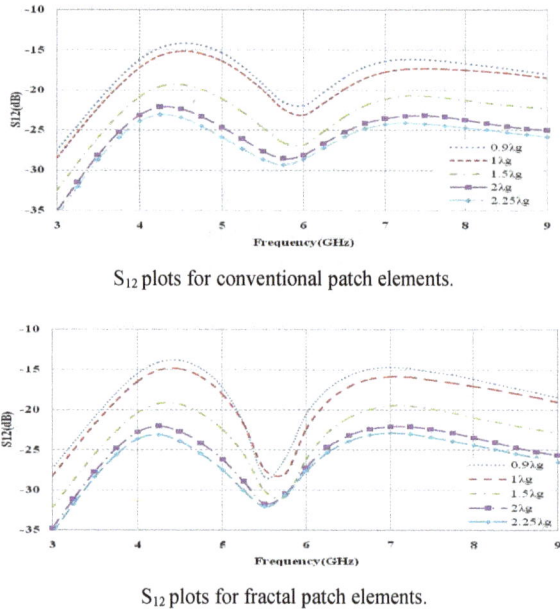

***Figure 8.*** *Comparison of mutual coupling effect in antenna array elements for different spacing.*

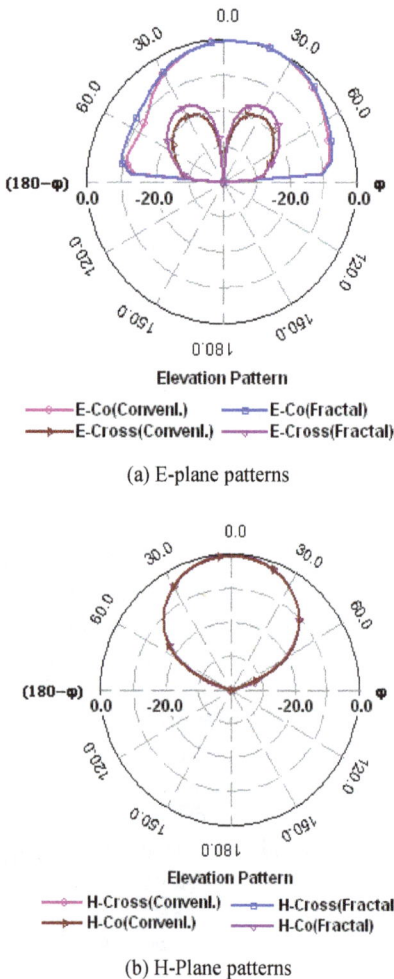

Gain vs. frequency plot for conventioanl antenna array.

Gain vs. frequency plot for fractal antenna array.

***Figure 10.*** *Gain vs. frequency plot of antenna array elements for different spacing between the elements.*

## 4. Conclusions

The fractal geometry of coplanar capacitively coupled probe fed microstrip antenna is presented. The size of antenna is reduced by increasing order of iteration. The fractal geometry offers similar performance as that of conventional patch. The effect of mutual coupling between the array elements has been presented for regular as well as fractal geometries. Fractal patches array exhibits less coupling effect between the elements with slightly increased gain. The increase in the gain is due to reduction in mutual coupling. Hence, fractal elements of capacitive coupled geometry may be used for the design of array to improve the overall performance of the antenna.

(a) E-plane patterns

(b) H-Plane patterns

***Figure 9.*** *Comparison of radiation patterns obtained from conventional and fractal array geometries for optimum spacing between the elements (0.9λ$_g$) at resonant frequency.*

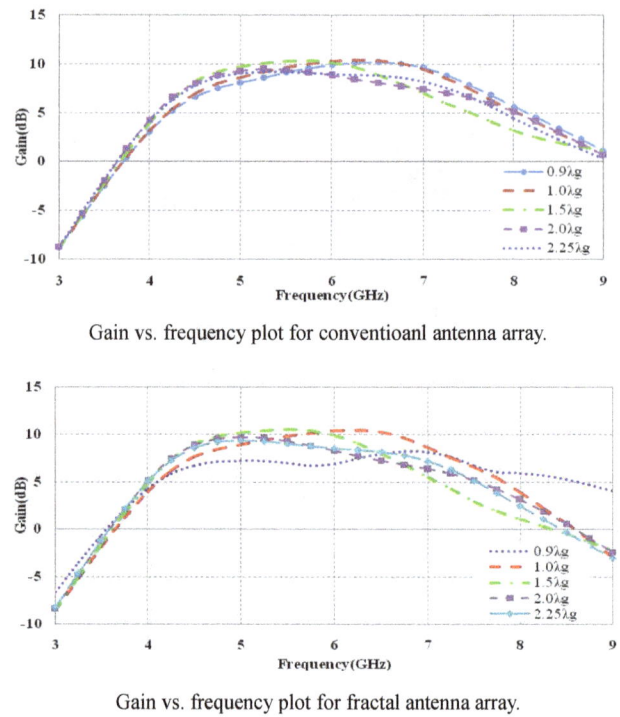

## References

[1]   C. A. Balanis, *Antenna Theory*, John Wiley & Sons, Hoboken, NJ, USA, 2$^{nd}$ Edition, 2004.

[2]   R. Garg, P. Bhartia, I. Bahl, and A. Ittipiboon, *Microstrip Antenna Design Handbook*, Artech House, Norwood, Mass, USA, 2001.

[3]   R. Kumar and P. Malathi, "On the design of fractal patch antenna and backscattering reduction," *International Journal of Recent Trends In Engineering*, vol. 2, pp. 4-6, 2009.

[4]   J. Bahl and P. Bhartia, *Microstrip Antennas*, Artech House, 1981.

[5]   R. Garg, "Progress in Microstrip Antennas," *IETE Technical Review*, vol. 18, pp. 85-98, 2001.

[6] S. A. Fares and F. Adachi, *Mobile and Wireless Communications Network Layer and Circuit Level Design*, Intech, 2010.

[7] N. Yousefzadeh and C. Ghobadi, "Consideration of mutual coupling in a micro strip patch array using fractal elements," *PIER*, vol. 6, pp. 41–49, 2006.

[8] N. Fourikis, *Advanced Array Systems, Applications and RF Technologies Phased Array Systems*, Academic Press, Ascot Park, Australia, 2000.

[9] A. A. Eldek, A. Z. Elsherbeni, and C. E. Smith, "Rectangular slot antenna with patch stub for ultra wideband application and phased array systems," *PIER*, vol. 53, pp. 227–237, 2005.

[10] P. S. Kooi, M. S. Leong, and T. S. Yeo, "A method of moments analysis of a micro strip hazed array in three-layered structures," *PIER*, vol. 31, pp. 155–179, 2001.

[11] W. L. Stutzman and G. A. Thiele, *Antenna Theory and Design*, John Wiley and Sons, 2nd Edition, 1998.

[12] V. G. Kasabegoudar, "Low profile suspended microstrip antennas for wideband applications," *Journal of Electromagnetic Waves and Applications,* vol. 25, pp. 1795-1806, 2011.

[13] V. G. Kasabegoudar, D. S. Upadhyay, and K. J. Vinoy, "Design studies of ultra-wideband micro strip antennas with a small capacitive feed," *International Journal of Antenna and Propagation*, pp. 1-8, vol. 2007.

[14] G. A. Mavridis, C. G. Christodoulou, and M. T. Chryssomallis, "Area miniaturization of a microstrip patch antenna and the effect on the quality factor Q," *IEEE Antennas and Propagation Society International Symposium*, pp. 5435-5438, 2007.

[15] K. L. Wong, *Compact and Broadband Microstrip Antennas*, John Wiley & Sons, 2002.

[16] R. O. Ouedraogo and E. J. Rothwell, "Metamaterial inspired patch antenna miniaturization technique," *IEEE Antennas and Propagation Society International Symposium*, pp. 1-4, 2010.

[17] M. Palandoken, A. Grede, and H. Henke, "Broadband microstrip antenna with left-handed metamaterials," *IEEE Trans. On Antennas and Propagation*, pp. 331-338, 2009.

[18] J. Guterman, A. Moreira, and C .Peixeiro, "Microstrip fractal antennas for multi standard terminals," *IEEE Antennas and Wireless Propagation Letters*, pp.351-354, 2004.

[19] R. V. Hariprasad, Y. Purushottam, V. C. Misra, and N. Ashok, "Microstrip fractal patch antenna for multiband communication," *IEEE Electronic Letters*, pp.1179-1180, 2000.

[20] H. T. Hui and M. E. Bialkowski, "Mutual coupling in antenna arrays," *International Journal of Antennas and Propagation*, 2010.

[21] C. Ludwig, "Mutual coupling, gain, and directivity of an array of two identical antennas," *IEEE Transactions on Antennas and Propagation*, vol. 24, no. 6, pp. 837–841, 1976.

[22] I. J. Gupta and A. A. Ksienski, "Effect of mutual coupling on the performance of adaptive arrays," *IEEE Transactions on Antennas and Propagation*, vol. 31, pp. 785–791, 1983.

[23] J. A. G. Malherbe, "Analysis of a linear antenna array including the effects of mutual coupling," *IEEE Transactions on Education*, vol. 32, pp. 29–34, 1989.

# 13

# A novel hybrid multiple access scheme in downlink for 4G wireless communications

**Poonam Singh[1], Saswat Chakrabarti[2]**

[1]Electronics & Comm. Engg. Dept., National Institute of Technology, Rourkela, India
[2]GSSST, Indian Institute of Technology, Kharagpur, India

**Email address:**

psingh@nitrkl.ac.in(P. Singh), saswat@ece.iitkgp.ernet.in(S. Chakrabarti)

**Abstract:** In this paper, we present a hybrid multiple access scheme to increase the data rate and to provide more flexibility for 4G wireless communications. The proposed scheme utilizes a combination of the multiple access techniques TDMA, CDMA and OFDMA. It uses variable time slots, spreading factors and number of subcarriers to provide flexible data rates for different classes of users. The performance analysis of the proposed scheme is carried out for different modulation schemes used in downlink and expressions for multiple access interference (MAI) and bit error rate (BER) are derived. The parameters for this scheme are selected depending on available bandwidth, required data rates, number of users, type of service provided etc. It is found that this scheme provides better scalability and can support more number of users.

**Keywords:** Multiple Access Scheme, FDMA, TDMA, CDMA, OFDMA

## 1. Introduction

An important goal of next generation wireless system is the convergence of multimedia services such as speech, audio, video, image and data. This implies that a future wireless terminal should be able to connect to different networks in order to support various services by guaranteeing high speed data. It is, therefore, very important to find a suitable transmission technique with high spectral efficiency and robustness to various distortions.

A variety of wireless, mobile Internet-access systems exist which provide broadband access to the Internet e.g. WLAN, WMAN etc. Orthogonal Frequency Division Multiple Access (OFDMA) is a multiple access scheme in which each user occupies different subcarriers. It offers dynamic user capacity and lower interference to adjacent cells [1] [2]. It has been adopted in WiMAX (IEEE 802.16e standard) [3] [4] and it is an attractive candidate for Beyond 3G systems due to its ability to limit intersymbol interference caused by multipath channels. The MC-CDMA and MC-DS-CDMA modulations combine the benefits of OFDM and CDMA to support high data rate transmission [5 - 9]. The hybrid TD-CDMA technique, which is selected

as a standard for TDD mode of 3rd Generation Partnership Project (3GPP) [10] [11], uses a combination of FDMA, TDMA and CDMA for multiple access and allows a flexible use of the limited transmission resources. Single-carrier FDMA (SC-FDMA) is a hybrid modulation scheme that combines the low PAPR of single-carrier systems with the multipath resistance and flexible subcarrier frequency allocation offered by OFDM and has been proposed for the uplink of 3GPP Long Term Evolution (LTE) [12].

In this paper, we present a hybrid multiple access scheme to increase the data rate and to provide more flexibility for 4G wireless communications e.g. LTE, LTE-Advanced and IEEE 802.16m. The proposed scheme including its system model is described in Section II, which utilizes a combination of the multiple access techniques TDMA, CDMA and OFDMA. It uses variable time slots, spreading factors and number of subcarriers to provide flexible data rates for different classes of users. The performance analysis of the proposed scheme is carried out for different modulation schemes used in downlink in Sections III and expressions for multiple access interference (MAI) and bit error rate (BER) are derived. Some simulation results are given in Section IV. The parameters for this scheme are selected depending on available bandwidth, required data rates, number of users, type of service provided etc. Finally, the conclusions are given in Section V. It is found that this scheme provides better scalability and can support more number of users than what is mentioned in LTE proposals.

## 2. A Hybrid Multiple Access Scheme (TD-MC-CDMA) in Synchronous Downlink

Though the multicarrier methods like OFDMA, MC-CDMA and MC-DS-CDMA can be used with TDM/TDMA for transmission, there is no explicit mention of TDMA application along with MC-CDMA in the existing literature. We would like to propose this combination for 4G applications to increase the scalability in the access. We are proposing variable spreading factors and number of time slots for different users depending on their data rates. This will provide a greater scalability, since the users with higher data rates will use lower spreading factor and will get more number of time slots. On the other hand, more number of users can be supported at low data rates with larger spreading factors. Analysis of OFDM and multicarrier spread spectrum modulation schemes have been done in many papers [13-20]. We have extended this analysis for the proposed TD-MC-CDMA system.

Fig.1 shows the structure of transmitter for the proposed scheme for synchronous downlink. We consider a system with K users having different data rates. The data rate $R_k$ of user k ($1 \leq k \leq K$) can be any one of the $N_d$ data rates supported by this scheme i.e. $R_{min} < R_k < R_{max}$. These $N_d$ data rates are integral multiples of the basic or minimum data rate $R_{min}$. The maximum data rate can be written as $R_{max} = N_{max} R_{min}$.

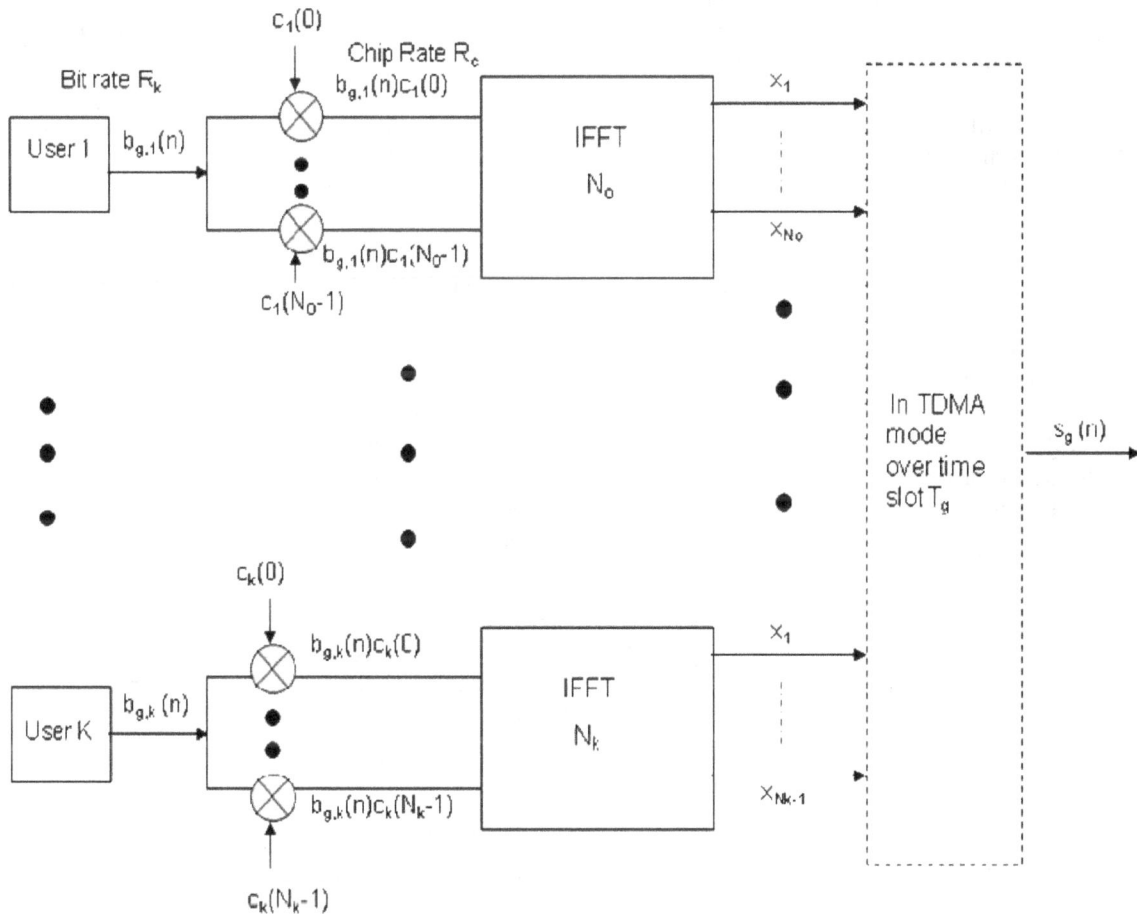

*Fig.1. Transmission Scheme for users of group 'g' in the downlink*

The complex valued data symbol $b_{g,k}$ of each user is multiplied with a spreading sequence $c_{g,k}$ of length L. The spreading factors are variable for different service classes, they are chosen depending on the data rates for each application. For lower data rates, a higher spreading factor and for higher data rates, a lower spreading factor is selected, so that the final chip rate $R_c$ for each user remains same. Thus, a large number of low rate users or a small number of high data rate users can be accommodated in a given bandwidth. The spreading factor Sf for one application can be selected from a set of spreading factors $Sf_{min}$ to $Sf_{max}$, ($Sf_{min} \leq Sf \leq Sf_{max}$). The complex valued data sequence for each user obtained after spreading is then modulated onto a number of subcarriers $N_k$.

The number of subcarriers is also variable and is equal to spreading factor for each user since each spread data is modulated on one subcarrier in this case. Thus each data symbol is spread over many subcarriers. As discussed in [8], the multicarrier CDMA schemes can be categorized into two groups: One (MC-CDMA) spreads the original data stream using a given spreading code and then modulates different subcarriers with each chip, i.e. spreading in the frequency domain. The other group (MC-DS-CDMA) spreads the serial to parallel converted data streams using a given spreading code and then modulates different carriers with each data stream i.e. spreading in the time domain similar to a conventional DS-CDMA scheme. In TD-MC-CDMA scheme also, the chips of spread data symbol are transmitted in frequency direction over several parallel subchannels or in time direction over several multicarrier symbols. We propose TD-MC-CDMA with spreading in frequency direction for the downlink and spreading in time direction for the uplink in order to optimize both the spectral efficiency and mobile power consumption. The orthogonality among codes is maintained in the downlink so frequency domain spreading can be used to get frequency diversity. However, in the uplink, the transmitted signal of each user is affected by different channel impulse responses. So the orthogonality of signals is destroyed increasing MAI [20]. In time domain spreading, all the chips are transmitted on each of the subcarriers, so synchronization problems are reduced. In both the cases, variable time slots, spreading factors and variable number of subcarriers are used. In this paper, we analyze the multiple access scheme proposed for downlink only.

In one frame duration $T_F$, there is $N_s$ number of time slots. The number of time slots $N_k$ allocated to a user k depends on its data rate $R_k$. For example, the number of time slots allocated to a user with audio signal having a smaller data rate will be less as compared to time slots allocated to users having video or multimedia transmissions with higher data rates. A number of users with same data rates can be grouped together as a group g and transmitted in one time slot. There can be maximum G number of groups, which is equal to the number of time slots in a frame $N_g$. In each OFDM symbol, a guard time of $T_{cp}$ is usually inserted to reduce intersymbol interference. The OFDM symbol duration including a guard interval is $T_{sym} = T + T_{cp}$, where T is the actual symbol duration. The maximum achievable data rate depends on the available channel bandwidth, number of time slots and the spreading factor.

The modulation scheme employed can be QPSK, 16-QAM or 64-QAM depending on the data rate of users and channel conditions. For audio data which require a lower bandwidth and better BER at a given SNR, lower modulation like QPSK is used, whereas for video signals requiring a higher bandwidth and comparatively lower BER requirements, higher modulation schemes like 16-QAM or 64-QAM is used. Adaptive modulation and coding can be used to improve the system efficiency.

Fig.2 shows the receiver of user k in the TD-MC-CDMA system for downlink. In the receiver, after removing the guard interval and taking inverse OFDM the received sequence is equalized to get the transmitted data. For inverse OFDM, $N_k$ point FFT is taken and then frequency domain equalization is performed. After channel estimation, the output of each subcarrier is equalized and then coherently combined over the parallel subcarrier components using same code as used at the transmitter i.e. despreading of signals is performed to recover the transmitted binary data. The demodulation and decoding used at the receiver should be corresponding to modulation and coding used at the receiver. All these information are passed to receiver in the beginning of every frame in form of a preamble.

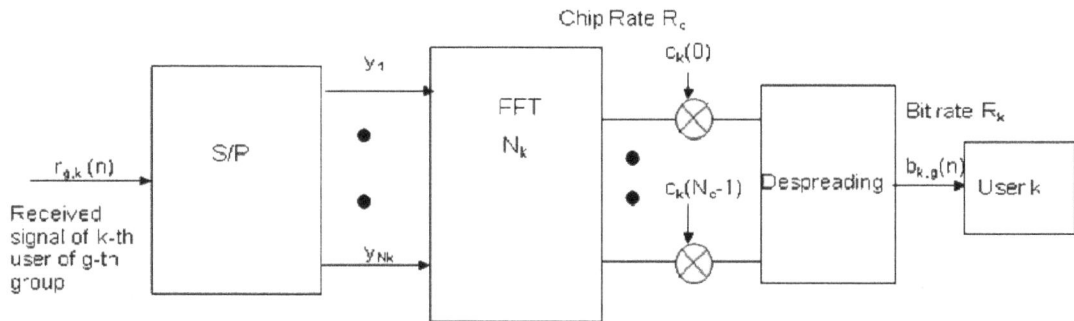

*Fig.2. Receiver of user k using TD-MC-CDMA for downlink*

### 2.1. Frame Structure for Downlink TD-MC-CDMA

The frame structure for TD-MC-CDMA system is shown in Fig.3; a radio frame with duration of $T_F$ is subdivided into $N_s$ main time slots (T1 to $TN_s$) of duration $T_g$ and a preamble $T_{pre}$. The preamble consists of time slots for downlink synchronization (DL Sync) and uplink synchronization (UL Sync). The burst structure of the main time slots consists of data blocks, pilot signals and a guard period of duration $T_p$. The frame duration can be written as

$$T_F = T_{pre} + N_s(T_g + T_p) \qquad (1)$$

In the time domain the period of one OFDM block is $T_{sym}$ including an effective block period of T and a cyclic prefix of $T_{cp}$. A time slot of $T_g$ is equivalent to $N_{os}$ OFDM symbols. $T_g$ can be written as

$$T_g = N_{os}T_{sym} = N_{os}(T + T_{cp}) \qquad (2)$$

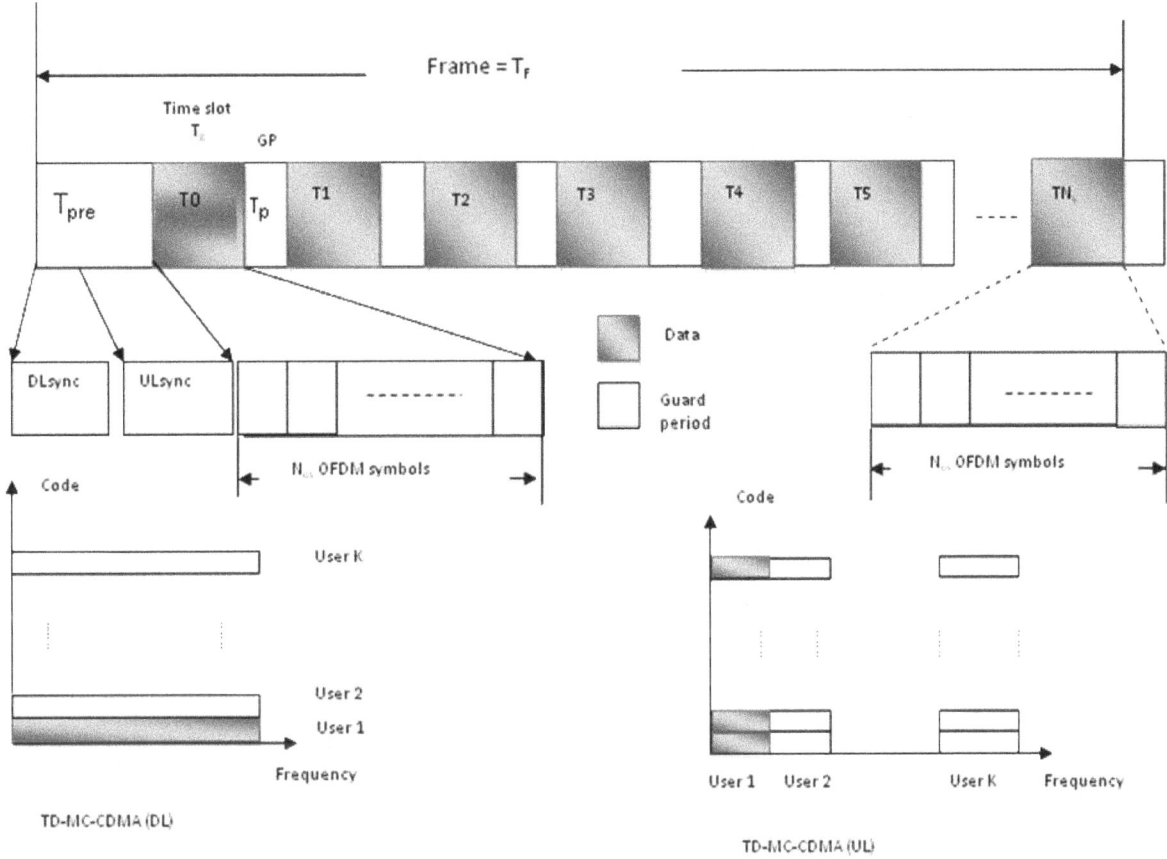

*Fig.3. The frame structure for downlink of a TD-MC-CDMA system*

# 3. Performance Analysis

We have taken the following assumptions for analysis:

a) There are G groups of K users each with data rates $R_k$ and binary spreading codes $c_k(t)$.

b) The transmitted binary data $b_k$ (k=1,2...K) for K users are independent and identically distributed (i.i.d.) random variables.

c) Users can be assigned more than one time slot, but for simplicity we assume one time slot per frame for each user.

d) Signal power of the data bits $P_s$ is same for all users.

e) $w_i = w_c + 2\pi m/T_b$ is the m-th subcarrier frequency and $w_c$ is the radio frequency

f) $\theta_k$ is the random carrier phase uniformly distributed over $[0,2\pi]$.

g) Walsh codes are used for spreading since they are orthogonal.

h) The modulation scheme employed is QPSK.

i) The number of subcarriers N is equal to spreading factor for each user since each chip is modulated on one subcarrier in the downlink.

j) The statistics of all the Rayleigh fading channels are assumed to be identical for different users.

The input information data symbols $b_{k,g}(n)$ of k-th user of g-th group are first modulated using QPSK. Then each symbol is multiplied with one chip of the spreading code $c_k(l)$ ( 1 = m = 1, 2,...N) with length $N_k$ and chip duration $T_c$ and modulated to one of the $N_k$ subcarriers as shown in Fig.1,. The transmitted signal for the k-th user in g-th group is given by

$$s_{k,g}(n) = \sum_{m=1}^{N_k} \sqrt{2P_s} b_{g,k}(n) c_{g,k}(m) \exp(jw_m t + \theta_{g,k,m}) \quad (4)$$

where $P_s$ is the signal power of each subcarrier and is assumed to be the same for all users, $w_m = w_c + 2\pi m/T_b$ is the m-th subcarrier frequency, $w_c$ is the radio frequency and $\theta_k$ is the random carrier phase uniformly distributed over $[0,2\pi]$.

The transmitted signal is the sum of K users' spread data modulated on $N_c$ subcarriers given by

$$s_g(t) = \sum_{k=1}^{K} \sum_{m=0}^{N_c-1} \sqrt{2P_s} b_{g,k}(n) c_{g,k}(m) \exp(jw_m t + \theta_{g,k,m}) \quad (5)$$

Each transmitted subcarrier signal passes through a frequency non-selective Rayleigh fading channel, and is also subjected to additive white Gaussian noise (AWGN) n(t), with a double-sided power spectral density $N_o/2$. For different users the statistics of all the Rayleigh fading channels are assumed to be identical. Furthermore, we assume that the channel fading is slow and remains unchanged over two consecutive symbols. The complex impulse response of the Rayleigh fading channel for the m-th subcarrier can be written as

$$h_m(t) = \beta_m e^{j\theta_m} \delta(t) \qquad (6)$$

where $\theta_m$ is a random phase introduced by the channel, it is modeled as uniformly distributed over the interval of $[0, 2\pi]$ and is assumed to be i.i.d. for each subcarrier and each user, $\beta_m$ is a Rayleigh random variable with power $E[\beta^2_m] = \sigma^2$ where $E[.]$ is the expectation operator.

Given the above channel characteristics, the received signal for k-th user can be written as

$$r_g(n) = h_g(n) * s_g(n) + w(n)$$

$$= \sum_{k=1}^{K} \sum_{m=0}^{N_c-1} \sqrt{2P_s} \beta_m s_k(n-n_k) \exp(jw_m t + \phi_m) + w(n)$$

$$= \sum_{k=1}^{K} \sum_{m=0}^{N_c-1} \sqrt{2P_s} \beta_m b_{g,k}(n-n_k) c_{g,k}(n-n_k) \exp(jw_m t + \phi_m) + w(n) \qquad (7)$$

where $n_k$ is the propagation delay, uniformly distributed over $[0, \pm T_c/2]$, w(n) is the additive white Gaussian noise (AWGN) with zero mean and a double-sided power spectral density of $N_o/2$, $\beta_m$ and $\varphi_m$ are the fading envelopes and random phase experienced by each user at subcarrier m. $\beta_m$, $\varphi_m$, $n_k$ and $b_{g,k}(t)$ are assumed to be independent and identically distributed (i.i.d.) for different k.

In this system, each chip modulates a different carrier frequency. Therefore, the chip duration is same as the bit duration. In multicarrier modulation, each subcarrier experiences frequency non-selective fading. Each subcarrier is modulated with a narrowband signal whose bandwidth is smaller than the channel coherence. A combination of IDFT and a cyclic prefix at the transmitter with the DFT at the receiver converts the frequency selective channel to separate flat fading channels. Therefore, in frequency-selective fading channels, each subcarrier experiences flat fading that may be jointly correlated.

Channel estimation is used to compensate for the amplitude and phase distortions associated with the received signal. To estimate the multiplicative channel response, pilot symbols are inserted among the transmitted data symbols. The receiver estimates the channel state information based on the received, known pilot symbols. At the receiver, the guard intervals of the received signals are removed and the resultant symbol sequence is converted to a modulated signal of each subcarrier using FFT. For channel estimation, the channel impulse response at each subcarrier is found by averaging the impulse response measured for each of the dedicated pilot symbols. Using the obtained channel impulse response, the output of each subcarrier is equalized and then coherently combined over the N parallel subcarrier components i.e. despreading the signals. Finally, the regenerated symbol sequences are parallel to serial converted to recover the transmitted binary data.

Since the subchannel bandwidth is smaller, fading per subchannel is flat and low complex detection techniques can be used. We consider a correlation receiver, assuming perfect channel estimation for the desired user. In the receiver, the received signal is copied and fed to $N_c$ subcarriers. Each subcarrier is demodulated and despread by multiplying with one chip of the spreading code. The combined signal is then passed through a matched filter followed by a maximum likelihood detector.

Let $x_m$ be the summation of K users' transmitted m-th chip during the i-th bit interval:

$$x_m = \sum_{k=1}^{K} b_i^k c_{i,m}^k \qquad m = 0,1,..N-1 \qquad (8)$$

where $c_{i,m}^k$ is the k-th user's m-th chip in the i-th bit interval. The sequence $x_m$ is parallel to serial converted and its IDFT is taken. The IDFT is

$$w_h = \frac{1}{N} \sum_{m=1}^{N} x_m \exp(j2\pi mh/N), \text{ h} = 0,1....N\text{-}1 \qquad (9)$$

The P/S conversion and zero-th order interpolation gives the continuous-time signal

$$w(t) = \sum_{h=0}^{N-1} w_h . q(t - hT_c) \qquad (10)$$

where q(t) is the unit rectangular pulse over a chip interval. Let the normalized frequency offset $\varepsilon$ be $f_o / \Delta f$, where $f_o$ is a frequency offset and $\Delta f = 1 / NT_c$. At the receiver input, the noiseless component of the received signal during the bit interval, impaired by the frequency offset, is

$$r(t) = \sum_{h=0}^{N-1} \sum_{m=0}^{N-1} x_m q(t - hT_c) \exp(j2\pi(m+\varepsilon)h/N) \qquad (11)$$

$$r_h = \frac{1}{T_c} \int_{hT_c}^{(h+1)T_c} r(t)q(t-hT_c)dt \qquad\qquad \text{h=0,1....N-1} \qquad (12)$$

The frequency offset produces ICI. Thus the h-th DFT input is

$$y_h = \frac{1}{N} \sum_{m=0}^{N-1} x_m \exp\left( j2\pi h \frac{m+\varepsilon}{N} \right)$$

$$= \frac{1}{N} \sum_{m=0}^{N-1} \sum_{k=0}^{K} b_i^k c^k_{i,m} \exp\left( j2\pi h \frac{m+\varepsilon}{N} \right) \qquad (13)$$

The N values of $y_h$ corrupted by AWGN noise and MAI are fed to the DFT. The DFT output is given by

$$z_g = \sum_{h=1}^{N} y_h \exp(-j2\pi hg/N)$$

$$= \sum_{h=0}^{N-1} \sum_{m=0}^{N-1} \sum_{k=0}^{K} b_i^k c^k_{i,m} \frac{1}{N} \exp\left( j2\pi(m+\varepsilon-g)h/N \right)$$

$$= \sum_{h=0}^{N-1} \sum_{m=0}^{N-1} \left[ \; b_i^d c^d_{i,m} + \sum_{k=0,k\neq d}^{K} b_i^k c^k_{i,m} \; \right] \frac{1}{N} \exp\left( j2\pi(m+\varepsilon-g)h/N \right) + \eta_g \qquad (14)$$

where $b^d_i$ and $c^d_{i,m}$ are the desired user's m-th chip in the i-th bit interval. $\eta_g$ is the Gaussian noise variable.

The DFT output is despread using the desired user's spreading sequence during the i-th bit interval and we obtain

$$\sum_{g=0}^{N-1} c^d_{i,g} z_g = b_i^d \sum_{h=0}^{N-1} \frac{1}{N} \exp(j2\pi\varepsilon h/N)$$

$$+b_i^d \sum_{g=0}^{N-1}\sum_{h=0}^{N-1}\sum_{\substack{m=0\\m\neq g}}^{N-1} c^d_{i,g} c^d_{i,m} \frac{1}{N} \exp\left( j2\pi(m+\varepsilon-g)h/N \right)$$

$$+\sum_{g=0}^{N-1}\sum_{h=0}^{N-1}\sum_{\substack{k=0\\k\neq d}}^{K-1} b_i^k c^d_{i,g} c^k_{i,g} \frac{1}{N} \exp\left( j2\pi\varepsilon h/N \right)$$

$$+\sum_{g=0}^{N-1}\sum_{h=0}^{N-1}\sum_{\substack{m=0\\m\neq g}}^{N-1}\sum_{\substack{k=0\\k\neq d}}^{K} b_i^k c^d_{i,g} c^k_{i,m} \frac{1}{N} \exp\left( j2\pi(m+\varepsilon-g)h/N \right) + \eta_g \qquad (15)$$

where the first, second, third and fourth terms at the right hand side represent the desired signal, self-interference, the MAI and the ICI respectively. Self interference is the interference from the desired user's own data in other chip intervals and is introduced due to non-zero frequency offset. The MAI is the interference from other users in the same chip interval. The ICI is the interference from other users in other chip intervals due to non-zero frequency offset. The estimated bit of the desired user can be obtained by using a QPSK demodulator followed by a hard limiter.

The variable of the MAI, ICI and noise for user d during the i-th bit interval can be written as

$$f_i^d = \sum_{g=0}^{N-1} c_{i,g}^d z_g = \sum_{g=0}^{N-1} \sum_{h=0}^{N-1} \sum_{m=0}^{N-1} c_{i,g}^d \sum_{\substack{k=0 \\ k \neq d}}^{K} b_i^k c_{i,m}^k \frac{1}{N} \exp(j2\pi(m-g+\varepsilon)h/N) + c_{i,g}^d \eta_g$$

$$= \sum_{g=0}^{N-1} \{\sum_{m=0}^{N-1} c_{i,g}^d u_m + n_g\} S_m$$

$$= \sum_{m=0}^{N-1} \mu_m S_m = \mu_0 S_0 + \sum_{m=1}^{N-1} \mu_m S_m \qquad (16)$$

Where $\mu_m$ is the MAI in the i-th bit and m-th chip interval and $S_m$ is the attenuation factor due to the relative carrier frequency offset $\varepsilon$ given by

$$\mu_m = \sum_{\substack{k=0 \\ k \neq d}}^{K-1} b_i^k c_{i,m}^k \qquad (17)$$

$$S_m = \frac{1}{N} \sum_{h=0}^{N-1} \exp\{j2\pi(m+\varepsilon)h/N\}$$

$$= \frac{\sin(\pi(m+\varepsilon))}{N \sin\left(\frac{\pi}{N}(m+\varepsilon)\right)} \exp(j\{\pi(1-\frac{1}{N})(m+\varepsilon)\}) \qquad (18)$$

The conditional pdf of the MAI, ICI and noise in frequency selective fading channels given the fading vector, $\vec{\alpha} = \{\alpha_0, ........\alpha_{N-1}\}$, where $\alpha_i$ is the fading parameter in the i-th subcarrier is obtained as weighted sum of nKN Gaussian pdfs

$$f(x) = \frac{1}{2^{nKN}} \sum_{i_0=0}^{nK} ..... \sum_{i_{N-1}=0}^{nK} \binom{nK}{i_0}, ..... \binom{nK}{i_{N-1}} \frac{1}{\sqrt{2\pi\sigma^2 \sum_{k=0}^{N-1} S_k^2}} \exp\left\{-\frac{\{x - A[K\sum_{k=0}^{N-1} \alpha_k S_k - 2(\sum_{k=0}^{N-1} i_k \alpha_k S_k / n)]\}^2}{2\sigma^2 \sum_{k=0}^{N-1} S_k^2}\right\} \qquad (19)$$

The variance is found to be

$$\sigma_I^2 = \sigma^2 \sum_{k=0}^{N-1} S_k^2 + \frac{K}{N} A^2 \sum_{k=0}^{N-1} \alpha_k^2 S_k^2 \qquad (20)$$

The pdf and variance without frequency offset for and

without fading can be obtained by replacing $S_0 = 1$, $S_k = 0$ and $\alpha_k = 1$ for all k as

$$f(x) = \frac{1}{2^{NK}} \sum_{l=0}^{NK} \binom{NK}{l} \frac{1}{\sqrt{2\pi\sigma^2}} \exp\{-[x - A(K - 2l/N)]^2 / 2\sigma^2)\} \qquad (21)$$

$$\sigma_I^2 = \sigma^2 + \frac{KA^2}{N} \qquad (22)$$

Equations 21 and 22 are derived assuming Gaussian approximations of MAI and ICI.

Then the Gaussian approximation of BER is obtained as

$$= \int_0^\infty Q\left(\sqrt{\sum_{i=0}^{N-1} \frac{S_0^2 \gamma_i(E_b/N_o)}{\sum_{k=0}^{N-1} S_k^2 + \frac{K}{n}(E_b/N_o)\sum_{j=0}^{N-1} \gamma_j S_j^2}}\right) f_\gamma(\gamma)d\gamma \qquad (23)$$

For flat fading, $\gamma_i = \gamma = \alpha^2$ for all i and the equation for BER reduces to

$$P_e = \int_0^\infty Q\left(\sqrt{\sum_{i=0}^{N-1} \frac{S_0^2 \gamma_i A^2/n}{\sigma^2 \sum_{k=0}^{N-1} S_k^2 + \frac{K}{n} A^2 \sum_{j=0}^{N-1} \gamma_j S_j^2}}\right) f_\gamma(\gamma)d\gamma$$

$$P_e = \int_0^\infty Q\left(\sqrt{\sum_{i=0}^{N-1} \frac{S_0^2 \gamma(E_b/N_o)}{(1 + \frac{K}{n}\gamma(E_b/N_o)(\sum_{j=0}^{N-1} S_j^2)}}\right) e^{-\gamma} d\gamma \qquad (24)$$

For AWGN channels, $\gamma_i = 1$ and equation (20) further reduces to

$$P_e = Q\left(\sqrt{\frac{S_0^2(2E_b/N_o)}{(1+\frac{K}{n}\frac{2E_b}{N_o})(\sum_{j=0}^{N-1}S_j^2)}}\right) \quad (25)$$

The MAI term can be removed for orthogonal spreading systems. However, the MAI is not completely removed in frequency selective fading channels even when orthogonal spreading sequences are employed. The fading in each carrier is not identical and may be correlated. At the receiver, the different amplitude in each chip may destroy the orthogonality of the sequences.

# 4. Simulation Results

Simulations are also conducted to evaluate error performance of the proposed multiple access scheme. Monte-Carlo simulation method is applied to estimate the bit error rate of the proposed scheme. The BER performance is analyzed and simulated for three different cases: low data rate (8 kbps) users having a high spreading factor of 32, high (10 Mbps) data rate users having a lower spreading factor of 4 and the average BER performance, where users at different data rates use different spreading

factors. It is found that the performance is better when transmission is at low data rates than at high data rates since at low rates, a higher spreading factor is used which reduces the MAI.

The simulation parameters are summarized in Table.1. In each frame, time slots are assigned to users independently in accordance with their data rates. The data rates vary from 8 kbps to 10 Mbps. The modulation used is QPSK and spreading factor is chosen to be 4, 8, 16 or 32 depending on the data rates. Walsh codes are used since they are orthogonal. The RF bandwidth is assumed to be 20 MHz and the total number of subcarriers is 1024 with a subcarrier spacing of 19.53 KHz. The ITU-R Vehicular Channel A model is used for simulation.

***Table.1.*** *Simulation parameters*

| Parameters | Values |
|---|---|
| Bandwidth | 80 MHz |
| Data Rate | 8 Kbps to 10 Mbps |
| Spreading Codes | Walsh Codes |
| Spreading Factor | 4, 8,16,32 |
| Modulation | QPSK |
| Number of subcarriers | 1024 |
| Subcarrier spacing | 78.125 KHz |

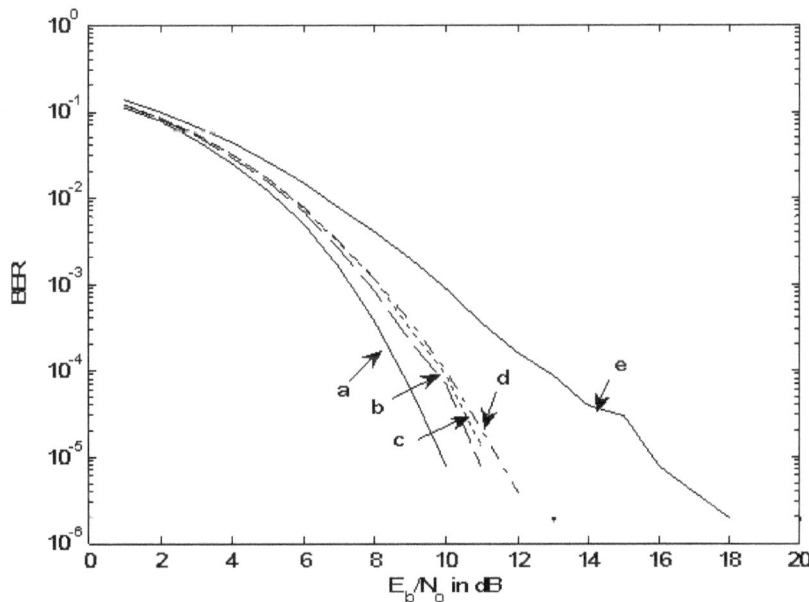

***Fig.4.*** *BER performance of the TD-MC-CDMA multiple access system in downlink for a data rate of 8 kbps (a) Simulation result for BER of 1 to 16 users in AWGN channel (b) Simulation result for BER of 1 user in fading channel (c) Simulation result for BER of 4 users in fading channel (d) Simulation result for BER of 16 users in fading channel and (e) Analytical result for BER of 16 users in fading channel.*

Fig.4 shows the BER performance of the TD-MC-CDMA scheme in synchronous downlink in AWGN and fading channels for a data rate of 8 kbps and a spreading factor of 32. It is found that the BER performance is same for any number of users in AWGN channel since the codes

are orthogonal and there is no interference. However, in a multipath fading channel, the performance degrades as the number of users increases due to multiple access interference. The required $E_b/N_o$ to get a BER of $10^{-4}$ is 13 dB for single user, 13.5 dB for 4 users and it is 14 dB for 16 users. Walsh codes are used which are orthogonal, but the MAI is not zero since each chip of the PN sequence experiences independent fading, which tends to destroy the orthogonality between spreading sequences. This increases

the MAI and degrades the BER performance. However, the degradation is very small as the number of active users increases. The simulation results are in good agreement with the predictions of analytical results.

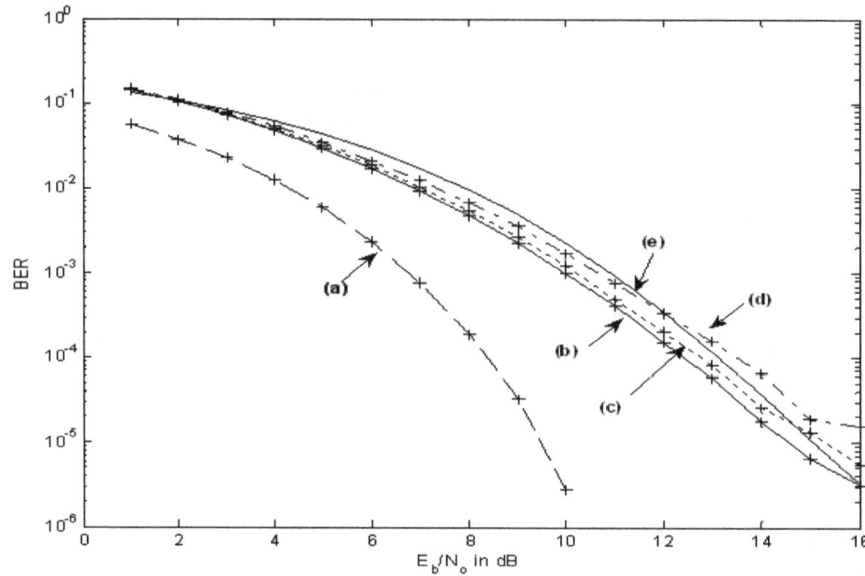

**Fig.5.** *BER performance of the TD-MC-CDMA multiple access system in downlink for a data rate of 10 Mbps (a) Simulation result for BER of 1 to 4 users in AWGN channel (b) Analytical result for BER of one user in fading channel (c) Simulation result for BER of 1 user in fading channel (d) Simulation result for BER of 2 users in fading channel and (e) Simulation result for BER of 4 users in fading channel.*

Fig.5 shows the BER performance of the TD-MC-CDMA scheme in synchronous downlink in AWGN and fading channels for a data rate of 10 Mbps and a spreading factor of 4. It is found that the BER performance is same for any number of users in AWGN channel since the codes are orthogonal and there is no interference. In a multipath fading channel, the required $E_b/N_o$ to get a BER of $10^{-3}$ is

17 dB for single user, 20 dB for 2 users and it is about 25 dB for 4 users. Since the spreading factor is less, the MAI is more in this case. The degradation in BER performance is also more as the number of users increases due to multiple access interference. The simulation results are in good agreement with the predictions of analytical results.

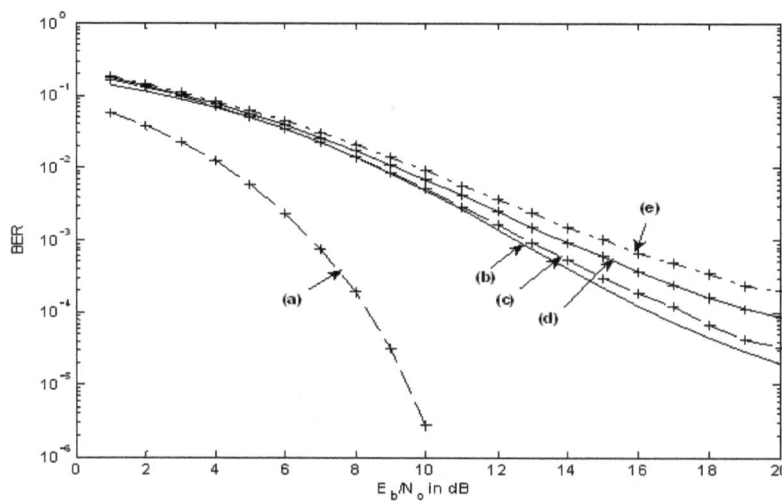

**Fig.6.** *Average BER performance of the TD-MC-CDMA multiple access system in downlink (a) Simulation result for BER of 1 to 16 users in AWGN channel (b) Analytical result for BER of 4 users in fading channel (c) Simulation result for BER of 4 users in fading channel (d) Simulation result for BER of 8 users in fading channel and (e) Simulation result for BER of 16 users in fading channel.*

Fig.6 shows the average BER performance of the TD-MC-CDMA scheme in synchronous downlink in AWGN and fading channels for a number of users having data rates from 8 kbps to 10 Mbps. It is found that the BER

performance degrades as the number of users increases due to multiple access interference. The required $E_b/N_o$ to get a BER of $10^{-4}$ is 13 dB for 4 users, 14 dB for 8 users and it is 16 dB for 16 users. However, the degradation is not very large as the number of active users increases. The simulation results are in good agreement with the predictions of analytical results.

# 5. Conclusions

TD-MC-CDMA scheme can be used for 4G cellular systems with spreading in frequency domain for downlink and spreading in time domain for uplink. 3GPP LTE and WiMAX Mobile use OFDMA for multiple access. The proposed scheme can give better diversity and flexibility for transmission of different classes of data e.g. audio, video, internet, ISDN, multimedia etc. having multiple data rates. The parameters e.g. spreading factor, number of time slots, number of subcarriers etc. are chosen according to the required data rate, the available bandwidth, the number of subscribers etc. The performances of proposed system for uplink and downlink channels have been studied. Expressions for multiple access interference and bit error rate are derived. The simulation results are in good agreement with the predictions of theoretical/ analytical results. Furthermore, simulations imply that MAI, that determines the error rate, can be approximated as Gaussian for many practical cases.

# References

[1] H.Sari and G. Karam, "Orthogonal Frequency Division Multiple Access and its Applications to CATV Networks," Eur. Trans. Comm., vol.45, pp 507-516, Nov.-Dec., 1998.

[2] S. H. Ali, K. Lee and V. C. M. Leung, "Dynamic Resource Allocation in OFDMA Wireless Metropolitan Area Networks", IEEE Wireless Communication, Feb, 2007.

[3] K. H. Teo, Z. Tao, and J. Zhang "The Mobile Broadband WiMAX Standard," IEEE Signal Processing Magazine, September, 2007.

[4] The WiMax Forum (www.wimaxforum.org)

[5] K. Fazel and S. Kaiser, Multi-Carrier and Spread Spectrum Systems, John Wiley and sons Ltd., 2003.

[6] L. Hanzo, B. J. Choi, Thomas Keller, and M. Mnster, *OFDM and MC-CDMA for Broadband Multi-User Communications, WLANs and Broadcasting*, John Wiley and Sons, 2003.

[7] L. Hanzo, L. L. Yang, E. L. Kuan, K. Yen, *Single and Multi Carrier DS-CDMA: Multiuser Detection, Space Time Spreading, Synchronization, Networking and Standards*, IEEE Press, 2005.

[8] S. Hara and R. Prasad, "Overview of multicarrier CDMA," IEEE Communication magazine, vol.35, pp.126-133, Dec., 1997.

[9] S. Hara and R. Prasad, "Design and Performance of Multicarrier CDMA System in Frequency-Selective Rayleigh Fading Channels," IEEE Transactions on Vehicular Technology, vol. 48, No. 5, pp.1584-1595, Sept. 1999.

[10] Tobias Weber, Johannes Schlee, Stefan Bahrenburg, Paul Walter Baier, Jürgen Mayer and Christoph Euscher, "A Hardware Demonstrator for TD-CDMA," IEEE Transactions On Vehicular Technology, Vol. 51, No. 5, September 2002.

[11] M. Haardt, A. Klein, R. Koehn, S. Oestreich, M. Purat, V. Sommer, T. Ulrich, *The TD-CDMA based UTRA TDD mode,"* IEEE Journal on Selected Areas in Communications, vol. 18 August 2000.

[12] 3GPP Specifications-Releases available at http://www.3gpp.org

[13] M. Chang and Y. T. Su, "Performance Analysis of Equalized OFDM Systems in Rayleigh Fading", IEEE Transactions on Wireless Communications, vol. 1, No.4, Oct. 2002.

[14] A. Chouly, A. Brajal, and S. Jourdan, "Orthogonal multicarrier techniques applied to direct sequence spread spectrum CDMA systems," Proc. IEEE GLOBECOM'93, pp.1723-1728, Nov., 1993.

[15] K. Fazel, "Performance of CDMA/OFDM for mobile communication system," Proc. IEEE ICUPC'93, pp. 975-979, Oct., 1993.

[16] Z. Hou and V. K. Dubey, "Bit Error Rate Probability of MC-CDMA System over Rayleigh Fading Channels", ISSSTA2004, Sydney, Australia, 30 Aug.-2 Sep. 2004.

[17] H. Ishikawa, M. Furudate, T. Ohseki and T. Suzuki, "Performance Analysis of Adaptive Downlink Modulation Using OFDM and MC-CDMA for Future Mobile Communication System", ISSSTA2004, Sydney, Australia, 30 Aug.-2 Sep. 2004.

[18] S. Kondo and L.B. Milstein, "Performance of multicarrier DS-CDMA systems," IEEE transactions on Communications, vol. 44, no.2, pp.238-246, Feb., 1996.

[19] E. A. Sourour and M. Nakagawa, "Performance of orthogonal multicarrier CDMA in a multipath fading channel," IEEE Transactions on Communications, vol. 44, no. 3, pp.356-367, March, 1996.

[20] S. Suwa, H. Atarashi, and M. Sawahashi, "Performance Comparison Between MC/DS-CDMA and MC-CDMA for Reverse Link Broadband Packet Wireless Access," IEEE Vehicular Technology Conference, 2002 Proceedings vol.4 pp. 2076-2080, Sept. 2002.

# 14

# Genetic algorithm based Finite State Markov Channel modeling

**Rakesh Ranjan[1, *], Dipen Bepari[2], Debjani Mitra[2]**

[1]Department of Electronics and Communication Engineering, National Institute of Technology (NIT), Patna, 800005, India
[2]Department of Electronics Engineering, Indian School of Mines (ISM), Dhanbad, 826004, India

**Email address:**
rakesh.r1804@gmail.com(R. Ranjan), dipen.jgec04@gmail.com(D. Bepari), debjani7@yahoo.com(D. Mitra)

**Abstract:** Statistical properties of the error sequences produced by fading channels with memory have a strong influence over the performance of high layer protocols and error control codes. Finite State Markov Channel (FSMC) models can represent the temporal correlations of these sequences efficiently and accurately. This paper proposes a simple genetic algorithm (GA) based search for the optimum state transition matrix for a block diagonal Markov model. The burst error statistics of the GA based FSMC model with respect to Autocorrelation Function and error free interval distribution of the original error sequence are presented to validate the proposed method. The superiority of the GA approach over the semi-hidden Markov model (SHMM) based Fritchman model is exhibited in significant improvement of closeness of match and in the usage of shorter length of error sequences. Another Baum-Welch algorithm (BWA) based GA search method has been proposed and compared with the BWA and SHMM methods for the same error sequence. Again the superiority of GA approaches is recognized, especially for the smaller error lengths.

**Keywords:** Genetic Algorithm, Finite State Markov Channel, Semi-Hidden Markov Model, Baum-Welch Algorithm, Autocorrelation Functions, Error-Free Interval Distributions

## 1. Introduction

The current generation networking and communication technology which has to cater to high speed multimedia traffic is strongly dependent on the identification and availability of accurate and tractable models for the mobile/wireless radio fading channels. Fading affects the overall network performance in the physical or media access control layers, and also the designing of efficient and reliable transceivers. The radio link is highly variable due to the statistical distribution of the environmental propagation parameters. Reliable and efficient channel modeling can thus be effective to develop network protocols that mitigate or exploit fading. There are two main approaches to capture the inherent memory in fading channels: the physical (analog) or waveform level and the digital or discrete channel modeling. Physical channel modeling is based on parameters such as received signal strength, noise/interference power, speed of mobile etc. They are more appropriate for design and testing of transmitter-receiver applications and parameter optimization. Discrete Channel Modeling on the other hand places

emphasis on the statistical properties of the bursty error sequence and without modeling the physical functionality of the channel, they can characterize the temporal correlation. The correlated error sequence of fading channels with memory is reproduced by representing the channel with a finite number of states: this is more commonly referred to as Finite State Markov Channel or FSMC modeling [1]. These models have a wide application in the design and performance evaluation of error control coding schemes and higher level wireless communication protocols. In this model, fading is approximated as a discrete-time Markov process with time discretized to a given interval, typically the symbol period. Specifically, the set of all possible fading gains is related to the received signal to noise ratio (SNR) and by partitioning the SNR into a finite number of levels, each interval is associated with a state of a Markov process. The channel varies over these states at each symbol period duration, according to a set of Markov transition probabilities. The FSMC model has evolved from the two states Gilbert Elliot channel [2, 3] representing a time varying binary symmetric channel. The crossover probabilities of the channel are determined by the current state of a discrete time stationary binary Markov process.

The source has two states: good or no errors and bad or burst errors. Fritchman [4] studied the Markov chains where the outputs are a deterministic function of the states based on the principle of semi-hidden Markov models. As an extension of the Hidden Markov Model (HMM), a hidden semi-Markov model (HSMM) is traditionally defined by allowing the underlying process to be a semi-Markov chain. Each state has a variable duration, which is associated with the number of observations produced while in the state [5]. Fritchman's model has received considerable attention in recent years because it is relatively easy to estimate the parameters of this model from burst error distributions. FSMC and other error modelling approaches and its features such as computational efficiency, etc have been excellently reviewed and discussed in literature [6-10]. HMMs are an important category of generative error models that can accurately represent the statistical patterns of bursty error data in fading channels [11]. The most important and difficult problem in HMMs is to estimate the model parameters that best explains the observations and training is usually performed by an iterative procedure following the Maximum likelihood criteria. Genetic Algorithms, on the other hand, are a powerful computational model having optimization capabilities that can encode a solution on a simple chromosome like data structure using techniques inspired by natural evolution such as inheritance, crossover, mutation and selection. GA's ability of global searching of better maxima without getting trapped into local maxima can offer better and computationally efficient solutions [12-16]. In this context, Genetic Algorithms (GA) for HMM training in general has already been identified to be a promising area mostly in applications related to automatic speech recognition [17], text/web information extraction [18-19], etc. Most of the GA and HMM related papers have used the chromosome structure as the complete set of $\Gamma = \{A, B, \Pi\}$, and the chromosomes have been evaluated for the objective function of log likelihood ratio. In the area of Discrete Channel Modeling, however, not much work has been reported on the applicability of Genetic Algorithms for reproducing error sequences following certain statistical characteristics. Zhao Zhi-Jin et al [20] have proposed a hybrid method of using (GA) and simulated annealing (SA) to train HMM for discrete channel modeling. A GA based equalization technique has been proposed [21] with much lower computational complexity for direct sequence ultra-wideband (DS-UWB) wireless communications. The Genetic Algorithm is associated with the RAKE receiver to combat the inter-symbol interference (ISI) due to the frequency selective nature of UWB channels. Another GA based novel technique has been developed to train a hidden Markov model (HMM) with for the cognitive radio channel [22]. This paper has used the idea of HMM and SHMM in proposing new GA based FSMC model which is slightly different from the papers encountered in the literature. The mean square error of the statistical properties of the error sequences has primarily been used as the fitness function in the GA for the search based estimation of the parameters of

the Markov model. Upon comparison of the validation with the analytical HMM/SHMM approach, the proposed GA methods are observed to have some significant improvement towards the closeness of match as well as in the length of training sequences required for a given accuracy.

The rest of the paper is organized as follows. Section 2 discusses the finite state channel model for discrete channel modeling. The GA based FSMC models have been proposed in section3. The proposed models are based on BWA and SHMM approaches. The simulation results for these two proposed technique have been provided in section 4. The results are compared with the BWA and SHMM methods for the same error sequence. The superiority of GA approaches is recognized, especially for the smaller error lengths and finally conclusions are summarized in Section 5.

## 2. Finite State Channel Model

Discrete or Finite State Channel models are characterized generally by conducting waveform level simulations and/or from measurement data. The output of waveform level simulation is a time series of bit errors that are long enough to represent the statistical properties of the discrete channel. Once a DCM is developed, it eliminates the necessity of further waveform level simulations at high sampling rates. From the modulator input at the transmitter to the output at the receiver, the blocks can be clubbed as a sequence of discrete symbols as shown in Figure 1.

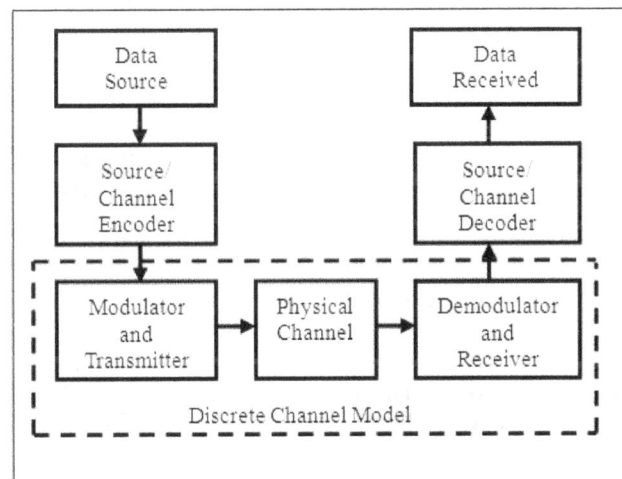

*Figure 1: Basic Communication system with the Discrete Channel Model.*

The relation between the output of the channel encoder at the transmitter and the input of the channel decoder at the receiver in the wireless fading propagation medium can be captured efficiently by using discrete-time Markov sequence to model the errors in the channel. In the sequence, the channel contains several different states and a set of transition probabilities and error probability matrix defining the inter-relationship between the states. Usually, we only know or can observe the input and output of a channel and therefore the error sequence, but the state sequence is not easily observed in a state channel and it is therefore called "hidden" and the Markov model is called a hidden Markov model (HMM). Both Gilbert and Fritchman are types of hidden Markov models that for large data sets need the powerful Baum Welch algorithm for computation of the HMM parameters [23]. Long error sequences of millions of symbols may be required for accurately estimating the small values of the state transition matrix. The variables in the algorithm are computed for each symbol in the given error sequence, resulting in slow convergence and high computational burden. To improve the efficiency and speed of estimation, several approaches have been adopted. A common one is based on the fact that for a general Markov model there is a statistically equivalent Fritchman like model with k good states and N-k bad states [24]. These models also referred as block diagonal Markov models are computationally more efficient as the variables in the estimation algorithms are computed at the beginning of each burst of errors rather than once every symbol. The motivation in this work was to work towards improving the efficiency of estimation of the HMM parameters still further with respect to better characterization of the original error vector and training with shorter length of error sequence.

# 3. The Proposed GA Based FSMC Model

Inspired from biology and based on natural genetics law of the survival of the fittest, GA has already proved to be very capable in many research areas and NP problems. The algorithm can evolve new and better solutions in the expectation of which an implementation of GA based FSMC models are being proposed here to search for the best model parameters for a block-diagonal Markov model. A three state Fritchman model with two good states and one bad state is considered, so that a known error observation matrix, say

$$B = \begin{bmatrix} 1 & 1 & 0 \\ 0 & 0 & 1 \end{bmatrix}$$ denotes the third state to be the 'bad' error

producing state. This model is semi hidden in the sense that if an error is produced, it is known that it has been generated by the third state. If however, no error is generated, the state cannot be identified. The implementation of this estimation algorithm based on block equivalent Markov model has been named as SHMM (semi-hidden Markov model) in the results of the current work. The state transition probability

matrix $B_{ij}$, of the 3-state channel model $\begin{bmatrix} A_{11} & A_{12} & A_{13} \\ A_{21} & A_{22} & A_{23} \\ A_{31} & A_{32} & A_{33} \end{bmatrix}$,

thus has basically six independent variables $(P_m = 1\%)$ to be estimated, as the sum of the rows of the $A_{ij}$ matrix is unity.

A master error sequence $E_M$ was made available from a waveform level simulation of an OFDM system. Another GA search method based on Baum-Welch algorithm has been proposed for 3-state channel model, with nine independent

variables $A_{11}, A_{12}, A_{22}, A_{23}, A_{31}, A_{33}, B_{11}, B_{12}, B_{13}$, with the sum of rows of state transition matrix $A_{ij}$ is unity and the sum of columns of error generation matrix $B_{ij}$ is also unity.

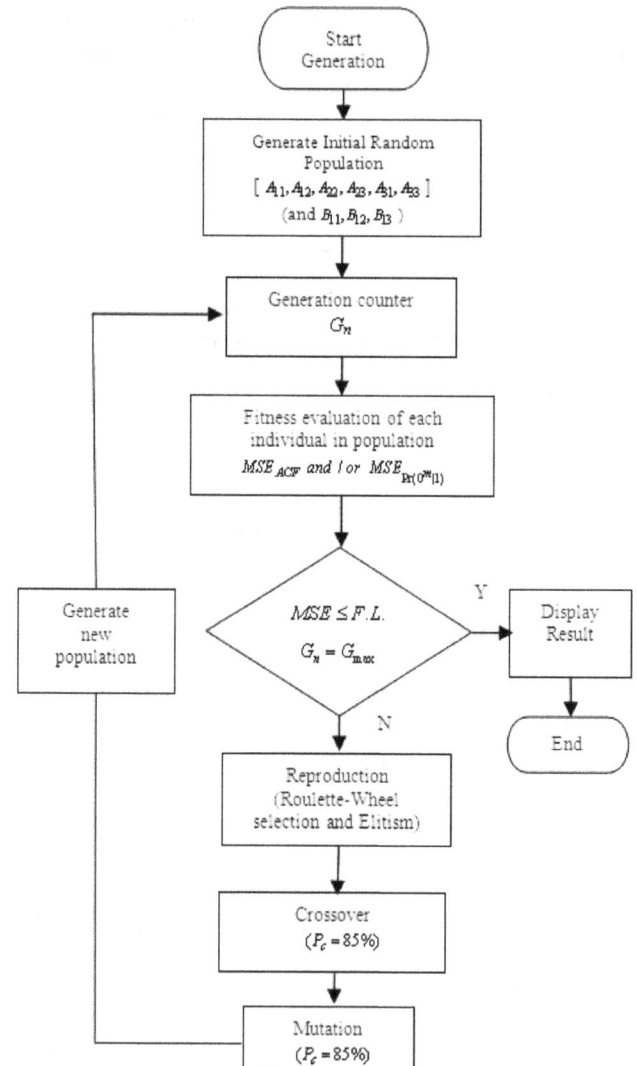

***Figure 2:*** *Block diagram of the proposed GA model*

The overall block diagram of the proposed GA based model has been illustrated in Figure 2. The GA was initialized with a suitable population size, wherein the independent variables were encoded into a string of real numbers ranging from 0 to 1 on each chromosome. Each chromosome was used to generate an error sequence and the Autocorrelation Function (ACF) comparisons of the original $E_M$ sequence and the GA generated sequence are used to evaluate the fitness function of the chromosomes. ACF comparison is one of the most popular techniques for performance evaluation of HMM and other FSCM for finding the closeness between the original and DCM generated data. The MSE (Mean Square Error) of the ACFs of the two error sequences are used as a measure of the fitness function of the chromosomes in the proposed GA based FSCM.

$$MSE_{ACF} = \frac{\sum_{i=1}^{N}(A_{Oi} - A_{Gi})}{N} \qquad (1)$$

where,

$N$ = total number of elements of Autocorrelation function,

$A_{Oi}$ = elements of ACFs of the original error sequences,

$A_{Gi}$ = elements of ACFs of the GA-generated error sequences.

The GA is made to minimize this error with a targeted value of $5 \times 10^{-7}$. This approach seems to be quite promising as seen from the results of the extensive simulations carried out. The distribution of error-free intervals is also a popular metric for validating the accuracy of DCMs. It is commonly denoted as, $\Pr(0^m | 1)$, where, $m$ is the length of intervals and $(0^m | 1)$ denotes the event of observing $m$ or more consecutive errors-free transmission followed by an error. Here both ACF and $\Pr(0^m | 1)$ have been used for validating the performance of our proposed GA method with the SHMM method. The roulette wheel selection method is capable of effectively eliminating the weaker chromosomes in preference to fitter solutions. The standard genetic operators of cross-over (two point crossovers have been

used) and mutation has been applied. The Elite count denoting the number of individuals that are assured to survive to the next generation is taken as four.

# 4. Results and Discussions

In this section the SHMM and BWA based GA approaches are discussed for channel modeling. Waveform level simulations of three different lengths, 50000, 10000, and 1000 bits of error have been used to estimate the transition probability matrix by the proposed GA approach and then the same has been compared with the popular SHMM technique for discrete channel modelling. The proposed GA algorithm was experimented upon to give a reasonably good match using a crossover fraction of 0.85 and an optimal population size of 30 for the different length of error sequences used. We can consider the smaller population size, but this may lead to missing of global fitness value. While larger population size, will results in increased simulation time. The variations in the values of MSE of ACFs with different population size, and with different crossover probabilities have been shown in Table 1 for N = 1000 and 10000. From the table, it is clear that population size of 30 and crossover fraction of 0.85 is most suitable for implementation of Genetic Algorithm for different length of error sequences.

**Table 1:** *Variations of MSE of ACFs with the population size, for a particular value of crossover fraction, for N (length of error sequence) = 1000, and 10000.*

| Crossover Fraction → | N = 1000 | | | | N = 10000 | | | |
|---|---|---|---|---|---|---|---|---|
| Population ↓ | 0.75 | 0.80 | 0.85 | 0.90 | 0.75 | 0.80 | 0.85 | 0.90 |
| 10 | 0.0016 | 0.0014 | 0.0008 | 0.0021 | 0.0440 | 0.0018 | 0.0010 | 0.0018 |
| 20 | 0.0013 | 0.0012 | 0.0014 | 0.0007 | 0.0341 | 0.0001 | 0.0330 | 0.0001 |
| 30 | **0.0015** | **0.0017** | **0.0012** | **0.0021** | **0.0023** | **0.0001** | **0.0001** | **0.0095** |
| 40 | 0.0013 | 0.0013 | 0.0014 | 0.0013 | 0.0214 | 0.0007 | 0.0246 | 0.0002 |
| 50 | 0.0009 | 0.0013 | 0.0009 | 0.0011 | 0.0026 | 0.0002 | 0.0001 | 0.0003 |
| 60 | 0.0014 | 0.0012 | 0.0007 | 0.0007 | 0.0271 | 0.0001 | 0.0002 | 0.0001 |
| 70 | 0.0011 | 0.0012 | 0.0005 | 0.0012 | 0.0001 | 0.0001 | 0.0379 | 0.0001 |

Figure 3 shows the ACF comparison of original training sequence error data along with that of the GA generated and SHMM generated error data, using a training sequence of length 50,000. Figure 4 presents the similar ACF comparisons for much shorter error sequences of length 10,000. From the figure, the superiority of the GA method over the SHMM method is clearly established as the former is seen to have better matching capability even in the face of a short length error sequence. This is indeed a significant result in context to discrete channel modeling. The statistical parameter $\Pr(0^m | 1)$, is also used to judge the performance comparison of GA over SHMM. Figure 5 shows the error-free interval distribution comparisons and again it is

conclusively observed that for lower length of error sequence, GA indeed has better performance in comparison to SHMM. The GA experimentation was repeated with several other error sequences of length 1000, generated arbitrarily from different state transition matrices and when compared to the SHMM method, in general it was found that for shorter sequences GA is capable of producing better match especially with respect to the error free interval distribution. Figure 5 shows one of the simulation results establishing this fact.

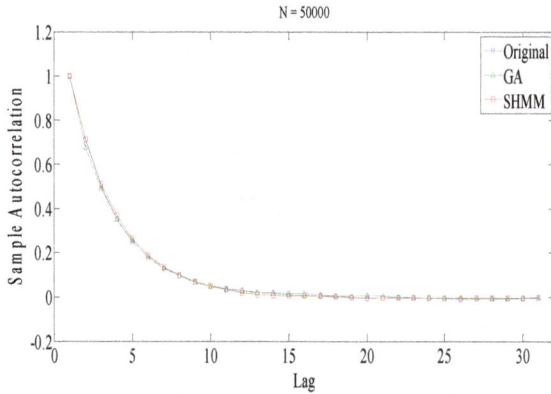

*Figure 3: ACF comparisons of GA-generated error data and SHMM-generated error sequence with the original channel error statistics for N = 50000.*

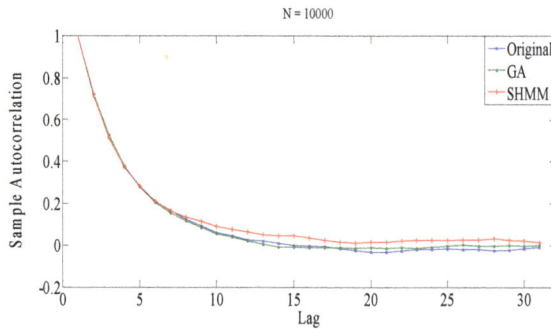

*Figure 4: ACF comparison of GA and SHMM methods for N = 10000.*

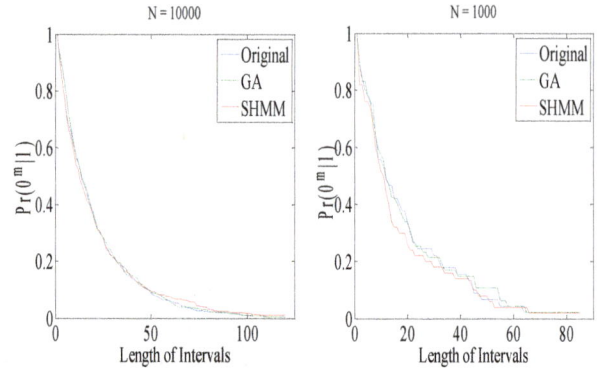

*Figure 5: Comparison of error-free interval distribution of the original error data with error-free interval distributions of GA-generated and SHMM-generated error sequences, for N = 10000 and 1000.*

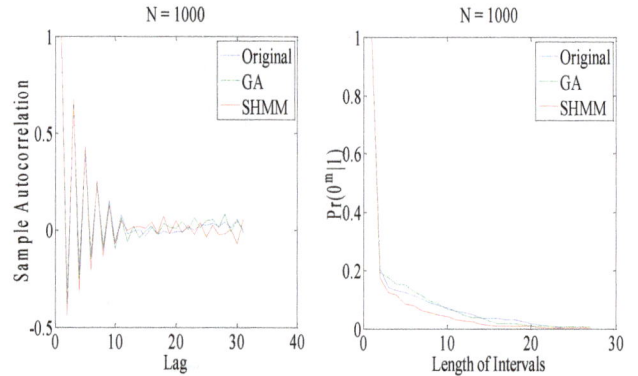

*Figure 6: ACFs and error-free interval distribution comparison for arbitrary error sequences with length of error sequence = 1000.*

The values of best fitness and mean fitness obtained during the GA experimentations for several generations have been plotted in the Figure 7 to depict the nature of the convergence in successive generations as the algorithm progresses towards the terminating criteria.

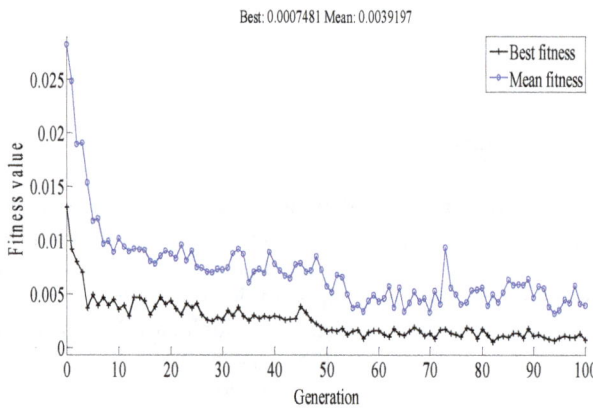

*Figure 7: A typical nature of the GA performance for a 1000 length error sequence.*

The MSE of the autocorrelation functions for different lengths of error sequences have been plotted in Figure 8 to compare the performance of the GA and SHMM methods. It

shows clearly that GA performs better in comparison to SHMM for lower lengths of error sequences up to about 30,000 while for larger length of error sequences both approaches have almost similar performance. Table 2 shows the values of MSEs of ACFs and probability of distribution of error-free intervals in testing the superiority of GA method over the SHMM method for different lengths of sequence.

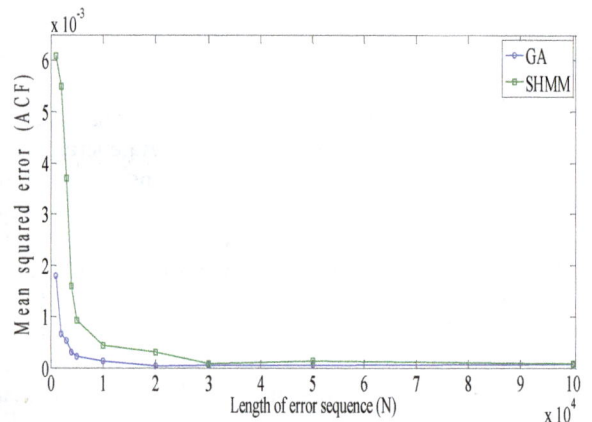

*Figure 8: MSE of ACF with the Length of error sequences (N) for GA and SHMM approaches.*

*Table 2:* MSEs of ACFs and Error-free interval distributions for different length sequences.

| Length of error sequence | MSE of ACF and $\Pr(0^m\mid 1)$ | | | | | |
| | 50,000 | | 10,000 | | 1000 | |
| | GA | SHMM | GA | SHMM | GA | SHMM |
| --- | --- | --- | --- | --- | --- | --- |
| $MSE_{ACF}$ | 0.000057 | 0.000063 | 0.00012 | 0.0011 | 0.0010 | 0.0020 |
| $MSE_{\Pr(0^m\mid 1)}$ | 0.0000081 | 0.000019 | 0.000088 | 0.00015 | 0.00034 | 0.0014 |

Another BWA based GA approach for the same error statistics has been provided next. The experiments show that GA formulations with the previous values of crossover fraction, population size, elite count, etc. can provide reasonably good fitting. Figure 9 shows the ACF comparison of original error data with the GA-generated and BWA-generated error data of lengths 50000 and 10000. The closeness of the curves establishes the superiority of GA approach over BWA method, especially for smaller lengths of error sequence.

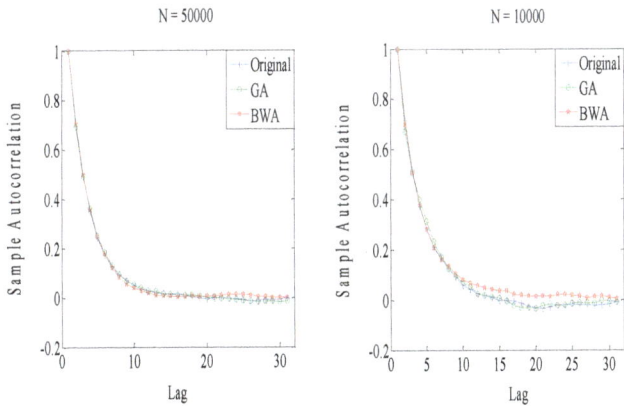

*Figure 9:* Comparison of Autocorrelation function of original error sequence with autocorrelation functions of GA-generated and SHMM-generated error sequences, for the same error data with lengths of 50000 and 10000.

Now the performances of both the proposed GA techniques have been compared with the BWA and SHMM techniques for the smaller lengths of error statistics (N = 1000). The autocorrelation function of the original error data has been compared with the ACFs of the BWA-generated, SHMM-generated and GA-generated error data in Figure 10. It illustrates that error sequences generated by both the proposed GA approaches have very close match with the original error data in comparison to the regenerated error data by the existing BWA and SHMM techniques. Figure 11 shows the similar performance comparison in terms of error-free interval distribution and again the

advantage of proposed GA approach for discrete channel modeling has been recognized. Therefore, for the shorter lengths of error data the proposed GA search techniques have significantly improved performance over the BWA and SHMM approaches.

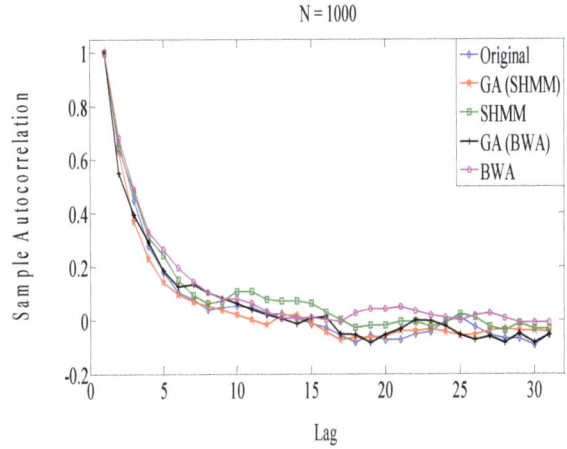

*Figure 10:* Comparison of ACF of original error data with ACFs of GA-generated and SHMM-generated, and BWA-generated error sequences, for length of error data = 1000.

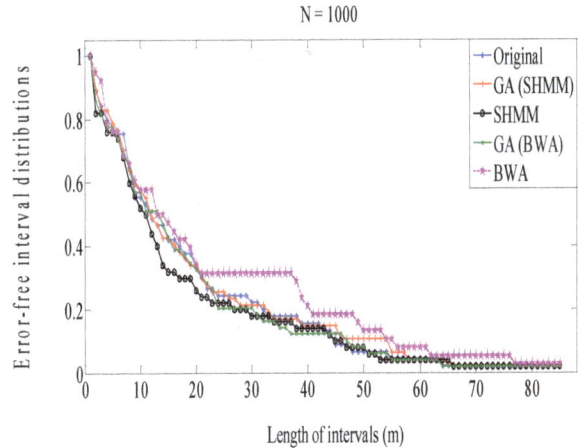

*Figure 11:* Comparison of $\Pr(0^m\mid 1)$ of original error data with $\Pr(0^m\mid 1)$ of the error data regenerated by proposed GAs, SHMM and BWA for length of error data = 1000.

## 5. Conclusion

Research studies have shown that the performance evaluation of high layer protocols, error control codes, interleavers, voice coders, actually become computationally very expensive using waveform level simulations. Efficiently designed accurate FSCMs that can reproduce the statistical properties of an error sequence can eliminate the necessity of further waveform level simulations at high sampling rates. This paper has proposed new GA based FSMC models for estimating the parameters of the state transition matrix of a block diagonal Markov model. Upon comparison of the validation with SHMM and BWA approach, the proposed method is observed to have some significant improvement towards the closeness of match.

Also, GA performs better in comparison to SHMM and BWA for lower lengths of error sequences, which is a significant result in context to discrete channel modeling. With HMM training, the variance of several random variables or estimated values of DCM is increased by decreasing the length of error sequence, while with the proposed GA method, these variations are very less. The training with shorter sequence is important for DCM, as it reduces the simulation run-time as well as computational burden of a particular training algorithm. Also in order to model the channel error-statistics dynamically/online, one cannot wait for the entire error sequence to be received at receiver and in that case small trace of error sequence plays a significant role in analysis and modeling of channel error-bursts.

# References

[1] H. S. Wang, and N. Moayeri, "Finite-State Markov Channel: A Useful Model for Radio Communication Channels," IEEE Transactions on Vehicular Technology, vol. 44, pp. 163–171, 1995.

[2] E. N. Gilbert, "Capacity of a burst-noise channel," Bell System Technical Journal, vol. 39, pp. 1253–1265, 1960.

[3] E. O. Elliott, "Estimates of error rates for codes on burst-noise channels," Bell System Technical Journal, vol. 42, pp. 1977–1997, 1963.

[4] B. D. Fritchman, "A binary channel characterization using partitioned Markov chains," IEEE Transactions on Information Theory, vol. IT-13, pp. 221–227, 1967.

[5] Shun-Zheng Yu, "Hidden semi-Markov models," Artificial Intelligence, vol. 174, pp. 215–243, 2010.

[6] P. Sadeghi, R. Kennedy, P. Rapajic, and R. Shams, "Finite-state Markov modeling of fading channels - A survey of principles and applications," IEEE Signal Processing Magazine, vol. 25, pp. 57-80, 2008.

[7] H. Bai, and M. Atiquzzaman, "Error modeling schemes for fading channels in wireless communications: A survey," IEEE Communications Surveys & Tutorials, vol. 5, pp. 2-9, 2003.

[8] C. Pimentel, and I. F. Blake, "Enumeration of Markov chains and burst error statistics for finite-state channel models," IEEE Transactions on Vehicular Technology, vol. 48, pp. 415–428, 1999.

[9] F. Babich, and G. Lombardi, "A Markov model for the mobile propagation channel," IEEE Transactions on Vehicular Technology, vol. 49, pp. 63-73, 2003.

[10] C. C. Tan, and N. C. Beaulieu, "On first-order Markov modeling for the Rayleigh fading channel," IEEE Transactions on Communications, vol. 48, pp. 2032-2040, 2000.

[11] N. Nefedov, "Generative Markov models for discrete channel modelling," The 8th IEEE International Symposium on Personal, Indoor and Mobile Radio Communications, vol. 1, pp. 7-11, 1997.

[12] D. Whitley, "A genetic algorithm tutorial," Statistics and computing, vol. 4, pp. 65-85, 2003.

[13] K. S. Tang, K. F. Man, S. Kwong, and Q. He, "Genetic algorithms and their applications," IEEE Signal Processing Magazine, vol. 13, pp. 22-37, 1996.

[14] T. V. Mathew, Genetic algorithm. http://www.civil.iitb.ac.in/tvm/2701_dga/2701-ga-notes/gad oc.pdf.

[15] S. Forrest, "Genetic algorithms: principles of natural selection applied to computation," Science, vol. 261, pp. 872-878, 1993.

[16] A. Chipperfield, P. Fleming, H. Pohlheim, and C. Fonseca, Genetic Algorithm Toolbox: User's Guide, For Use with MATLAB, Version 1.2. http://www.pohlheim.com/Papers/tr_gatbx12/ChipperfieldFl emingPohlheimFonseca_tr_GATbx_v12.pdf.

[17] S. O. C. Morales, Y. P. Maldonado, and F. T. Romero, "Improvement on Automatic Speech Recognition Using Micro-genetic Algorithm," in 11th Mexican International Conference on Artificial Intelligence, pp. 95-99, 2012.

[18] R. Li, Jia-heng Zheng, and Chun-qin Pei, "Text Information Extraction Based on Genetic Algorithm and Hidden Markov Model," in First International Workshop on Education Technology and Computer Science, vol. 1, pp. 334-338, 2009.

[19] J. Xiao, L. Zou, and C. Li, "Optimization of Hidden Markov Model by a Genetic Algorithm for Web Information Extraction," in International Conference on Intelligent Systems and Knowledge Engineering, 2007.

[20] Z. Zhi-Jin, Z. Shi-Lian, X. Chun-Yun, and K. Xian-Zheng, "Discrete channel modeling based on genetic algorithm and simulated annealing for training hidden Markov model," Chinese Physics, vol. 16, pp. 1619-1623, 2007.

[21] N. Surajudeen-Bakinde, Xu Zhu, J. Gao, and A. K. Nandi, "Genetic algorithm based equalization for direct sequence Ultra-Wideband communications systems," In Proceedings of the 2009 IEEE wireless communications and networking conference, pp. 145-149, 2009.

[22] T. W. Rondeau, C. J. Rieser, T. M. Gallagher, and C. W. Bostian, "Online modeling of wireless channels with hidden Markov models and channel impulse responses for cognitive radios, In proceeding of 2004 IEEE International Microwave Symposium Digest, vol. 2, pp. 739-742, 2004.

[23] W. H. Tranter, K. S. Shanmugan, T. S. Rappaport, and K. L. Kosbar, Principles of Communication Systems Simulation with Wireless Applications. Prentice Hall, NJ, Professional Technical Reference, 2004.

[24] W. Turin, Performance Analysis and Modeling of Digital Transmission Systems. Kluwer Academic/Plenum Publishers, 2004.

# 15

# Optimal resource allocation for LTE uplink scheduling in smart grid communications

**Jian Li, Yifeng He, Yun Tie, Ling Guan**

Department of Electrical and Computer Engineering, Ryerson University, Toronto, Canada

**Email address:**

jian.li@ryerson.ca(J. Li), yhe@ee.ryerson.ca(Y. He), ytie@ee.ryerson.ca(Y. Tie), lguan@ee.ryerson.ca(L. Guan)

**Abstract:** The success of the smart grid majorly depends on the advanced communication architectures. An advanced smart grid network should satisfy the future demands of the electric systems in terms of reliability and latency. The latest 4th-generation (4G) wireless technology, the 3rd Generation Partnership Project (3GPP) Long Term Evolution (LTE), is a promising choice for smart grid wide area networks (WAN), due to its higher data rates, lower latency and larger coverage. However, LTE is not a dedicated technology invented for smart grid, and it does not provide Quality of Service (QoS) guarantee to the smart grid applications. In this paper, we propose an optimal LTE uplink scheduling scheme to provide scheduling timeguarantee at the LTE base station for different class of traffic, with a minimal number of total resource blocks. A lightweight heuristic algorithm is proposed to obtain the optimal allocation of resource blocks for each class of traffic. In the simulation, we compare the proposed optimal scheduling scheme and two existing scheduling schemes, the Large-Metric-First scheduling scheme and the Guaranteed Bit Rate (GBR) /Non-GBR scheduling scheme. The comparison results demonstrate that the proposed optimal scheduling can use less resource blocks to satisfy the scheduling time requirements than the other two existing scheduling schemes.

**Keywords:** LTE, Uplink, Scheduling, Smart Grid Communications, Quality of Service

## 1. Introduction

The smart grid is a modern electric system which uses sensors, automation, computers and other application-specific devices to control and monitor the grid system. Currently, the constant improvements of smart grid technology have made a significant progress on flexibility, security, reliability and efficiency of the electricity system. Meanwhile, the advanced systems and devices generate a largevolume of traffic flows and placeanenormous challenge on real-time communications. Therefore, an advanced and efficient smart grid communication network should be able to satisfy the future demands in terms of reliability and latency. Smart grid communication architecture consists of three interconnected networks: Wide Area Network (WAN), Neighborhood Area Network (NAN) and Home Area Network (HAN) [1]. The WAN plays animportant role in the smart grid communication network, which connects various NANs, and forms a connected, integrated and robust smart grid system. The performance of WAN directly affects the systemmonitoring and controlling, or even whole electric system operations. In this paper, we study the uplink scheduling problem in the smart grid WAN.

In many countries, data fiber network, synchronous optical network (SONET), supervisory control and data acquisition (SCADA) network, etc., are being deployed in current smart grid WANs [2]. However, the smart grid WAN is still unsatisfactory in some aspects, such as lack of common backhaul medium for data communication. Furthermore, future demands including Plug-in Hybrid Electric Vehicle (PHEV) and Community Energy Storage (CES) require reliable two-way communications and interactivities that traditional network systems cannot provide [3,4]. Therefore, the traditional communication architecture should be urgently upgraded, or even replaced by a more advanced communication technology.

There are several kinds of technologies available for smart grid WAN, such as Long Term Evolution (LTE), Worldwide Interoperability for Microwave Access (WiMAX) [5], etc.Compared with other technologies, the latest 4th-generation (4G) wireless technology, the 3rd Generation Partnership Project (3GPP) LTE is a promising option for smart grid because of its higher data rates, lower latency and larger coverage. The 3GPP LTE Release 8 shows that LTE provides up to 300Mbps download rate and 75Mbps upload rate [6]. The specification also defines Orthogonal Frequency Division Multiple Access (OFDMA) as the access technique for the downlink and Single Carrier FDMA (SC-FDMA) for the uplink. OFDMA has advantages of robustness against multi-path fading, higher spectral efficiency and bandwidth scalability; and SC-FDMA makes user equipments (UEs) energy-saving. The additional crucial technique applied in LTE is Multiple-Input-Multiple-Output (MIMO) that uses multiple transmitters and receivers to achieve higher bit rates and improved coverage [7]. Fig.1 shows the smart grid connectivity supported by LTE.

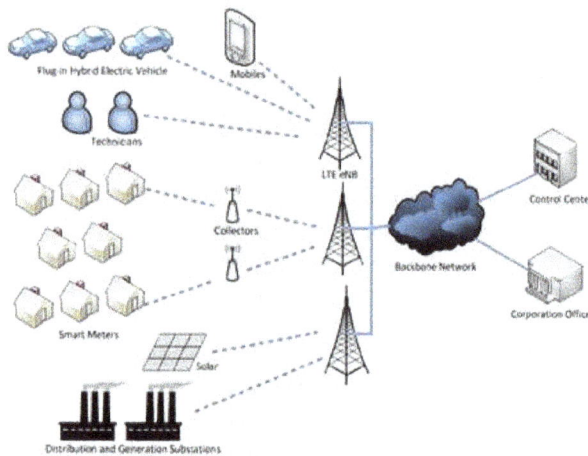

*Figure 1.The smart grid connectivity supported by LTE*

However, LTE is not a dedicated technology invented for smart grid. Smart grid applications have outstanding Quality of Service (QoS) requirements. Smart grid applications need more stringent latency requirements in WAN than other public applications such as web. An excessive delay of the criticaldata may delay the power restoration, which may lead to severe economic and social consequences. Reducing latency or end-to-end delay becomes one of the major challenges in smart grid communication networks.

In this paper, we investigate the LTE uplink scheduling problem in a smart grid WAN. Ourcontribution is that we optimize the allocation of resource blocks (RBs) in the LTE base station, named evolved NodeB (eNB) in LTE, to provide scheduling timeguarantee to different classes of smart grid traffic.

This paper is organized as follows. We describe the related works in Section 2. In Section 3, we study a queuing model and an LTE uplink scheduling model. Based on the models, we theoretically analyze the relationship between the scheduling time and the resource blocks to be allocated. Section 4 presents the problem formulation and heuristic algorithm forthe LTE uplink scheduler. The experiment results are presentedin Section 5, and the conclusions are drawn in Section 6.

# 2. Related Work

Since the LTE technical standard does not define a unique scheduling algorithm, the LTE scheduling has attracted significant attention from researchers. In [8], YuzheXu et al. investigated the latency performance of LTE network in smart grid, and proposed a new LTE scheduler that firstly allocates resources for smart grid. In [9], Yasir et al. presented a Bandwidth and QoS Aware (BQA) LTE uplink scheduler, which maximizes the cell throughput by giving priority to user equipments (UEs) with better channel conditions and maintains some level of fairness by providing resources to UEs with adverse channel conditions. In [10], Oscar et al.proposed two novel LTE uplink resource allocation algorithms for multiclass services, which adapt dynamically to the number of requests in the system, assigning resources as fair as possible.

# 3. System Models

We study a queuing model and LTE uplink scheduling model in this section. The queuing model illustrates a method to obtain the scheduling time for the LTE uplink scheduler. Section 3.2 shows the LTE uplink scheduling model, which presents the mechanism of LTE uplink scheduling and the relationship between the scheduling time and LTE scheduling resources.

## 3.1. Queuing Model

According to the requirements, the data in the smart grid can be classified into different classes. For example, data on remote workforce is classified into low-priority class while the control data from the control center and exception messages such as the outage notifications are classified into the critical class. The queuing model consisting of several queues and one scheduler is involved in our study. The queuing model in LTE scheduler is shown in Fig.2. Suppose that the scheduler provides $C$ classes of traffic with different priorities. A smaller classnumber corresponds to a higher priority. The traffic of class-$c(\forall c=1, 2,..., C)$is characterized by four parameters: 1) the arrivals of the class-$c$requests, which are modeled as a Poisson process with average arrival rate $\lambda_c$ requests/second; 2) the average request size$F_c$Kbytes/request which is specified by the size of every request; 3) the upper bound of scheduling time $\tau_c$in seconds; and 4) the possibility $\rho_c$that an arriving request belongs to class-$c$.

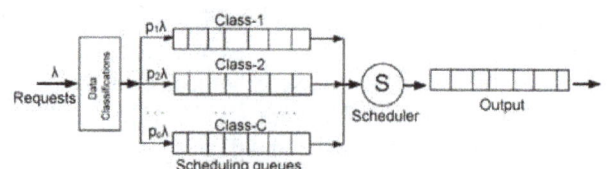

*Figure 2.Queuing model*

In order to simplify the queuing model, we assume that the model consists of $C$ queues connecting to a scheduler, and each queueis used to hold the traffic of the corresponding class. Requests can be served immediately at the scheduling rate $S_{total}$ by the scheduler. In this paper, we employ the preemptive priority service scheme.

In the queuing model, the scheduling rate for class-$c$ is denoted as $S^{(c)}$. We have $S_{total} = \sum_{c=1}^{C} S^{(c)}$. The average size of requests of class-$c$ is $F_c$. Thus, the scheduling time for class-$c$ traffic flows is assumed to be Poisson distributed with a mean timeof $F_c/S^{(c)}$. In accordance with the composition property of the Poisson process, the arrivals of task requests in class-$c$ follow a Poisson process with arrival rate $\lambda_c = \rho_c \lambda$ and the total arrivals of all requests follow a Poisson process with average arrival rate $\lambda = \sum_{c=1}^{C} \lambda_c$. In a preemptive priority M/M/1 queuing system, the mean scheduling time for class-$c$ data flowis given by[11]:

$$T_{sch}^{(c)} = \frac{F_c/S^{(c)}}{1-\beta^{(c-1)}} + \frac{\sum_{j=1}^{c}(\rho_c \lambda F_j^2/S^{(j)^2})}{(1-\beta^{(c-1)})(1-\beta^{(c)})} \quad (1)$$

where $\beta^{(c)} = \sum_{j=1}^{c}(\rho_j \lambda F_j)/(S^{(j)})$. To ensure the schedule queue stable, $\beta^{(C)} = \sum_{j=1}^{C}(\rho_j \lambda F_j)/(S^{(j)}) < 1$ should be satisfied.

### 3.2. LTE Uplink Scheduling Model

The LTE uplink scheduleris located at the base station in LTE. The minimum transmission unit of LTE scheduleris known as a resource block. The radio resource that is available in the uplink LTE system is defined in both frequency and time domains. In the frequency domain, each RB consists of 12 consecutive subcarriers, and in the time domain it is made up of one time slot of 0.5ms duration. Each 1ms Transmission Time Interval (TTI) consists of 2 slots, and eachsub-frame is defined as 10 TTIs. At each TTI, multiple RBs can be assigned to a number of users with different classes; each resource block, however can be assigned to at most one user.

The LTE scheduler has $B$ MHz bandwidth, divided into $N$ RBs. We assume that the scheduler is capable of assigning RBs arbitrarily to all users, and each RB $n$ has a bandwidth of $B/N$. Let $n = \{1, 2..., N\}$ denotes the RB indexset. For simplicity, we suppose that uniform power allocation across all subcarrier.

We define a variable $x_c$ to indicate the number of resource blocks assigned to class-$c$ traffic flows. Assuming that the throughput usage can achieve the Shannon rate limit [12], the maximum uplink channel throughput of class-$c$ for uplink direction according to Shannon-Hartley theory can be expressed as [13]:

$$S^{(c)} = x_c \frac{B}{N} \log_2(1 + SINR_{c,n}) \quad (2)$$

where $SINR_{c,n}$ is the average Signal to Interference and Noise Ratio (SINR) for the RB $n$ at the transmitter. The LTE standard provides reporting mechanisms (Channel State Information and Buffer Status Reporting) to providethe packet scheduler with valuable information

about the cellular environment that can assist in increasing the scheduling operation in the uplink [14].

Therefore, the total scheduling rate for the LTE uplink scheduling is given by:

$$S_{total} = \sum_{c=1}^{C} x_c \frac{B}{N} \log_2(1 + SINR_{c,n}) \quad (3)$$

Based on the above analysis, we can formulate the mean scheduling time for processing class-$c$ traffic as follows:

$$T^{(c)} = \frac{F_c/(x_c \frac{B}{N} \log_2(1 + SINR_{c,n}))}{1 - \sum_{j=1}^{c-1} \frac{\lambda_j F_j}{x_j \frac{B}{N} \log_2(1 + SINR_{j,n})}} + \frac{\frac{\sum_{j=1}^{c}(\lambda_j F_j^2/(x_j \frac{B}{N} \log_2(1+SINR_{c,n}))^2)}{1-\sum_{j=1}^{c-1} \frac{\lambda_j F_j}{x_j \frac{B}{N} \log_2(1+SINR_{j,n})}}}{1 - \sum_{j=1}^{c} \frac{\lambda_j F_j}{x_j \frac{B}{N} \log_2(1+SINR_{j,n})}} \quad (4)$$

$$\forall c = 1, 2, ..., C,$$

The totalschedulingtime for processing all requests is formulated as follows:

$$T_{total} = \sum_{c=1}^{C} \frac{\lambda_c}{\lambda} T^{(c)} \quad (5)$$

## 4. Optimal Resource Allocation

In this section, we formulate the resource allocation minimization problem depend on the queuing model and LTE uplink scheduling model, and propose a heuristic scheme to obtain the solution of the problem.

### 4.1. Problem Formulation

We formulate the allocated resource blocks minimization problem based on the queuing model and the LTE uplink scheduling model, aiming to minimize the total number of the allocated resource blocks while satisfying the scheduling time constraint for each class of traffic. The problem of resourceblock minimization can be written as follows:

$$\text{Maximize}_{\{x_c\}} x_c \quad (6a)$$

Subject to

$$x_c \leq N, \quad \forall c = 1,2,...,C, \quad (6b)$$

$$\sum_{c=1}^{C} x_c \leq N, \quad (6c)$$

$$\sum_{j=1}^{c} \frac{\rho_j \lambda F_j}{S^{(c-1)}} < 1, \quad (6d)$$

$$\lambda_c F_c < S^{(c)}, \quad \forall c = 1, 2, ..., C, \quad (6e)$$

$$T^{(c)} < \tau_c, \quad \forall c = 1, 2, ..., C, \quad (6f)$$

where(6b) indicates that the resource blocks assigned to class-$c$ must be less than or equal to the total number of resource blocks. Equation (6c) declares that the number of allocated resourceblock also cannot beyond the total number of resource blocks. Equation (6d) presents the inequality to ensure the queuing model stability. Equation (6e) shows the relationship between the arrival rate and the scheduling rate for class-$c$. $\tau_c$in(6f) is the upper bound of the scheduling time for class-$c$ service. It is pre-defined according to the QoS requirements for different classes.

### 4.2. Heuristic Scheme

Although global searching to find the solution of the minimum resource blocks could be feasible, such method is inefficient. Therefore, we propose a heuristic scheme to obtain the optimal solution.

---

**Algorithm 1** Proposed LTE uplink scheduling algorithm

---

1: Let $N$ be the set of resource blocks
2: Let $n$be the index of resource blocks
3: **for**$n = 1$ to $N$**do**
4: Calculate SINR value of each resource block
5: **end for**
6: $n = 0$
7: **for**$c = 1$ to $C$**do**
8:$x \leftarrow 1$
9: **while**$n < N$**do**
10: Calculate $T_x^{(c)}$
11: **if**$T_x^{(c)} < \tau_c$ and $\sum_{i=1}^c (\rho_i \lambda F_i)/(S^{(J)}) < 1$**then**
12:**if**$n + x_c < N$**then**
13:$x_c \leftarrow x$
14: Assign $x_c$resource blocks for the class-$c$data flow
15: $n \leftarrow n + x_c$
16: **else**
17: $x_c \leftarrow (N - n)$
18: Assign $x_c$ resource blocks for the class-$c$data flow
19: Stop allocation until next TTI
20: **end if**
21: **else**
22: $x \leftarrow x + 1$
23: **end if**
24: **end while**
25: **end for**

---

The proposed heuristic scheme dynamically decides the number of resource blocks allocated to each class in each TTI. Algorithm 1 describes the proposed heuristic scheme in details. In each TTI, the eNBscheduler captures the Buffer Status Report and Channel State Information from the user equipments and calculates SINR values. The principle of the allocation is to assign RBs in sequential order from the higher-priority classes to lower-priority classes. The initial number of resource blocks to be allocated for class-$c$ $x$ is set to 1. Next, the scheduler calculates the scheduling time $T_x^{(c)}$ using (4) and compares it with the scheduling time requirement $\tau_c$. If the value of calculation is larger than the requirement, the value $x$ is increased by 1. The scheduler keeps increasing the resource blocks, until the scheduling time $T_x^{(c)}$ meet the requirement. When the calculated value becomes smaller than $\tau_c$, $x_c$is determined and the scheduler allocates $x_c$ resource blocks to the class-$c$ traffic flows. Then the scheduler begins to process the next classtraffic flows. Finally, once the allocationis performed, the system updates all the relevant parameters.

Note that the system adjusts itself in order to match the QoS target. The proposed allocation scheme aims to allocate minimum RBs. Our proposed heuristic can guarantee each class of traffic is allocated the minimum RBs. If any of the class obtains one less resourceblock, this class of traffic cannot satisfy the scheduling time requirements.

## 5. Simulations

In this section, we perform LTE uplink scheduling heuristic algorithm simulations to evaluate the networkperformance in terms of the resources allocation and the scheduling time.

### 5.1. Simulation Setting

Table 1 summarizes the parameter settings of smart grid traffic. All traffic is divided into three classes. Class-1 traffic has the highest priority, including the exception messages and alarms. Class-2 contains the control messages which is not as critical as class-1. The normal operation traffic is classified into class-3. The total arrival rate of the incoming trafficis set in the range of 100-300 requests/second. The other simulation configurations can be seen in Table 2. We evaluate the number of RBs allocated for different classes in different scheduling algorithms as well as the performance of the algorithms in terms of scheduling time.

*Table 1.Parameter Settings of Smart Grid Traffic*

| Service class | 1 | 2 | 3 |
|---|---|---|---|
| Percentage of arrival rate | 20% | 30% | 50% |
| Average requests size (bytes) | 30k | 50k | 700k |
| Upper bound of scheduling time (sec) | 0.001 | 0.003 | 0.007 |

*Table 2. Major Simulation Parameters of LTE*

| Parameter | Setting |
|---|---|
| System bandwidth | 10MHz |
| Number of RBs | 50 |
| Number of subcarriers per RB | 12 |
| RB bandwidth | 180kHz |
| Transmission time interval | 1ms |
| Transmission power | 125mW |
| Noise power per Hz | 160dBm |
| Traffic arrival model | Poisson |

### 5.1.1. Comparison with Existing Algorithms

We compare the performances among three scheduling schemes: 1) our proposed scheduling scheme, 2) a Large-Metric-First scheduling scheme [8], and 3) the Guaranteed Bit Rate (GBR)/Non-GBR scheduling scheme [15]. The Large-Metric-First scheduling schemeis determined by a utility function for UEs, which is given by $\lambda = W_P + P_{PF}$, where $W_P$is the weight for UEs in the smart grid communication network and $P_{PF}$is given by traditional

LTE scheduling proportional fair (PF) algorithm. The GBR/Non-GBR scheduling represents a guaranteed minimum bit rate requested by an application. In LTE, the GBR bearers and non-GBR bearers can be provided. Of these, the GBR bearers are typically used for applications such as exception messages and control messages, with an associated GBR value; higher bit rates can be allowed if resources are available. Non-GBR bearers do not guarantee any particular bit rate, which usually are used for the normal operation applications. All simulations have been conducted with the parameters described in Section 3.4.3.

### 5.1.2. Simulation Results

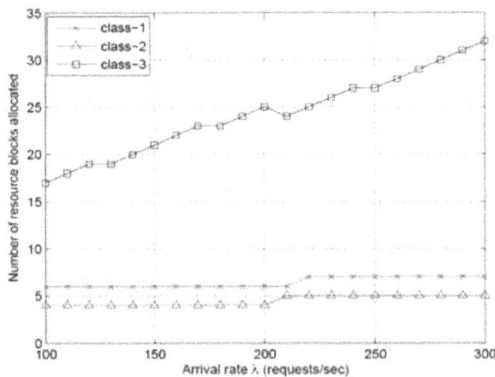

**Figure 3.** Number of resource blocks for different classes in the proposed scheduling scheme

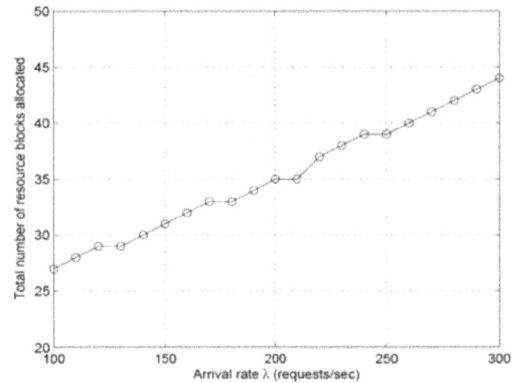

**Figure 4.** Total number of resource blocks in the proposed scheduling scheme

Fig.3 shows the number of resource blocks for different classes in the proposed LTE scheduling scheme. With the increasing of the arriving rate, the number of allocated resource blocks is dynamically adjusted to satisfy the scheduling time requirements. For classes with small traffic volume, class-1 and class-2, the change is slowly while class-3 has a higher increasing because of its largetraffic volume.

Fig.4 shows the total number of resource blocks allocated in the proposed scheduling scheme is increased. In Fig.5 and Fig.6, we let the number of total resource blocks be the value shown in Fig.4, and perform resource allocations using the three scheduling scheme, respectively.

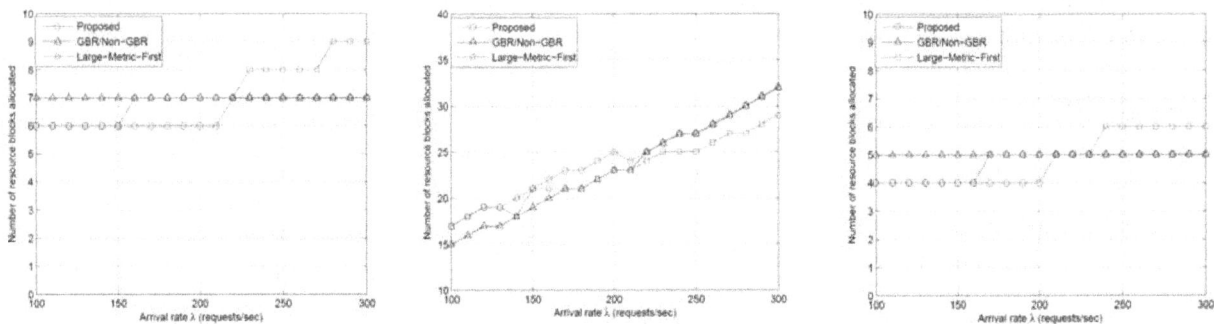

**Figure 5.** The number of allocated resource blocks for different classes

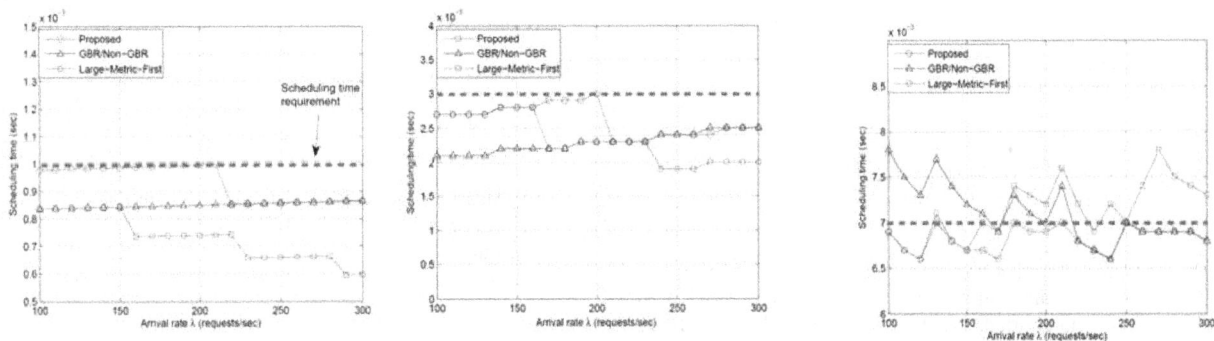

**Figure 6.** The scheduling time for different classes

Fig.5 shows the number of allocated resource blocks for different classes among the three scheduling schemes. The total available resource blocks for an arrival rate are the same among the three schemes. For example, when $\lambda = 150$ request/s, our proposed algorithm uses 31 resource blocks, then the other two algorithms also have 31 resource blocks available in the same arrival rate. Fig.5(a), Fig.5(b) and Fig.5(c) show the number of resource blocks for three classes, respectively. We can see that more resource blocks are assigned for class-1 and class-2 in Large Metric-First algorithm than our proposed algorithm. That is because the Large-Metric-First scheduling scheme is more focused on the traffic with higher priorities. In the GBR/Non-GBR scheduling scheme, class-1 and class-2 traffic flowsare assigned to GBR bearers and the values do not dynamically change with the increasing of arrival rate.

We can see that such kind of scheduling is inflexible. If the smart grid system encounters an emergency situation, for example, the additional volume of exceptional messages, alarms and control trafficare added in the networks, the scheduling time performance will get much worse.

Fig.6 shows the scheduling time for different classes using the same amount of resource blocks indicated in Fig.4. The grey dash lines represent the scheduling time requirements for different classes (see Table 1). The Large-Metric-First scheduling and GBR/Non-GBR scheduling allocates more resource blocks in class-1 and class-2, and less resource blocks in class-3. All these three algorithms satisfy the scheduling time requirements for class-1 and class,but for class-3 traffic flow, the other two algorithms cannot satisfy the requirement because less resource blocks are left for class-3. The Large-Metric-First scheduling has a better performance in class-1 and class-2 while sacrifices the scheduling time in class-3. The GBR/Non-GBR can partly satisfy the requirement for class-3 when arrival rate is larger than 220 requests/sec.

# 6. Conclusion

In this paper, we proposed an optimal scheduling algorithm for LTE in smart grid and evaluated the scheduling time performance in the smart grid network environment. The experiment results show that the scheduling time in the proposed scheme outperforms Large-Metric-First scheduling scheme and GBR/Non-GBR scheduling scheme.

# References

[1]   L. B. Le and T. Le-Ngoc, "QoS provisioning for OPFMA-based wireless network infrastructure in smart grid," in proc. of IEEE Electrical and Computer Engineering 2011 24th Canadian Conference on CCECE, Niagara Falls, May 2011.

[2]   Y. Kim and M. Thottan, "SGTP: Smart Grid Transport Protocol for secure delivery periodic real time data," Bell Labs Technical Journal, vol. 16, no. 3, pp. 83-99, December 2011.

[3]   O. C. Onar, M. Starke, G. P. Andrews and R. Jackson, "Modeling, controls, and applications of community energy storage systems with used EV/PHEV batteries," in proc. of IEEE Transportation Electrification Conference and Expo (ITEC), Dearborn, June 2012.

[4]   B. P. Roberts and C. Sandberg, "The role of energy storage in development of smart grids," in Proceedings of the IEEE, vol. 99, no. 6, pp. 1139-1144, June 2011.

[5]   P. Cheng, L. Wang, B. Zhen and S. Wang, "Feasibility study of applying LTE to smart grid," in proc. ofSmart Grid Modeling and Simulation (SGMS), 2011 IEEE First International Workshop, Brussels, Oct 2011.

[6]   S. Abeta, "Toward LTE commercial launch and future plan for LTE Enhancements (LTE-Advanced)," in proc. of Communication Systems (ICCS), 2010 IEEE International Conference, Singapore, Nov 2010.

[7]   A. Ghosh and R. Ratasuk, Essentials of LTE and LTE-A, Cambridge University Press, 2011.

[8]   Y. Xu and C. Fischione, "Real-time scheduling in LTE for smart grids," in proc.of IEEE Communications Control and Signal Processing (ISCCSP), 2012 5th International Symposium, Rome, May 2012.

[9]   S. N. K. Marwat, T. Weerawardane, Y. Zaki, C. Goerg and A. Timm-Giel, "Performance evaluation of bandwidth and QoS aware LTE uplink scheduler," in 10th International Conference, WWIC 2012, Santorini, Greece, June 6-8, 2012. Proceedings, June 2012.

[10]  O. Delgado and B. Jaumard, "Scheduling and resource allocation in LTE uplink with a delay requirement," in proc. or IEEECommunication Networks and Services Research Conference (CNSR), 2010 Eighth Annual, Montreal, May 2010.

[11]  D. Gross and C. M. Harris, Fundamentals of queuing theory, New York: Wiley, 1998.

[12]  M. M. Tantawy, A. S. T. Eldien and R. M. Zaki, "A novel cross-layer scheduling algorithm for Long Term-Evolution (LTE) wireless system," Canadian Journal on Multimedia and Wireless Networks, vol. 2, no. 4, pp. 57-62, December 2011.

[13]  "ETSI TR 136 942; LTE; evolved universal terrestrial radio access(E-UTRA); radio frequency (RF) system scenarios (3GPP TR 36.942 version 9.3.0 Release 9)," 3GPP, 2012

[14]  .H. Safa and K. Tohme, "LTE uplink scheduling algorithms: Performance and Challenges," in proc. of IEEETelecommunications (ICT), 2012 19th International Conference, Jounieh, April 2012.

[15]  M. Alasti, B. Neekzad, J. Hui and R. Vannithamby, "Quality of service in WiMAX and LTE networks [Topics in Wireless Communications]," Communications Magazine, IEEE, vol. 48, no. 5, pp. 104-111, May 2010.

# 16

# Optimization of core network router for telecommunication exchange

**Diponkar Paul***, **Subrata Kumar Sarkar, Rajib Mondal**

World University of Bangladesh

**Email address:**
dipo0001@ntu.edu.sg (D. Paul)

**Abstract:** The operation of Core Router is to restrict Network Broadcast to the LAN, to act as default gateway, to move data between different networks and to advertise loop free Path. The technology of Wifi, WiMAX (Worldwide Interoperability for Microwave Access), GPRS (General Packet Radio Services), EDGE (Enhanced Data Rates For GSM Evolution), EV-DO (Evolution Data Optimized) which are used for remote data access. A Router must be able to support multiple telecommunications interfaces of the highest speed in use in the core Internet and must be able to forward IP packets at full speed on all of them. It must also support the routing protocols being used in the core. In telephone system, Core Routers installed on the network are used as carriers to carry data from traffic sources to sinks. The optimization formulation based on Telecommunication network to obtain an optimal decision for the Core Router on whether to accept packets from a traffic source. This decision is made to maximize the reward of data delivery while the quality-of-service performance is guaranteed. From the performance evaluation, a Core Router with network optimization can achieve the highest reward while the maximum packet-blocking probability requirements met. The Core Router is a cell-site access platform specifically designed to optimize, aggregate and transport mixed-generation Radio Access Network (RAN) traffic. It is used at a cell site as part of an IP-RAN or Cell Site DCN solution. An IP RAN solution in which the Core Router extends IP connectivity to the cell site and Base Transceiver Station (BTS), through a Fast Ethernet interface to the BTS, the router provides bandwidth-efficient IP transport of voice and data bearer traffic, as well as maintenance, control and signaling traffic over the IP using traditional circuits.

**Keywords:** IIG, DNS, Router

## 1. Introduction

A router is a device that forwards data packets between computer networks, creating an overlay internet work. A router is connected to two or more data lines from different networks. When a data packet comes in on one of the lines, the router reads the address information in the packet to determine its ultimate destination. Then, using information in its routing table or routing policy, it directs the packets to the next network on its journey .Routers perform the "traffic directing" functions on the internet. A data packet is typically forwarded form one router to another through the networks that constitute the internet work until it gets to its destination node. The most familiar type of routers are home and small office routers that simply pass data, such as web pages and email, between the home computers and the owner's cable or DSL modem, which connects to the internet through an ISP. More sophisticated routers, such as enterprise routers, connect large business or ISP networks up to the powerful core routers that forward data at high speed along the optical fiber lines of the internet backbone. When multiple routers are used in interconnected network, the exchange information about destination addresses, using a dynamic routing protocol. Each router builds up a table listing the preferred routes between any two systems on the interconnected networks. A router has interfaces for different physical types of network connections (such as copper cables, fiber optic, or wireless transmission).It also contains firmware for different networking protocol standards. Each network interface uses this specialized computer software to enable data packets to be forwarded form one protocol transmission system to another. Routers may also be used connect two or more logical groups of computer devices known as subnets, each with a different sub-network address. The subnets address recorded in the router do not necessarily map directly to the physical interface connections. A router has to stages of operation called planes. A router records a routing table listing what routes should be used to forward a data packet, and through which physical interface connection. It does this using internal pre-configured address, called static routes. A typical home or small office router showing the ADSL telephone line and Ethernet network cable connections. The router forwards data packets between incoming and outgoing interface connections. It routes it to the correct network type using information that the packets header contains. It uses data recorded in the routing table control plane. Routers may provide connectivity within enterprises, between enterprises and the internet, and between internet service providers (ISP) networks. Smaller routers usually provide connectivity for typical home and office networks. Other networking solutions may be provided by a backbone wireless system (WDS), which avoids the costs of introducing networking cables into buildings. A screenshot of the luCI web interface used by open wrt. Access routers, including 'small office/home office'(SOHO) models, are located at customer sites such as branch offices that do not need hierarchical routing of there own. Typically, they are optimized for low cost. Some SOHO routers are capable of routing alternative free Linux-based firmwares like Tomato, Open Wrt or DD-WRT. Distribution routers aggregate traffic from multiple access routers, either at the same site, or to collect the data streams from multiple sites to a major enterprise location. Distribution routers are open responsible for enforcing quality of service across a WAN interface connections, and substantial onboard data processing routines [1]. They may also provide connectivity to groups of file servers or other external networks. External networks must be carefully considered as part of the overall security strategy. Separate from the router may be a firewall or VPN handling device, or the router may including Cisco systems 'PIX and ASA5500 series, junipers Net screen, Watch guard's Firebox, Barracuda's variety of mail-oriented devices, and many others. In enterprises, a core router may be provided a "collapsed backbone" interconnecting the distribution tier routers buildings of a campus, or large enterprise locations. They tend to be optimized for high bandwidth. Routers intended for ISP and major enterprise connectivity usually exchange routing information using the border gateway protocol (BGP).RFC 4098 standard defines the types of BGP-protocol routers according to the routers functions. Edge router also called a provider Edge router is placed at the edge of an ISP network. The router uses external BGP to EBGP protocol routers in other ISPs, or a large enterprise autonomous system. Subscriber edge router also called a customer Edge router, is located at the edge of the subscriber's network, it also uses EBGP protocol to its provider's autonomous system. It is typically used in an organization. Inter-provider border router. Interconnecting ISPs is a BGP-protocol router that maitains BGP session with other BGP protocol routers in ISP autonomous systems. A core router resides within an Autonomous System as a back

bone to carry traffic between edge routers. In the ISPs Autonomous system, a router uses internal BGP protocol to communicate with other ISP edge routers, or the ISPs internet provider border routers. The internet no longer has a clearly identifiable backbone, unlike its predecessor networks. The major ISPs system routers make up what could be considered to be the current internet backbone core. ISPs operate all four types of the BGP-protocol routers described here. An ISP "core" router is used to interconnect its edge and border routers. Core routers may also have specialized functions in virtual private networks based on a combination of BGP and Multi protocol Label Switching protocols. Port forwarding Routers are also used for port forwarding between private internet connected servers [2]. Voice/Data/Fax/Video processing Routers Commonly referred to as access servers or gateways, these devices are used to route and process voice, data, video, and fax traffic on the internet .Since 2005,most long distance phone calls have been processed as IP traffic (VOIP) through a voice gateway. Voice traffic that the traditional cable networks once carried. Use of access server type routers expanded with the advent of the internet, first with dial up access, and another resurgence with voice phone service. Router is a kind of network equipment that connects many networks or network segments. A mobile network is composed of one or more routers and mobile or fixed nodes. All packets destined to or out of the mobile network should pass through the mobile router which manages the connectivity of the network. The mobile router can translate data and information between different networks or segments to make them understand each other[3]. The mobile router has the responsibility to manage the mobility of the whole network. While the mobile network moves, the mobile router has to find out the point of attachment and update the location information of every node which is attached to the mobile network. Generally connections of heterogeneous networks .The MWR 1941-DC Mobile Wireless Edge Router is a networking platform optimized for being used in mobile wireless networks. It is specifically designed to be used at the cell site edge as a part of an IP Radio Access Network (IPRAN) or Cell Site Data Communications Network (DCN).It offers high performance at a low cost while meeting the critical requirements for deployment in cell sites, including small size, high availability, and DC input power flexibility. Further more it can generate revenue from new cell-site IP-based services and enable rapid deployment of next-generation mobile services. The router comprises high-performance architecture, driven by a powerful MIPS RISC processor coupled with an optional ATM network processing engine. The Cisco MWR-1941-DC- provides Abis Optimization as part of CDMA 1xRTT IP RAN or Cell Site DCN solution. In an IP RAN solution, the MWR 1941-DC extends IP connectivity to the cell site and Base Transceiver Station (BTS). Through a Fast Ethernet interface to the BTS, the router provides bandwidth-efficient IP transport of voice and data bearer traffic, as well as maintenance, control, and signaling traffic over IP using the leased line or backhaul network. The router also supports standards-based Internet Engineering Task Force (IETF) Internet protocols over the RAN transport network. In computer networking, a gateway is a router on a TCP/IP network that serves as an access point to another network. A default gateway is the node on the computer network that the network software uses when an IP address does not match any other routes in the routing table. In home computing configurations, an ISP often provides a physical device which both connects local hardware to the Internet and serves as a gateway. In organizational systems a gateway is a node that routes the traffic from a workstation to another network segment. The default gateway commonly connects the internal networks and the outside network (Internet). In such a situation, the gateway node could also act as a proxy server and a firewall. The gateway is also associated with both a router, which uses headers and forwarding tables to determine where packets are sent, and a switch, which provides the actual path for the packet in and out of the gateway. In other words, a default gateway provides an entry point and an exit point in a network.

## 2. Methodology

The Internet is a global system of interconnected computer networks that use the standard Internet Protocol Suite (TCP/IP) to serve billions of users worldwide. It is a network of networks that consists of millions of private, public, academic, business, and government networks, of local to global scope, that are linked by a broad array of electronic, wireless and optical networking technologies. The Internet carries a vast range of information resources and services, such as the inter-linked hypertext documents of the World Wide Web (WWW) and the infrastructure to support electronic mail.

A firewall can either be software-based or hardware-based and is used to help keep a network secure. Its primary objective is to control the incoming and outgoing network traffic by analyzing the data packets and determining whether it should be allowed through or not, based on a predetermined rule set. A network's firewall builds a brigade between an internal network that is assumed to be secure and trusted, and another network, usually an external (inter)network, such as the Internet, that is not assumed to be secure and trusted. A fiber media converter is a simple networking device that makes it possible to connect two dissimilar media types such as twisted pair with fiber optic cabling. MRTG is free software for monitoring and measuring the traffic load on networks links [3] It allows the user to see traffic load on a network over time in graphical form. MRTG uses the Simple Network Management Protocol (SNMP) to send requests with two object identifiers (OIDs) to a device. The device, which must be SNMP-enabled, will have a management information base (MIB) to look up the OIDs specified. After collecting the information it will

send back the raw data encapsulated in an SNMP protocol. MRTG records this data in a log on the client along with previously recorded data for the device. The software then creates an HTML document from the logs, containing a list of graphs detailing traffic for the selected device.

Ping is a computer network administration utility used to test the reach ability of a host on an Internet Protocol (IP) network and to measure the round-trip time for messages sent from the originating host to a destination computer. Ping operates by sending Internet Control Message Protocol (ICMP) echo request packets to the target host and waiting for an ICMP response. In the process it measures the time from transmission to reception (round-trip time) and records any packet loss [4]. The results of the test are printed in the form of a statistical summary of the response packets received, including the minimum, maximum, and the mean round-trip times, and sometimes the standard deviation of the mean. The Ping process are given bellow the source host generates an ICMP protocol data unit. The ICMP PDU is encapsulated in an IP datagram, with the source and destination IP addresses in the IP header. At this point the datagram is most properly referred to as an ICMP ECHO datagram, but we will call it an IP datagram from here on since that's what it looks like to the networks it is sent over. The source host notes the local time on its clock as it transmits the IP datagram towards the destination. Each host that receives the IP datagram checks the destination address to see if it matches their own address or is the all hosts address (all 1's in the host field of the IP address).If the destination IP address in the IP datagram does not match the local host's address, the IP datagram is forwarded to the network where the IP address resides. The destination host receives the IP datagram, finds a match between itself and the destination address in the IP datagram. The destination host notes the ICMP ECHO information in the IP datagram performs any necessary work then destroys the original IP/ICMP ECHO datagram. The destination host creates an ICMP ECHO REPLY, encapsulates it in an IP datagram placing it's own IP.

Address in the source IP address field, and the original sender's IP address in the destination field of the IP datagram. The new IP datagram is routed back to the originator of the PING. The host receives it, notes the time on the clock and finally prints PING output information, including the elapsed time. The process above is repeated until all requested ICMP ECHO packets have been sent and their responses have been received or the default 2-second timeout expired. The default 2-second timeout is local to the host initiating the PING and is NOT the Time-To-Live value in the datagram. This error message indicates that the requested host name cannot be resolved to its IP address; check that the name is entered correctly and that the DNS servers can resolve it.

Trace route is a computer network diagnostic tool for displaying the route (path) and measuring transit delays of packets across an Internet Protocol (IP) network. MTR (My trace route, originally called Matt's trace route) is computer software which combines the functionality of the trace route and ping programs in a single network diagnostic tool. MTR probes routers on the route path by limiting the num-

ber of hops individual packets may traverse, and listening to responses of their expiry. It will regularly repeat this process, usually once per second, and keep track of the response times of the hops along the path.

```
Command Prompt                                                    _ □ X
Microsoft Windows XP [Version 5.1.2600]
(C) Copyright 1985-2001 Microsoft Corp.

C:\Documents and Settings\user>ping mediacollege.com

Pinging mediacollege.com [66.246.3.197] with 32 bytes of data:

Reply from 66.246.3.197: bytes=32 time=280ms ITL=46
Reply from 66.246.3.197: bytes=32 time=279ms ITL=46
Reply from 66.246.3.197: bytes=32 time=279ms ITL=46
Reply from 66.246.3.197: bytes=32 time=279ms ITL=46

Ping statistics for 66.246.3.197:
    Packets: Sent = 4, Received = 4, Lost = 0 (0% loss),
Approximate round trip times in milli-seconds:
    Minimum = 279ms, Maximum = 280ms, Average = 279ms

C:\Documents and Settings\user>_
```

```
Command Prompt                                                    _ □ X
C:\>tracert mediacollege.com

Tracing route to mediacollege.com [66.246.3.197]
over a maximum of 30 hops:

  1    <10 ms    <10 ms    <10 ms  192.168.1.1
  2    240 ms    421 ms     70 ms  219-88-164-1.jetstream.xtra.co.nz [219.88.164.1]
  3     20 ms     30 ms     30 ms  210.55.205.123
  4      *         *         *     Request timed out.
  5     30 ms     30 ms     40 ms  202.50.245.197
  6     30 ms     40 ms     40 ms  g2-0-3.tkbr3.global-gateway.net.nz [202.37.245.140]
  7     30 ms     30 ms     40 ms  so-1-2-1-0.akbr3.global-gateway.net.nz [202.50.116.161]
  8    160 ms    161 ms    160 ms  p1-3.sjbr1.global-gateway.net.nz [202.50.116.178]
  9    160 ms    171 ms    160 ms  so-1-3-0-0.pabr3.global-gateway.net.nz [202.37.245.230]
 10    160 ms    161 ms    170 ms  pao1-br1-g2-1-101.gnaps.net [198.32.176.165]
 11    180 ms    181 ms    180 ms  lax1-br1-p2-1.gnaps.net [199.232.44.5]
 12    170 ms    170 ms    171 ms  lax1-br1-ge-0-1-0.gnaps.net [199.232.44.50]
 13    240 ms    241 ms    240 ms  nyc-n20-ge2-2-0.gnaps.net [199.232.44.21]
 14    240 ms    251 ms    250 ms  ash-n20-ge1-0-0.gnaps.net [199.232.131.36]
 15    241 ms    240 ms    250 ms  0503.ge-0-0-0.gbr1.ash.nac.net [207.99.39.157]
 16    251 ms    260 ms    250 ms  0.so-2-2-0.gbr2.nwr.nac.net [209.123.11.29]
 17    250 ms    260 ms    261 ms  0.so-0-3-0.gbr1.oct.nac.net [209.123.11.233]
 18    250 ms    260 ms    261 ms  209.123.182.243
 19    250 ms    260 ms    261 ms  sol.yourhost.co.nz [66.246.3.197]

Trace complete.

C:\>
```

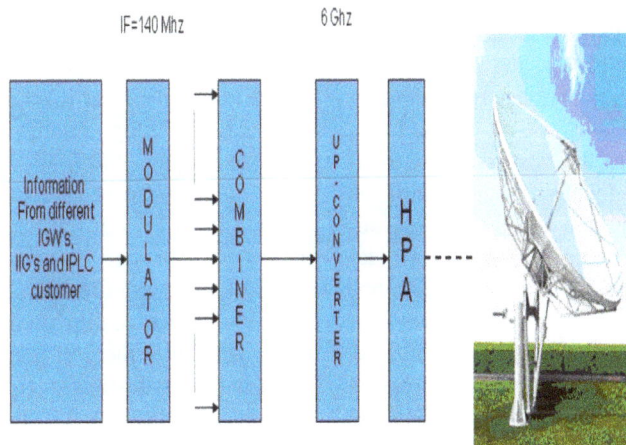

The Internet work Protocol identifies hosts with a 32-bit number called IP address or a host address. To avoid confusion with MAC addresses, which are machine or station addresses, the term IP address, will be used to designate this kind of address. IP addresses are written as four dot-separated decimal numbers between 0-255.IP addresses must be unique among all connected machines (are any hosts that you can get over a network or connected set of networks, including your local area network, remote offices joined by the company's wide-area network, or even the entire Internet community).The Internet Protocol moves data between the hosts in the form of datagram's. Each datagram is delivered to the address contained in the destination address of the datagram's header. The Destination Address is a standard 32-bit IP address that contains sufficient information to uniquely identify a network and a specific host on that network [8]. If your network is connected to the Internet, you have to get a range of IP addresses assigned to your machines through a central network administration authority. The IP address uniqueness requirement differs from the MAC addresses. IP addresses are unique only on connected networks, but machine MAC addresses are unique in the world, independent of any connectivity. Part of the reason for the difference in the uniqueness requirement is that IP addresses are 32-bits, while MAC addresses are 48-bits, so mapping every possible MAC address into an IP address requires some overlap. Of course, not every machine on an Ethernet is running IP protocols, so the many-to-one mapping isn't as bad as the numbers might indicate. There are a variety of reasons why the IP address is only 32 bits, while the MAC address is 48 bits, most of which are historical. Since the network and data link layer use different addressing schemes, some system is needed to convert or map the IP addresses to the MAC addresses.[9] Transport-layer services and user processes use IP addresses to identify hosts, but packets that go out on the network need MAC addresses. The Address Resolution Protocol (ARP) is used to convert the 32-bit IP address of a host into its 48-bit MAC address. When a host wants to map an IP address to a MAC address, it broadcasts an ARP request on the network, asking for the host using the IP address to respond. The host that sees its own IP addresses in the request returns its MAC address to the sender. With a MAC address, the sending host can transmit a packet on the Ethernet and know that the receiving host will recognize it. The standard structure of an IP address can be locally modified by using host address bits as additional network address bits. Essentially, the dividing line between network address bits and host bits is moved, creating additional networks [5], but reducing the maximum number of hosts that can belong to each network. These newly designed network bits define a network within the larger network, called a subnet. Sub netting allows decentralized management of host addressing. With the standard addressing scheme, a single administrator is responsible for managing host addresses for the entire network. By sub netting, the administrator can delegate address assignment to smaller organizations within the overall organization. Sub netting can also be used to overcome hardware differences and distance limitations. IP routers can link dissimilar physical networks together, but only if each physical network has its own unique network address. Sub netting divides a single network address into many unique subnet addresses, so that each physical network can have its own unique address. A subnet is defined by applying a bit mask, the subnet mask, to the IP address. If a bit is on the mask, that equivalent bit in the address is interpreted as a network bit. If the bit in the mask is off, the bit belongs to the host part of the address. The subnet is only known locally. To the rest of the Internet, the address is still interpreted as a standard IP address. The IP address and the routing table direct a datagram to a specific physical network, but when the data travels across a network, it must obey the physical layer protocol used by that network. The physical networks that underlay the TCP/IP network do not understand IP addressing. Physical networks have their own addressing schemes and there are as many different addressing schemes as there are different types of physical networks. One task of the network access protocols is to map IP addresses to physical network addresses. In figure, when an ARP request is sent, all fields in the layout are used except the Recipient Hardware Address (which the request is trying to identify). In an ARP reply, all the fields are used. The fields in the ARP request and reply can have several values. The ARP software maintains a table of translations between IP addresses and Ethernet addresses. This table is built dynamically. When ARP receives a request to translate an IP address, it checks for the address in its table. If the address is found, it returns the Ethernet address in its table. If the address is not found in the table, ARP broadcast a packet to every host on the Ethernet. The packet contains the IP address for which an Ethernet address is sought. If a receiving host identifies the IP address as its own, it responds by sending its Ethernet address back to the requesting host. The response is then cached in the ARP table [7]. The arp -a command display all the contents of the ARP table. It is a distributed database system that doesn't bog down as the database grows. It guarantees that new host information will be disseminated to the rest of the network as it is needed to those who are interested. If a DNS server receives a request for information about a host for which it has no information, it passes on the request to an authoritative server (is any server responsible for maintaining accurate information about the domain which is being queried). When the authoritative server answers, the local server saves (caches) the answer for future use. The next time the local server receives a request for this information, it answers the request itself [6]. The ability to control host information from an authoritative source and to automatically disseminate accurate information makes DNS superior to the host table, even for small networks not connected to the Internet. We see that at first telephone call from user Telephone set gp to the Local Exchange (LE). Local Exchange than pass to the call TANDEM.TANDEM is the central switch of all Local Exchange. TANDEM than just pass to the call TAX. TAX is located in the big city. If the call is local tax than pass to the call to the TANDEM and if the call is international TAX than pass to the call International Gateway (IGW).The international call from IGW is passed by Satellite Earth Station or Submarine Cable. The bellow Figure illustrated how to International call exchange by Submarine Cable: As defined in the national telecommunications policy 1998 and international long distance telecommunications services (ILDTS) policy 2007, all mobile operators is to interconnect through Interconnection Exchange (ICX) s and all international calls to be handled by International Gateway (IGW) which is to be connected to the mobile and fixed operators through the ICXs. The Interconnection Exchange (ICX) will receive all calls from the mobile and fixed operators whenever the call is made to other network and will pass it to the destination network if the call is local, and will pass to the IGWs if the call is international. ICX will also deliver calls received from IGWs where the call is destined. Below illustrate the structure of interconnection between different interfaces. South East Asia –Middle East – Western Europe ( SEA-ME –WE 4 ) is an optical fiber submarine communications cable system that carries telecommunications between Singapore , Malaysia, Thailand, Bangladesh ,India , Sri Lanka , Pakistan, United Arab Emirates, Saudi Arabia, Sudan, Egypt, Italy, Tunisia, Algeria and France. The cable is approximately 18,800Km long, and provides the primary internet backbone between South East Asia, the Indian subcontinent, the Middle East and Europe. SEA-ME-WE 4 are used to carry telephone, internet, multimedia and various broadband data applications. The  SEA-ME-WE 3 and the SEA-ME-WE 4 cable systems are intended to provide redundancy for each other .The two cable systems  are complementary, but  separate , and 4 is not intended to replace 3. SEA-ME-WE 3 are far longer at 39,000 km (compare to SEA-ME –WE 4's 18,800km) and extend from Japan and Australia along the bottom of the Eurasian landmass to Ireland and Germany. SEA-ME-WE 4 has a faster rate of data transmission at 1.28 Tbit/s against SEA-ME-WE 3'S 0.96 Tbit/s. SEA-ME-WE 3 provides connectivity to a greater number of countries over a greater distance, but SEA-ME-WE 4 provides far higher data transmission speeds intended to accommodate increasing demand for high speed internet access in developing countries. The cable uses dense wavelength – division multiplexing (DWDM), allowing for increased communications capacity per fiber and also facilitates bidirectional communication within a single fiber. DWDM does this by multiplexing different wavelengths of laser light on a single optical fiber. Two fiber pair able to carry 64 carriers at 10 Gbit/s each. This enables Terabit per second speeds along the SEA-ME-WE 4 cable, with a total capacity of 1.28 Tbit/s. While the ISDN TDM switching feature can switch any type of traffic, the main application for the feature is video traffic. This scenario, which was tested for this document, uses ISDN video endpoints for TDM switching. The ISDN PRI to the ISDN network uses E1 interface 0/0/0 with the configuration of 10 B channels. The video endpoints use EM-4BRI-NT/TE BRI interfaces on an EVM-HD-8FXS/DID, slots 2/0/16, 2/0/17, and 2/0/18.The EVM-HD has a 50-way amphenol Champ RJ-21 connector. The connector connects to a Black Box JPM2194A special patch panel. A male-to-female 50-way cable connects the EVM ports to the patch pane [6]. When we have to connect two switches then we make TG or TGs between them. TG

contains 5E1s then there will be total of 160(32*5) circuits between these two switches. One more important thing, when we connect two switches then we have to make at least 2 Signaling Links between them. Signaling Link will be always made on a 16th circuit of an E1.In above example we have total 5 E1s, so we will use 16th circuit of any two E1s for making two Signaling Links. In these two E1s we will be able to use 30 circuits of each E1 for voice & data. But in remaining 3 E1s we won't use 16th circuit for Signaling & we will have 31 circuits for carrying voice & data. Synchronization will be done for each E1 so 0th interval will be used in all the 5E1s.Maximum 16 Signaling Links can be made between two switches. Digital Circuit Multiplication Equipment (DCME) performs voice compression over TDM and IP networks to reduce bandwidth requirements for microwave, wire line and costly satellite links by up to 16:1, without causing degradation in voice quality. DCME voice trunking gateways employ voice detection and silence suppression techniques to enable enterprises, cellular operators and carriers to cut operating costs and open up more lines of communications using existing bandwidth capacity.

## 3. Receive Path

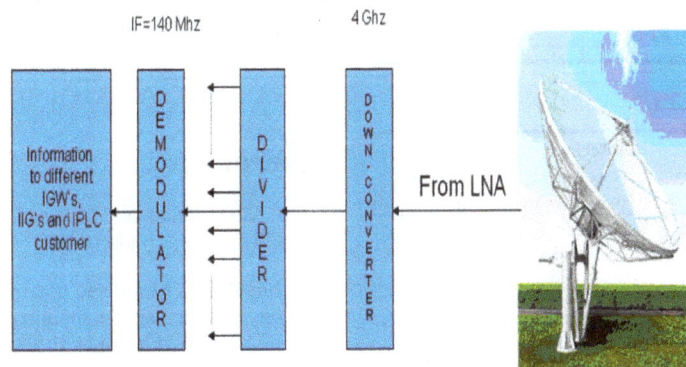

### 3.1. Compression Equipment

- Digital circuit multiplication equipment (DCME).
- Low Rate Encoder (LRE).

### 3.2. Advantages

The advantages of satellite communication over terrestrial communication are:
- The coverage area of a satellite greatly exceeds that of a terrestrial system.
- Transmission cost of a satellite is independent of the distance from the center of the coverage area.
- Satellite to Satellite communication is very precise.
- Higher Bandwidths are available for use.

### 3.3. Disadvantages

The disadvantages of satellite communication:
- Launching satellites into orbit is costly.
- Satellite bandwidth is gradually becoming used up.
- There is a larger propagation delay in satellite communication than in terrestrial communication. A network with three Routers and three hosts, connected to the Internet through Router1.

Hosts and addresses:
- PC1 10.1.1.100, default gateway 10.1.1.1
- PC2 172.16.1.100, default gateway 172.16.1.1
- PC3 192.168.1.100, default gateway 192.168.1.96

Router1:
Interface 1  5.5.5.2 (public ip)
- Interface 2  10.1.1.1

Router2:
- Interface 1  10.1.1.2
- Interface 2  172.16.1.1

Router3:
- Interface 1    10.1.1.3
- Interface 2    192.168.1.96

Network mask in all networks: 255.255.255.0 (/24 in CIDR notation).

If the routers do not use a Routing Information Protocol to discover which network each router is connected to, then the routing table of each router must be set up.

*Router 1*

| Network ID | Network mask | Gateway | Interface (examples; may vary) | Cost (decreases the TTL) |
|---|---|---|---|---|
| 0.0.0.0 (default route) | 0.0.0.0 | Assigned by ISP (e.g. 5.5.5.1) | eth0 (Ethernet 1st adapter) | 10 |
| 10.1.1.0 | 255.255.255.0 | 10.1.1.1 | eth1 (Ethernet 2nd adapter) | 10 |
| 172.16.1.0 | 255.255.255.0 | 10.1.1.2 | eth1 (Ethernet 2nd adapter) | 10 |
| 192.168.1.0 | 255.255.255.0 | 10.1.1.3 | eth1 (Ethernet 2nd adapter) | 10 |

*Router 2*

| Network ID | Network mask | Gateway | Interface (examples; may vary) | Cost (decreases the TTL) |
|---|---|---|---|---|
| 0.0.0.0 (default route) | 0.0.0.0 | 10.1.1.1 | eth0 (Ethernet 1st adapter) | 10 |
| 172.16.1.0 | 255.255.255.0 | 172.16.1.1 | eth1 (Ethernet 2nd adapter) | 10 |

*Router 3*

| Network ID | Network mask | Gateway | Interface (examples; may vary) | Cost (decreases the TTL) |
|---|---|---|---|---|
| 0.0.0.0 (default route) | 0.0.0.0 | 10.1.1.1 | eth0 (Ethernet 1st adapter) | 10 |
| 192.168.1.0 | 255.255.255.0 | 192.168.1.96 | eth1 (Ethernet 2nd adapter) | 10 |

Router2 manages its attached networks and default gateway, router 3 does the same, router 1 manages all routes within the internal networks. Accessing internal resources If PC2 (172.16.1.100) needs to access PC3 (192.168.1.100), since PC2 has no route to 192.168.1.100 it will send packets for PC3 to its default gateway (router2). Router2 also has no route to PC3, and it will forward the packets to its default gateway (router1). Router1 has a route for this network (192.168.1.0/24) so router1 will forward the packets to router3, which will deliver the packets to PC3; reply packets will follow the same route to PC2.Accessing external resources If any of the computers try to access a webpage on the Internet, like http://en.wikipedia.org/, the destination will first be resolved to an IP address by using DNS-resolving. The IP-address could be 91.198.174.2. Here none of the internal routers know the route to that host, so they will forward the packet through router1's gateway or default route. Every router on the packet's way to the destination will check whether the packet's destination IP-

address matches any known network routes. If a router finds a match, it will forward the packet through that route but if not, it will send the packet to its own default gateway. Each router encountered on the way will store the packet ID and where it came from so that it can pass the request back to previous sender. The packet contains source and destination, not all router hops. At last the packet will arrive back to router1, which will check for matching packet ID and route it accordingly through router2 or router3 or directly to PC1 (which was connected in the same network segment as router1.

# 4. Conclusion

The Core Network Router is very essential for maintenance of Telephone Exchange System. Routing information is exchanged only upon the establishment of new neighbor adjacencies. To find a solution to the simultaneous routing, frequency planning and power allocation problem in a telecommunication network with fixed relay infrastructure and conclude that the major benefit of relays is to make the system more equitable while extending coverage. By using the Core Router, operators can simplify and optimize their current network with a compact, high-performance, and modular cell-site access platform, reduce operating costs and enhance profit opportunities. This report can be a guideline for proper operation, maintenance, monitoring and troubleshooting of Telecommunication Network System.

# References

[1]    I. Mohammad, and M. Imad, Handbook of Sensor Networks, CRC Press, London, 2005.

[2]    Jun-Zhao, "Mobile ad hoc networking: an essential technology for pervasive computing", International Conferences on Info-tech and Info-net, Proceedings, 2001, pp. 316-321.

[3]    W. Heinzelman, A. Chandrakasan, and H. Balakrishnan, "Energy- efficient communication protocol for wireless microsensor networks", Proc. 33rd Hawaii Int. Conf. Syst. Sci. (HICSS'00),                                        2000. http://wwl.microchip.com/downloads/en/devicedoc/41211b. pdf,    Microchip    Technology    Incorporated,    2006. http://www.labcenter.co.uk, Labcenter electronics, 2006.

[4]    Suri, S., Waldvogel, M., Warkhede, P.R.: Profile-Based Routing: A New Framework for MPLS Traffic Engineering. In: Quality of Future Internet Services. LNCS, vol. 2156, Springer Verlag, Heidelberg (2001).

[5]    Yilmaz, S., Matta, I.: On the Scalability-Performance Tradeoffs in MPLS and IP Routing. In: Proceedings of SPIE ITCOM (May (2002).

[6]    Ott, T., Bogovic, T., Carpenter, T., Krishnan, K.R., Shallcross, D.: Algorithms for Flow Allocation for Multi Protocol Label Switching. MPLS International Conference (October 2000).

# 17

# Meeting the challenges for wireless sensor network deployment in buildings

**Costas Daskalakis, Nikos Sakkas, Maria Kouveletsou**[*]

Applied Industrial Technologies Ltd., Gerakas, Attiki, Greece

**Email address:**

daskalakis@apintech.com (C. Daskalakis), sakkas@apintech.com (N. Sakkas), kouveletsou@apintech.com (M. Kouveletsou)

**Abstract:** Wireless sensor networks (WSNs) in buildings are faced with transmission issues, much more severe than those of outdoor applications. Next to the transmission effective range, battery lifetime is also of a high importance, as it can significantly affect network performance and maintenance requirements. In this paper we present an architectural concept, in fact a dynamic routing protocol, for the setup of a building WSN. Three key goals have underpinned the protocol design; ability to cost efficiently address transmission distance within buildings, acceptable battery longevity, typically up to a year, and no data loss. Experimental data have been collected over a period of several months and have demonstrated the much enhanced performance of the network, when compared to the performance before the protocol implementation.

**Keywords:** Wireless Networks, Dynamic Routing, Relaying

## 1. Introduction

In the framework of an EU research program [1], an advanced energy monitoring and control system was planned for installation in a three floor, plus basement, building. This system would be based on a wireless sensor network (WSN). The WSN would be deployed with the aim to provide for energy management as well as for a joint and real time view on energy consumption and the respective indoor environment quality. Indoor installations of wireless networks present many challenges, primarily in terms of transmission effective ranges as well battery lifetime and, consequently, network, maintenance free, operation.

The pilot building had an overall height around 12m and an average surface area of about 640 sq. m. The building volume was approximately 1000 cubic meters. Load bearing elements had been constructed from reinforced concrete, and had a typical thickness 25- 30 cm; bricks, synthetic panels, etc, had been used as non load bearing structural elements. Cooling was based on heat pumps; because of the passive architectural elements used, cooling needs were minimal; ventilation was purely natural and no mechanical means had been installed.

The project was planned for completion by early 2013. At the moment of the paper preparation (2012), work is still in progress; a number or wireless nodes have been installed in the building, and sensors have been linked to them; sensor-boards deployed till now monitor electric energy as well as environmental parameters and occupancy; thermal energy and water temperature sensors nodes will follow in the next months.

The network planned for installation in this building will be gradually deployed all across it, and will monitor approximately 25- 40 electric and thermal energy consumption, storage and production points. Thermal and PV panels have been deployed on the building roof as energy producers. In the basement, a hot water boiler and a battery pack provide for thermal and electric, respectively, energy storage. The network will also monitor environmental conditions such as temperature, humidity, $CO_2$, in order to provide for a correlation between the energy used in particular building spaces (offices) and the quality of the environment.

The nodes that have been currently deployed are located in all three floors as well as the basement and the terrace. The network gateway, which collects all the sensor data, is on the top floor. The gateway node communicates with an application that then sends the data to a web server where they are published real time. Network nodes have distances ranging from 2 m to 25 m from this gateway node.

## 2. Background and Motivation

The network rationale has been multipurpose; to will allow monitoring devices, user and building performance, calculating respective performance indicators and issuing advice for the building user. To also support several control operations of the renewable energy infrastructure.

The network was based on wireless nodes [2] programmed on TinyOS [3, 4]. A number of known architectural approaches, such as single-hop transmission and the Collection Tree Protocol (CTP) [5], implemented for the TinyOS, were initially used, which however did not perform well in terms of the above criteria. The problems manifested as significant data loss due to frequent instances of communication loss, as well as a rapid deterioration of battery performance.

To address these issues, a dynamic routing architecture has been designed and the wireless sensor network has been respectively programmed. Similar routing techniques have been introduced [6] which do not have the drawbacks of CTP. Another type of routing protocols, known as position based routing [7] exist, but they usually include a Global Position System (GPS) module, which is not acceptable due to increased energy consumption and specific operating conditions (open space).

The routing was somehow designed with the nature of the application context in mind; energy management in buildings. Because of this particular context, it has been safe to assume that there would, by definition, be a number of points were electric energy would be monitored. These wireless nodes, obviously battery independent, would then assume a special, relaying role in the transmission architecture. Therefore, although several of the concepts used can be relevant also in outdoor environments, where network nodes may not be easily sourced by electricity, the architecture and routing relevance is far greater in the case of indoor applications, which assumes that certain nodes may be mains powered.

## 3. The Deployed Wireless Network

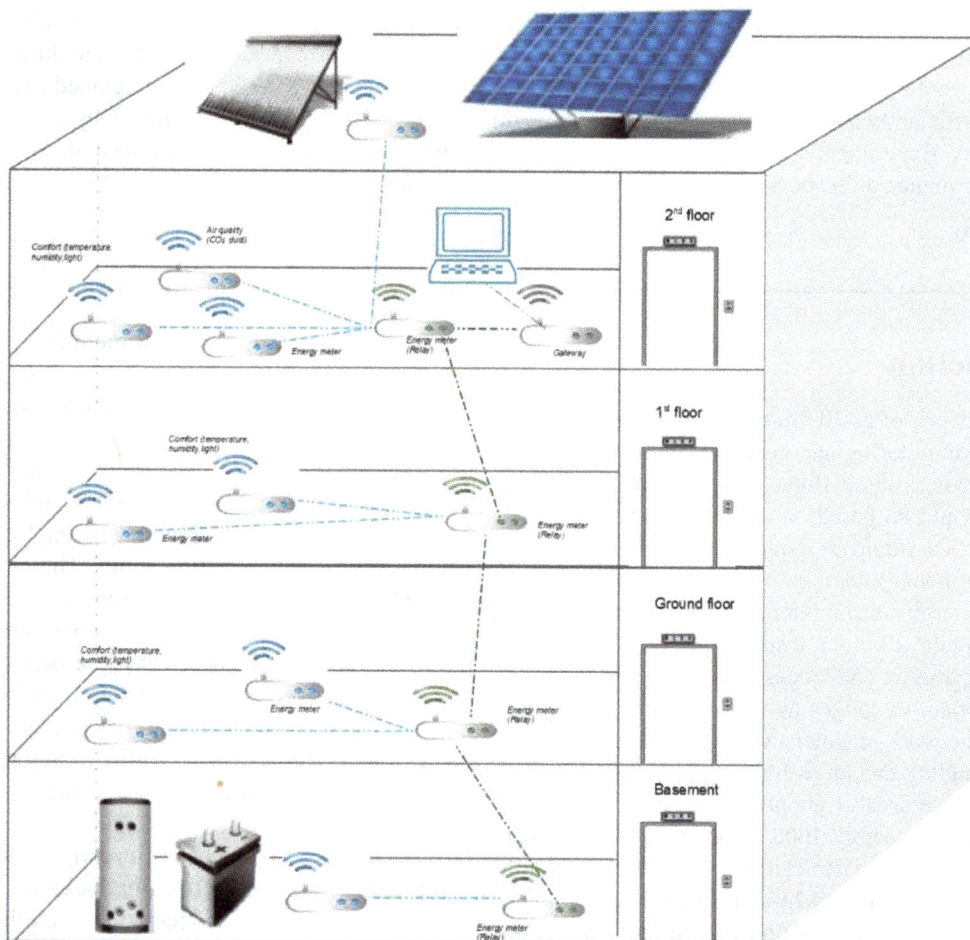

**Figure 1.** *WSN Topology.*

The wireless nodes have been based on the TinyOS technology and have been provided by MaxFor (Korea) and SowNet (Netherlands). In the former case, the operation frequency was that of 2.4GHz; in the latter 800MHz. Deployed nodes have been equipped with either an on board or an external antenna. Each node has been connected to one or more sensors; sensor boards for the environmental parameters typically included temperature, humidity (provider Sensirion) and visible light intensity (provider Hamamatsu). Another sensor board deployed provided measurements of CO, dust (provider NIDS Co.), and $CO_2$ (provider SOHA Tech). As to the electric sensors deployed these have been developed on the ADE chip (Analog Devices ADE7753 chip). The current sensing element was of a shunt type; the sensors were capable of measuring one phase current up to 20 A. Further sensors are now being developed on the same chip for higher currents as well as three phase currents. A special type of a, so called, occupancy sensor board has also been deployed in several spaces (rooms, etc.). This board is based on the Murata PIR (pyro-electric infrared) sensor, typically used for human presence detection. The occupancy board, however, does not just sense presence. It has been designed as to discern between humans entering and exiting spaces and, in this way, provides real time values for space occupancy. Such values can then be used to develop a new generation of actual-data based energy efficiency indicators. Space energy consumed can then be related with actual occupancy, i.e., the serviced population, and not just with space surface or theoretic populations, as typically the case. Fig. 1 below broadly illustrates the WSN topology.

# 4. Problems Encountered

Two types of problems were encountered from the very first trials. First, despite the relatively small distances (maximum 25m) because of the indoor nature of the installation the nodes would often loose communication with the base gateway. Second, batteries, in several cases, were ex-

hausted in a matter of weeks. To address these issues a number of routing provisions were made [8]. These will be briefly discussed below.

## 4.1. Addressing the Limited Battery Life

Initially, the single-hop transmission was attempted, where nodes transmitted the data directly to the gateway. This approach is very simple in concept and very effective in terms of battery lifetime, but is the least flexible and doesn't allow for nodes to be deployed far from the gateway. The shortcomings of the approach and its inability to challenge the required distances, very soon became apparent. As a next option, the well established CTP was used. In this approach, nodes were constantly listening for incoming requests, which they then forwarded to the next listener and so on, till the data reached the gateway. This approach is flexible, results, however, to a fast degradation of the network performance; listening, unfortunately, comes at an energy cost that appeared detrimental in terms of battery lifetime [9].

Fig. 2 shows the performance of a node with an external antenna, with a sampling rate of 30 sec. The battery was exhausted in as little 10 days! A voltage below 2.5V will not transmit reliable data.

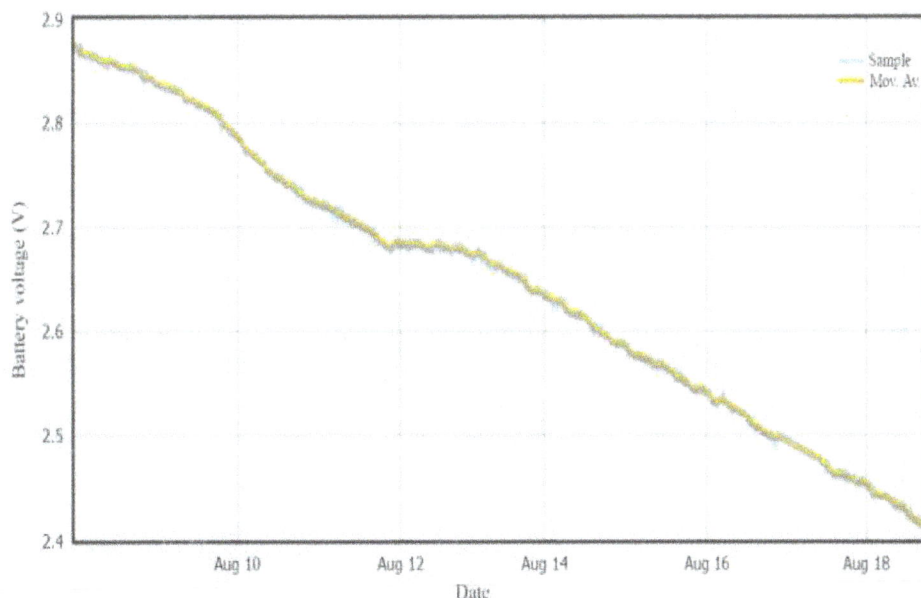

Figure 2. Battery performance for a node with external antenna, with a sampling rate of 30 seconds.

The approach that yielded the most promising results in addressing the limited battery lifetime issue was the adoption of an energy efficient, communication protocol. This protocol required that radios remained switched off and were active only during the transmission of the data to the gateway. Obviously, we had to completely depart from any radio "listening" concept as this resulted to a rapid decrease of the network performance [10].

Initially we opted for a fixed configuration; nodes re-

mained in sleep mode till the moment came to wake up, sense and transmit. This was done always towards the gateway node, although it was obvious that it would eventually turn out impossible for a node at the basement to directly reach the gateway in the third floor. Even in our early setup on the upper two floors, nodes were often loosing communication with the gateway; thus, this was an issue that had to be addressed within the "effective distance" set of problems. Battery wise, however the new communication protocol, where nodes were most of the time in the energy optimal state of sleep, increased the performance by an order of magnitude.

Fig. 3 shows a node deployed in an office. Now the battery remains in operation after two months, well above the critical operation voltage (~2.55 V) even during stressful conditions. On Fig. 4 the temperature variation in the period is displayed.

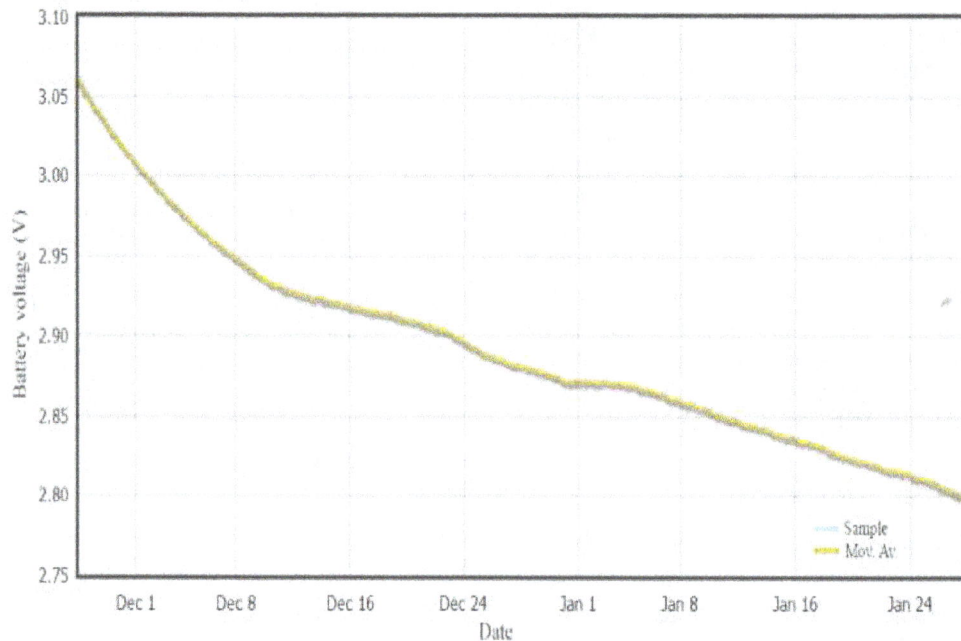

*Figure 3: Battery remains in operation after two months.*

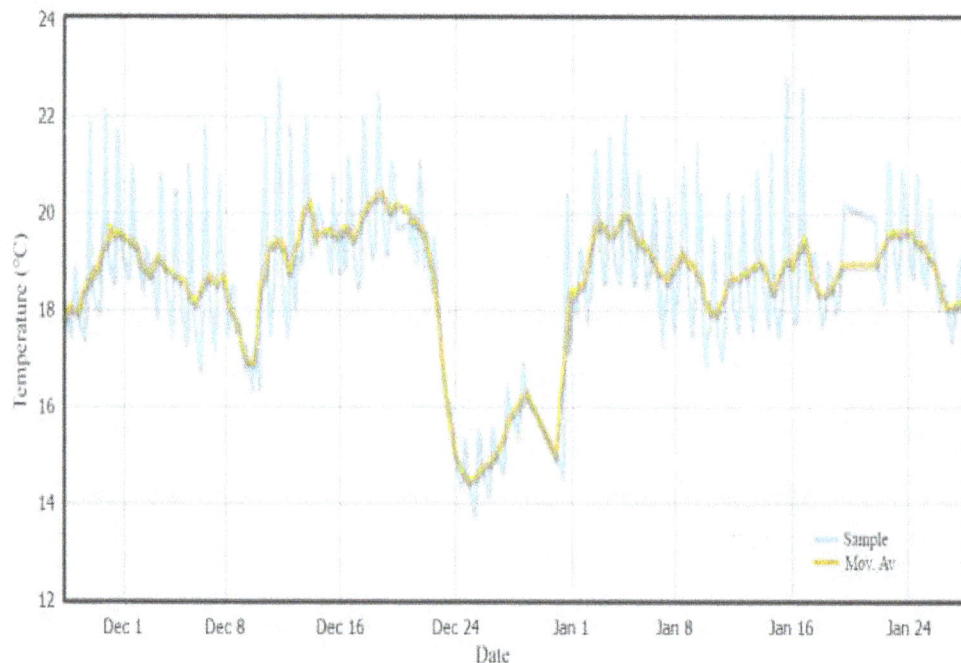

*Figure 4. Temperature variation for the same period.*

Obviously the sampling rate has a significant impact on such a configuration. The less often sensors are sampled, the less the radio needs to wake up. We did implement a number of sampling rates (between 30 and 300 sec) which fully justified this view. This, in addition with a possible removal of the unnecessary and energy consuming external antenna and its substitution with an on-board, more energy efficient variant, would bring the battery longevity close to a year.

### 4.2. Data Loss Issues

As said above, CTP was not found to have a good energy performance; because of this, it was very early abandoned. It does however have an ability of flexible meshing, via which data are able to automatically find a way to the gateway, provided no distance between two nodes is far too big to communicate the data across.

In our network implementation and despite the relatively small distances (maximum 25m) it was not always possible to reach directly the gateway; even more we noticed that this was not always an on/ off situation; often the node lost communication to the gateway which it resumed however at later moment. In the meanwhile, however, data read from the sensors during the non communication period, were lost. The second reason of data loss incidents was related to Internet connectivity issues. As the network serves a web application and transfers its data to a web server database, every loss of internet connectivity resulted also to data loss.

Finally, data loss could arise due to a failure of the PC hosting the gateway, or due to a power disruption, exceeding its UPS capacity. This is the case illustrated in fig. 3 and fig. 4, for several days between end of July and beginning of August. This was a power failure of the gateway PC in a period where offices were shut due to vacations.

Thus, we had to address two important situations. First, to provide a solution for those nodes who, due to distance, would never be able to directly reach the gateway, without resorting to the energy inefficiencies of CTP [11]; second to provide a solution to prevent data loss, in all situations as those described above.

### 4.2.1 Relaying

By the very purpose of our network, it will always include a number of electric meter nodes, measuring electric energy and power on a number of locations (devices, cabinets, etc.).

Fig. 5 illustrates a prototype electric energy sensor (the Analog Devices energy sensing chip is on the back side). Such sensors were installed in a cabinet to monitor space consumption or near devices to monitor, for example, cooling heat pumps, heaters, washing machines, burners etc.

**Figure 5.** *Electric sensor used as a relay node.*

Such nodes do not need any battery and do not have any respective energy limitations; instead of programming

those to be, by default, in sleep mode, they can be, at no performance cost, on the alert. Thus, we can use these nodes as, so called, data relaying nodes. They offer a good, energy wise, pathway for data transmission, for those remote nodes, not directly reaching the gateway. Making use of such nodes and assigning them this special relaying role will solve the effective distance problem of the remote nodes. Of course, similar may apply in environments were no mains fed energy sensors are used. Any sensor board, which can in fact connect to the mains, is a candidate for assuming a relaying role. However, convenience dictates that should energy metering be part of the set-up, these wireless meters are the best and most convenient candidates for relaying data.

### 4.2.2 Data Routing and Buffering

Before presenting the dynamic routing protocol we will illustrate the key network functionality in a more static and preconfigured setup. In fact, this was the first step to address the key issues of our concern. As soon as the functionality was proven, then the dynamic, ad-hoc features of the network were implemented, resulting to a flexible and completely dynamic routing protocol.

A first requirement was to embed in the design some contingency. If a specific path could not, for any reason, transmit the data, then a second option should be tried. Fig. 6 illustrates two such alternative paths programmed for a sensor node; one (primary) via a relaying node and one (secondary) directly to the gateway. It is again emphasized that these two paths, are, for now, hardwired; later they will result as part of the dynamic routing protocol.

Here is how the setup works: The node is most of the time in a sleep mode, until its timer wakes it up (wake-up mode) to sense the data it has been assigned to. After the data are sensed the node seeks to transmit them via its primary path, leading, in this example, to a relay node. The relay node may truly receive them and dispatch an acknowledgment, mission complete, signal to our node, which then enters again its sleep mode. If however the relay node, for any reason, will not acknowledge data reception then the node will try its second transmission path, in this example, by sending directly to the gateway. A similar acknowledgment process is executed. It the gateway confirms data reception then the mote goes safely in its sleep mode. If, however, again the mote does not receive the acknowledgment then it will start buffering the data to its local memory. Buffering offers a means to avoid data loss. As we will see, as soon as the problem, whatever it may be, is removed, the mote will offload all its locally saved data to either of its target motes, starting from the default one (the relaying node).

Overall a node may be in one of three states; sleep, wake-up and buffering. When in a wake- up mode, then it will be trying to deliver its sensor data via one of the two possible routes. When in its buffering mode it will be storing locally its data before again entering its sleep mode. In its next wake up, it will seek to off load all the past data.

When this happens, and an acknowledgment signal is eventually received, then operation will be back to normal, the local memory will be flushed and the mote will again go in its sleep mode. The local memory size and the sam-pling rate will define how long the node may store its data locally, with no data loss. In our case this has ranged between several months and a year!

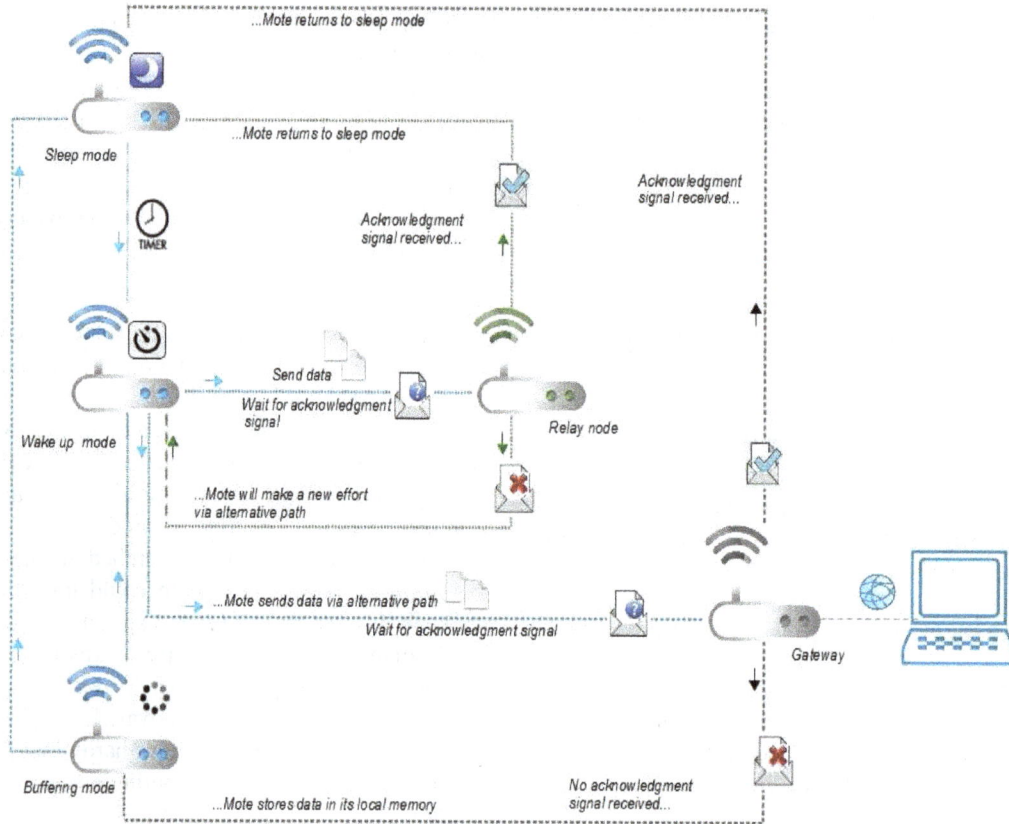

*Figure 6.* Sensor node; 3 modes (sleep, wake, buffer), 2 paths.

By providing these alternative paths we were able to re-program two of the nodes deployed, which experienced transmission problems and data loss. Fig. 7 and Fig. 8, and in particular the horizontal segments illustrate such cases of data loss due to loss of contact.

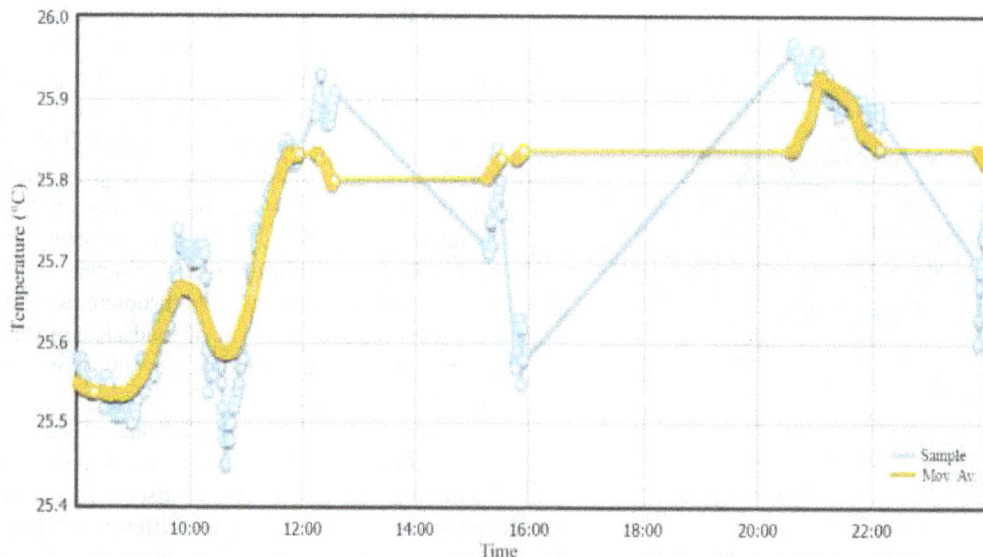

*Figure 7.* Significant data loss due to loss of contact and no buffering implemented.

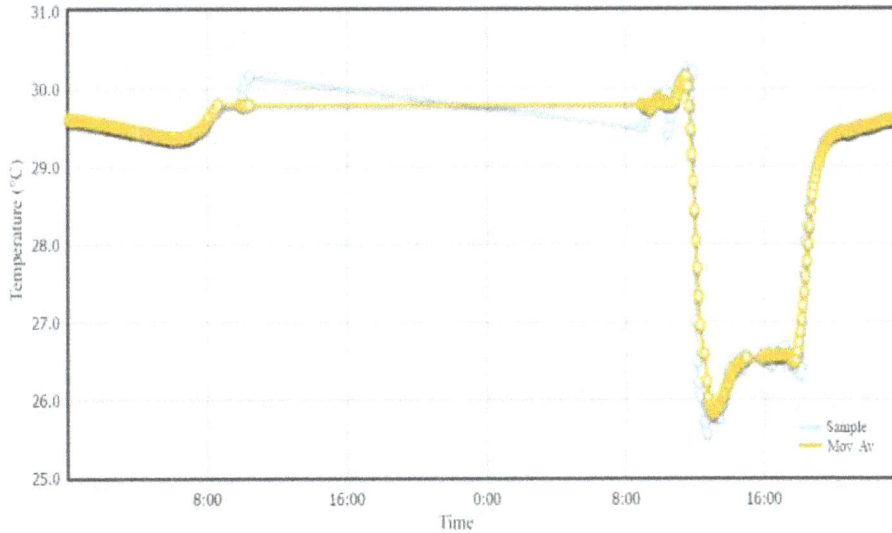

*Figure 8. Data loss from noon 11 July till noon 12 July; connection restored thereafter.*

Their preferred transmission path was now via energy meter relaying nodes that were located more closely to them. As explained above, a buffering concept was also programmed to address data loss.

Through this buffering provision, all data loss that was observed previously to this implementation was completely resolved and the robustness of the network considerably increased.

Fig. 9 shows the performance of a sensor node with buffering implemented. Horizontal segments, indicative of a communication loss, are no more the case.

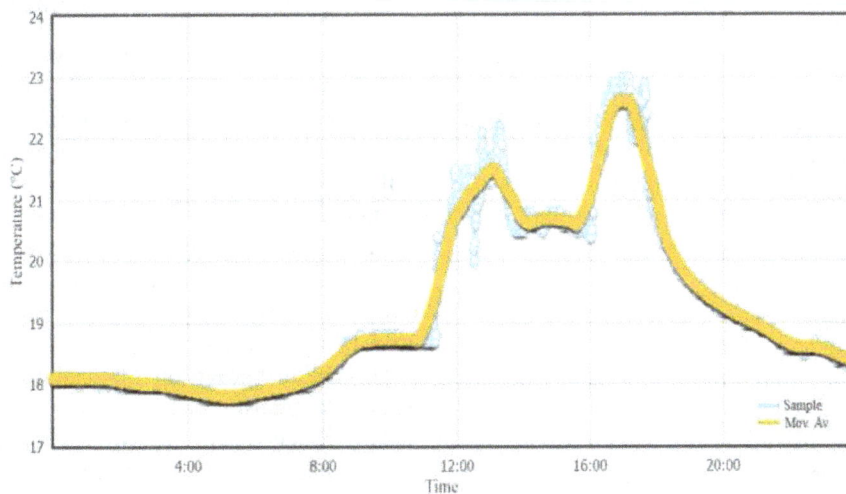

*Figure 9. Sensor node with buffering implemented; no data loss experienced.*

### 4.3. Time Stamping the Data Samples

When read, sensor samples need to be time-stamped. Nodes do not have any real time clock (RTC) to time-stamp the samples precisely. At the single-hop case, where data are transmitted directly to the gateway and no data buffering is taking place, we can safely assume that the time of the sampling is the time the data reaches the gateway and the PC. All PCs obviously are equipped with RTCs. At the current case, where relaying nodes are used and buffering might occur at any time, delaying the data in reaching the gateway, this assumption cannot stand.

Every node, since it includes a microprocessor, has an internal clock which can be used to measure the time elapsed since the boot-up of the node. Although it doesn't provide an absolute, real time, it can provide a relative time. In TinyOS, there is an implementation of TimeSync [12], where the local time of each node is synchronized with the RTC of the PC and the gateway connected to it. This method seems to be very poor energy-wise, as it needs a radio communication in short intervals to synchronize the clocks. Also, depending on the network topology, it might take several minutes until all the nodes are synchronized [13]. A simple concept of TimeOffset was introduced to timestamp the data samples. This method uses only the local clock of each node to help determine the real time of the sampling, without any need of time synchronization.

When a data sample at a node occurs, this sample is time

stamped with the local time of the node. Then, the sample is transmitted to the next node with the difference between the sampling time and the transmitted time. If this transmission happens immediately, then the time-offset is 0. If for some reason the transmission is not successful and is retried at a later time, then the time-offset will be that time difference. The next node will receive the sample and will timestamp it with its own local time, keeping the previously time-offset from the previous node too.

Again, the sample will be transmitted to the next node with the time offset between the time received and the time transmitted plus the previous time offset. Eventually, the sample will reach the gateway and the PC with a total time offset between the sample time and the receive time. Subtracting this time offset from the local real time of the PC, will result to a very precise and absolute time-stamp of the sample.

# 5. Dynamic Routing

We will now investigate how the above relaying, buffering, etc., principles were integrated in a dynamic routing protocol that allowed the network to fully self configure its communication path [14]. It is interesting to note that compared to the above static routing configuration, in the dynamic routing the requirement for the two, alternative, fixed paths, was not any more in place. We will explain this automatic configuration process, by looking in how every new node was accommodated in the network.

Every new node entering the network needs first to identify a parent. Before a transmission of a sample, the node checks if there is a parent defined. If there is no parent defined, the node will broadcast a "request for parent" message and will wait one second for responses. All nodes in radio range, which succeed in receiving the "request for parent" message and have a parent different from the requesting node, will respond by sending back a "request for parent response" message with a random delay (0-500ms). The random delay was introduced to minimize the possibility of colliding response messages.

The node will now construct a list including the nodeID and the RSSI value, for every candidate parent. RSSI corresponds to the received signal strength indicator and is a measure of the power present in a received radio signal. From this list, the best candidate is chosen as parent, by means of the best RSSI and is assigned as the node parentID parameter.

Every node which is assigned as a parent will construct a childlist table. This table consists of pairs [childID → parentID]. When a node sends a sample to its parent it includes in the message the respective [childID → parentID] pair.

## 5.1. An Example of Dynamic Routing

The following example demonstrates how the childlist table is constructed.

Fig. 10 illustrates a network comprising 9 nodes, among

which a gateway (Node 0), connected to a PC. The network has been gradually set up according to the above presented process. Let us now see how the data transmission will occur and what exactly will be, every time, transmitted. Let us assume Node 7 as the start node.

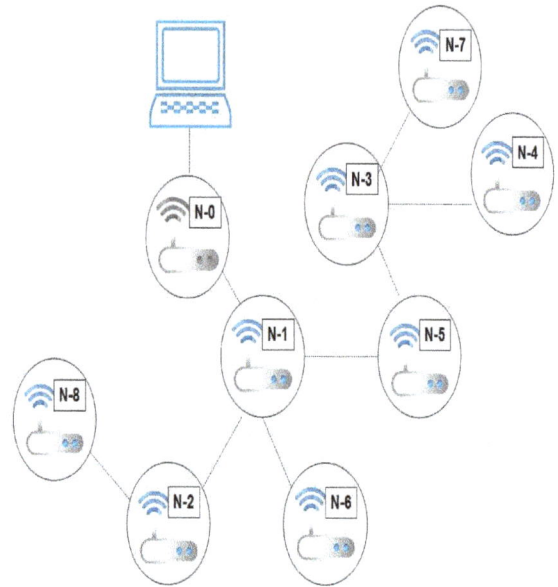

***Figure 10.*** *WSN topology; dynamic routing.*

When node 7 will wake up, it will transmit to its parent node 3 its sample along the pair [7→3]. This pair is stored on the node 3 childlist. Node 3 will forward this sample to its own parent, node 5. Then, node 5 will also store to its childlist the same pair [7→3]. Likewise, node 5 will forward the sample to its parent, node 1, along the pair [7→3], which again updates the childlist of node 1. Finally, node 1 will forward the sample to node 0. The same process is applied to every node, when it transmits its sample to its parent, in order to reach the gateway.

This childlist provision is crucial in order to support cases, where the gateway needs to send a message to a specific node or when a node needs to communicate directly with another node. There are certain cases where data from a given node need be transferred to another node to support some control action carried out at that point. One of the purposes of the dynamic routing presented here is to allow for such a clear and unambiguous transmission path between any node pair.

With regard to the above example, Table 1 below shows the childlist table of every node, constructed in the above mentioned way.

What if some parent - child relationship ceases, for any reason, to exist? This may happen during a network change/ reconfiguration, which will trigger a new parent identification routine and may result to a new Childlist structure. It is now necessary to update the Childlist, so that nodes are aware of the change. Here is how this happens: When a child-parent pair is, for any reason updated, then this update will be broadcast through the tree. The gateway will receive the update and will realize the change (its Childlist has changed! and in particular a parent-child relationship has changed!). It will then broadcast a message communicating to all nodes the change in its Childlist structure. The gateway will take no action in cases on new nodes. It will update its Childlist but there is now no need to broadcast anything. The above process applies to changes in the [child, parent] relationship and not when a new node is introduced.

*Table 1. Parent- Childlist table.*

| NodeID | ParentID | Childlist | |
|---|---|---|---|
|  |  | ChildID | ParentID |
| 7 | 3 | - | - |
| 4 | 3 | - | - |
| 8 | 2 | - | - |
| 3 | 5 | 7 | 3 |
|  |  | 4 | 3 |
| 2 | 1 | 8 | 2 |
|  |  | 3 | 5 |
| 5 | 1 | 7 | 3 |
|  |  | 4 | 3 |
| 6 | 1 | - | - |
| 1 | 0 | 8 | 2 |
|  |  | 7 | 3 |
|  |  | 4 | 3 |
|  |  | 3 | 5 |
|  |  | 2 | 1 |
|  |  | 5 | 1 |
|  |  | 6 | 1 |
| 0 | - | 8 | 2 |
|  |  | 7 | 3 |
|  |  | 4 | 3 |
|  |  | 3 | 5 |
|  |  | 2 | 1 |
|  |  | 5 | 1 |
|  |  | 6 | 1 |
|  |  | 1 | 0 |

# 6. Conclusions

Building level WSN applications are faced with significant challenges as regards transmission range, battery longevity and data loss.

An architectural concept has been implemented on a TinyOS WSN that has allowed to successfully addressing such requirements, by eliminating data loss and securing battery lifetimes up to a year. This overcomes the problems associated with the CTP meshing protocol, which has not been found to be an optimal approach in the building context, because of its energy inefficient characteristics.

Besides the static, node level, concepts, such as relaying and data buffering, the architecture includes also a dynamic routing protocol that allows the network to self configure, define its optimal transmission paths and assume instant corrective actions whenever a node enters or exits the network. Ad hoc communications can be thus optimally achieved, not only between a node and the gateway but also reversely, between the gateway and a node as well as between any pair of network nodes. This has been found a very useful feature for building applications where data from a given node need be transferred to another node to be used in decentralized control operations.

# References

[1] EnergyWarden, 2010, European Commission, 23 Jun 2012 http://energywarden.net.

[2] Karl H. and Willig A. (2007) Protocols and Architectures for Wireless Sensor Networks. England: John Wiley & Sons.

[3] Levis P. and Gay D. (2009) TinyOS Programming. United Kingdom: Cambridge University Press.

[4] TinyOS, 2011, University of California Berkeley, 23 Jun 2012 http://www.tinyos.net.

[5] Gnawali O., Fonseca R., Jamieson K., Moss D., Levis P., Collection tree protocol, Proceedings of the 7th ACM Conference on Embedded Networked Sensor Systems (2009) 126-127.

[6] D. B. Johnson and D. A. Maltz, Dynamic Source Routing in Ad-Hoc Wireless Networks, Mobile Computing (1996).

[7] Sinchan R. and Chiranjib P., Geographic Adaptive Fidelity and Geographic Energy Aware Routing in Ad Hoc Routing, Special Issue of IJCCT Vol.1 Issue 2, 3, 4; 2010 for International Conference [ACCTA-2010], 3-5 August 2010.

[8] Gelenbe E., Lent R., Power-aware ad hoc cognitive packet networks, Ad Hoc Networks 2(3) (2004) 205-216.

[9] Wattenhofer R., Li L., Bahl P., Wang Y.-M., Distributed topology control for power efficient operation in multihop wireless ad hoc networks, Proceedings of INFOCOM 2011, Twentieth Annual Joint Conference of the IEEE Computer and Communications Societies, IEEE 3 (2001) 1388-1397.

[10] Mhatre V., Rosenberg C., Design guidelines for wireless sensor networks: communication, clustering and aggregation, Ad Hoc Networks 2(1) (2004) 45-63.

[11] Qiu W., Skafidas E., Hao P., Enhanced tree routing for wireless sensor networks, Ad Hoc Networks 7(3) (2009) 638-650.

[12] Maroti M., Kusy B., Simon G., Ledeczi A., The flooding time synchronization protocol, Proceedings of the 2nd international conference on Embedded networked sensor systems (2004) 39-49.

[13] Sundararaman B., Buy U., Kshemkalayani A.D., Clock synchronization for wireless sensor networks: a survey, Ad Hoc Networks 3(3) (2005) 281-323.

[14] L. Subramanian and R. H. Katz, An Architecture for Building Self Configurable Systems, In Proceedings of the Seventh Annual ACM/IEEE International Conference on Mobile Computing and Networking 2001, pp. 70-84.

# 18

# On the embedded intelligent remote monitoring and control system of workshop based on wireless sensor networks

## YUE Xiangyu

School of Management and Engineering, Nanjing University; Nanjing Jiangsu; 210008; PR China

**Email address:**

yuexiangyu168@gmail.com

**Abstract:** The embedded intelligent remote monitoring and control system of workshop based on wireless sensor networks sets the sensor technology, embedded technology, network communication technology, data processing technology, Beidou positioning technology, sensing information technology of image and weather, geographic information technology and remote sensing technology in one, forming a digital information management system that can provide a full range of electronic remote monitoring and control for the workshop. Its wireless sensor network is a self-organizing network that is constructed from a large number of sensor nodes, which sets such three technologies of sensor, micro-electromechanical system and network in one, taking the perception, collection and processing of the information of the perceive objects in the network coverage as its aim and transfering it to data processing center to provide a basis for the remote monitoring and control of the workshop. This embedded intelligent remote monitoring and control system of workshop has many advantages: high safety, low cost, intelligence, timely alarm, energy conservation, good real-time control, wide monitoring range, strong adaptability, and so on. This monitoring system can be applied to not only plant monitoring but also other fields, such as environmental monitoring, industrial control, intelligent city, intelligent home, etc, so it has important practical significance and valuable practical value for exerting network advantage and making artificial intelligence promote social progress.

**Keywords:** Embedded, Wireless Sensor Networks, Remote Monitoring and Control, Beidou Positioning System, Microprocessor, Wireless Communication

The recent years' emergence of wireless sensor network has become a frontier international multidisciplinary research hotspot. This wireless sensor network consists of several single points with computing, wireless communication, sensing or control capability and has several functions such as signal acquisition, real-time monitoring, Information transmission, collaborative processing, information service and etc. [1] It expands network technology and realizes the real-time monitoring of environmental conditions and facilities. Becoming the core of the internet of things, sensor network is mainly used in the senor layer of the internet of things and for the purpose of tracking, monitoring, decision-making support.

Embedded System refers to a specialized computer application system based on the computer technology, with hardware and software portability and strict requirements for reliability, cost, volume, power consumption and other functions. Embedded system consists of hardware and software. Hardware includes embedded microprocessor and peripheral equipments; while the software includes embedded operating system and specific application programs. The embedded technology is widely used intelligent control, monitoring and management.

Measurement and Control Technology refers to the monitoring and control of the characteristics of a certain thing. [2] Remote Control refers that measurement and control personnel uses computer network connects controlled equipments through remote of computer such as the equipment inquiry, configuration, modification and etc. It can realize the seamless connection between office automation and industrial automation. Through B/S model, clients only need to install a web browser to download the program from the web server to achieve installation and operation so as to realize remote measurement and control.

The rapid development of computer, communication and microelectronics technology result huge changes in the field of measurement and control system. Networking and informationization have become the developmental direction of measurement and control technology. Sensor network, embedded technology, and comprehensive utilization of industrial measurement and control system technology realize Web-based remote measurement and control system, which greatly improves the real-time, security, maintainability of measurement and control, strengthens the centralized monitoring and unified scheduling and management optimization. What author researches in this article is a web-based embedded intelligent workshop remote measurement and control system through which successfully designs and realizes intelligent home furnishing systems.

# 1. The Overall Structure of Embedded Intelligent Workshop Remote Measurement and Control System

Embedded intelligent workshop remote measurement and control system mainly consists of three parts: the embedded measurement and control unit, local server, remote management mainframe computer. The basic architecture of the remote measurement and control system is shown in Figure 1.

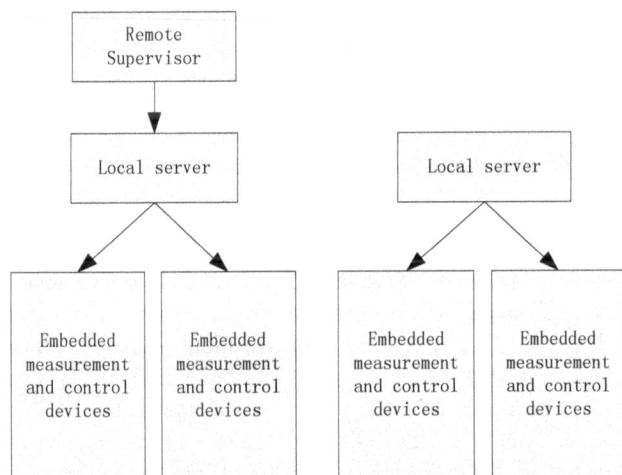

*Figure 1: The basic framework of embedded intelligent workshop remote measurement and control system [3]*

The embedded measurement and control units at the bottom layer are mainly distributed in the measurement and control node points including the central processor, sensor, actuator, network interface, BDS module, GPRS module, which is mainly for environmental parameter acquisition in workshop and then compared with set point. When the set value range is exceeded, the system alarm immediately alerts and starts the actuator to obtain the normal parameters and simultaneously transfer the testing date to the local server through local protocol for manager display. The BDS module can clearly position the faulted equipment. As wireless transceiver module, GPRS module can transmit data with high quality and realizes the realistic feasibility of the application of embedded system in remote measurement

and control system.

As a middle bridge, the local server not only needs to receive monitoring parameters uploaded by measurement and control units, but also save them to the database, regularly makes inquiries and analyze the status of measurement and control units. If error is found, clients will be immediately notified with short messages. At the same time, the server will regularly feedback the date of database to remote manager in order to let clients timely monitor the status of measurement and control equipments and let manager realize the control on measurement and control equipments.

Placed on the top layer, Remote Manager can analyze inquiry and manage measurement and control units, receive all kinds of information uploaded from the measurement and control server. It can send order to control and deal with each measurement and control unit at the bottom layer in accordance with IP address to achieve a unified management of equipments. System implements wireless connection operation through GPRS, and users can achieve real-time control on them at any time.

Embedded intelligent workshop remote measurement and control system collects the parameters of each node through pre-sensor module and information receiving module and sends them to the embedded system for storage and processing, and simultaneously transmits the monitored information to the measurement and control server through which displays it on the manager so as to achieve the control of the measurement and control equipment. This embedded intelligent remote measurement and control system can link a small device to the Internet and timely monitor the operation of each device. The network monitoring is more flexible with relatively low cost for construction and maintenance and it can save large amounts of data with data storage methods and high system integration. It breaks time and geographical constraints. As long as network is available and user is authorized, the measurement and control task can freely accord with needs and achieve plug-and-play.

# 2. The Key Technologies of Embedded Intelligent Workshop Remote Measurement and Control System

## 2.1. Embedded System Design

The measurement and control equipment unit of embedded intelligent workshop remote measurement and control system equals each node point of wireless sensor network for information collection, transferring optical signal, chemical signals and other signals into electrical signals, and transmitting them to the microcontroller for processing. The measurement and control equipment of embedded intelligent workshop remote measurement and control system is an intelligent measurement and control system by using the R&D of embedded technology, and the core of which is central microprocessor, equipped with SDRAM, NAND FLASH, RJ-45 network interface, 4 lines of touch screen interface, serial interface, LCD, USB interface, SD memory card interface, A/D and D/A converter. Through SP3243ECA chip with UART1 and UART2, the

TTL level is changed into RS232 level, which realizes the information exchange between BDS and GPRS. The hardware structure of embedded intelligent workshop remote measurement and control system is shown in Figure 2.

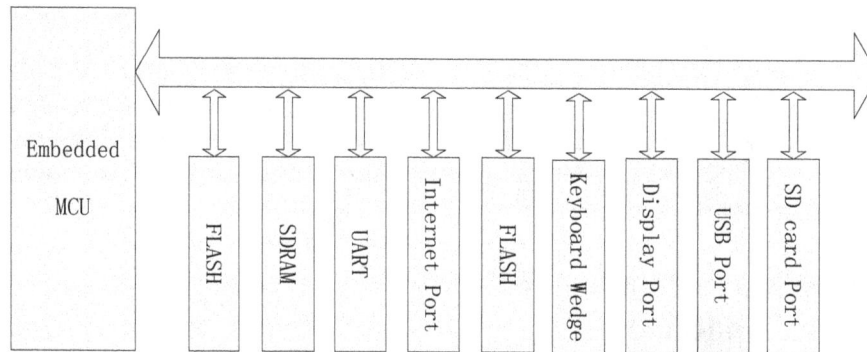

**Figure 2:** *The hardware structure of embedded intelligent workshop remote measurement and control system [3]*

The software of embedded intelligent workshop remote measurement and control system mainly includes startup programs, drivers, operating system, application program and etc. The open source code of Linux operating system can tailor kernel and operate on the hardware platforms like ARM with high efficiency [3], which has an advantage of equipping powerful network functions and excellent file system support functions. Therefore, we can use the Linux operating system. After the procedure of tailoring and cross compiler, an ARM executable file is formed and then it will be downloaded to FLASH through serial and network port. The application program of measurement and control system mainly includes data acquisition module, LCD module, keyboard control module, network service module, communication service module, control execution module and etc. In order to construct a better human-computer interface, we can transplant the QT/Embedded graphical user interface support system to the Linux [4] to develop appropriate graphical user interface.

### 2.2. Web Application Technology

The embedded intelligent workshop remote measurement and control system adopts B/S model. In order to directly access to the intelligent measurement and control equipment unit from the measurement and control server or measurement and control manager, it is possible to install a server with web browsing function on the intelligent measurement and control equipment unit. Here, we adopt Boa as web server on the Linux platform, which will be convenient for system stability and efficient operation in the target system, and or use and maintenance with high reliability. Boa and a normal web server share similarity in the capability of receiving the request from clients, analyzing request, responding to request, and returning the result of request to client and other tasks. The difference between Boa and normal web server is that the HTTP request for processing of Boa enjoys high speed and efficiency, and therefore it will have a high value in the

application in the embedded system.

The main steps of Boa program transplantation are shown as follows: (1) Downloading Boa source code for ARM-Linux and Unzipping it (2) Generating and modifying Makefile file, and changing CC=gcc to CC=/usr/local/arm/2.95.3/bin arm-linux-gcc, and CPP=gcc-E to CPP=/usr/local/arm/2.95.3/bin/ arm-linux-gcc-E; (3) The binary file compiled by Boa server will be downloaded to the FLASH, and then web server will have its functions. [5]

### 2.3. Data Communication Technology

The communication equipment of embedded intelligent workshop remote measurement and control system mainly includes wireless network, BDS, GPRS and custom protocol communication and etc.

The selected date of this communication system design is transmitted and collected through the MC35 of GPRS module produced by the subsidiary of Siemens. The MC35 module supports GSM900 and GSM1800, a dual-band network, with a receiving rate of 86.20Kb/s, sending rate of 21.5Kb/s, and it is easy to be integrated. Through TTL232 level, it is converted and connected with the interface of UART1 of ARM2440 processor.[3]

The communication system design adopts GPS and Beidou Double-Star, a dual-mode navigation positioning technology that can position the intelligent measurement and control equipment units, researched and developed by China. The system uses Beidou Double-Star terminal to achieve fast positioning, and transmits the position information to the central station through GSM and Double-Star short message to user to achieve zoom alarm. The Beidou Double-Star Positioning System can not only achieve fast positioning to provide users all-weather, real-time positioning service with an positioning accuracy equivalent to GPS, a short message transmitting 120 Chinese characters information; and accuracy timing that can reach an accuracy of 20ns.

The communication of intelligent measurement and control equipment units, local server, and remote manger is realized through TCP/IP protocols. Due to limited resources of embedded systems and huge quantity of TCP/IP protocols, reasonable tailoring of TCP/IP protocols is particularly important. The simplification of TCP/IP protocols should follow two principles: Firstly, the characteristics of connection oriented cannot be changed; secondly, the simplification should fit the applied protocols at the application layer. [6] Local server and remote manager realize the communication with intelligent measurement and control equipment units through its dynamic web pages. Then, the receiving and sending orders are realized.

# 3. The Application of Embedded Intelligent Workshop Remote Measurement and Control System

With the development of sensing technology, semiconductor technology and embedded technology [7], wireless sensor networks have obtain rapid development and research results. Its application expends from single military field to diversified fields, such as environment monitoring, industrial control, intelligent city, intelligent home furnishing and other fields where practical results are received. [3] What the author describes here is an intelligent workshop remote measurement and control system platform based on sensor network and ARM in order to achieve monitoring and control on the fire alarm, rainwater parameters and other equipment. The personnel for monitoring and measuring can remotely control the equipment scatted in different positions with no need to go to the scene.

## 3.1. Selection and Design of Embedded Measurement and Control Equipment Unit

The embedded intelligent workshop remote measurement and control equipment units are allocated in the different positions within the workshop to collect the information of workshop site information, and to control them. The information collected mainly includes workshop temperature, humidity, smoke, carbon dioxide, other environmental factors and the electrical states of the work. And the system should measure and control these parameters and realize the communication of the whole network as well as the display function of the web server.

The wireless network of embedded measurement and control system is mainly used to measure temperature, humidity and smoke. The humidity sensor is installed on the switch of the window to control it based on the collected humidity information such as whether it rains. Here we use HM1500[8] as the humidity acquisition unit because its humidity measurement range is 0%~100% with a relatively wide range. Also it works well even if soaked, depends little on temperature and has good linearity. Integration of temperature sensor and air conditioning can achieve

automatic temperature control; fire measurement sensor uses NIS-09C[9] because it has a function of high sensitivity in smoke detection and a fire alarm function.

When the measurement and control system detects that the workshop humidity is higher than the prescribed range, the actuator will close the window; otherwise, there will be no operation. When the detected temperature is higher or lower than the specified temperature range, air conditioning will be started automatically to adjust the temperature of the workshop; when smoke is detected, switches of the extinguishers and windows will be controlled and the alarm will be alerted, the monitoring and measuring personnel will be notified by short messages to effectively deal with the fire.

The remote control switch of measurement and control system is functionalized through a control circuit. The remote control circuit for apparatus is shown in Figure 3:

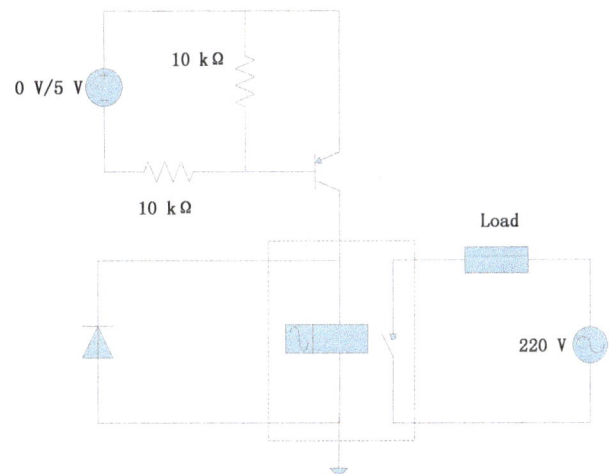

**Figure 3:** *the control circuit of [3] with remote control electrical appliance switch embedded intelligent measurement and control system*

The output signal of the microcontroller of the embedded intelligent measurement and control system will be functioned as the input signal of relay after driven by transistors. If the output of the microcontroller is low level, the transistor will reach saturation and it will energize the relay for actuation and load will be energized as well; if the microcontroller outputs is high level, the transistor will reach scanty and the relay will be cut off. Therefore, remote manager can control the switches of the electrical appliance.

## 3.2. The Design of Embedded Measurement and Control System Software

This plan is implemented through embedded measurement and control system. The monitoring of temperature, humidity, smoke and the control driving program of buzzer and short messages needed to be compiled and configured on the Linux cross development platform. At the same time, redundant drivers should be cut off and drivers that are useful to this platform should be retained. Finally, they will be downloaded to FLASH. However, the software of other measurement and control server needs no change and can be used directly.

In addition, the middleware technology [10] can be used to make corresponding configuration, which makes remote measurement and control platform widely used in fields. In different situations, the data transmission at the higher layer and application programs need no change. The only thing need to be done is to change sensor and corresponding driver.

## 4. Conclusion

Based on the wireless sensor network, the embedded intelligent workshop remote measurement and control system platform is a normal web remote measurement and control structure. Its structure and key technology includes embedded technology, communication technology, web server technology and etc. Based on the wireless sensor network, this embedded intelligent workshop remote measurement and control system can not only applied in workshop monitoring, but also in other environmental monitoring, industrial control, intelligent city, intelligent home furnishing and other fields. Therefore, this will provide significant practical value to bring network advantage into full play and to the realization of artificial intelligence to promote social progress.

## Brief Introduction to the Author

YueXiangyu (1992- ), who is a male, Han nationality, was born in Weifang, Shandong, with BA, Nanjing University, mainly engaged in electrical information and automation, aesthetics and other aspects of learning and research, specializing in modern industrial embedded control systems, networked control systems, intelligent control. He presided over one national college student science and technology innovation project, taking part in the research on one "Eleventh Five-Year Plan" education and science key project of the Education Ministry, and one soft science research project of Shandong Province. He has 12 science and technology papers published in Chinese and oversea academic journals. He has won the 1st scholarship of China, the 1st scholarship of people, the 1st and the 2nd prizes of China Education Robotics Competition, the 1st prize of China Mathematical Modeling Contest, the top award of scientific research achievement of Nanjing University, two China science and technology patents and the honorary title, "three-good-student" of Jiangsu Province. In addition, as a representative of Nanjing University, he went to the National University of Singapore to participate in an academic exchange program. Address: School of Management and Engineering, Xianlin Campus of Nanjing University, 163, Xianlin Avenue, Qixia District, Nanjing, Jiangsu Province, PR China; Zip: 210008.

## References

[1]   XU Shu-kail, XIE Xiao-rong, XIN Yao-zhong. Present application situation and development tendency of synchronous phasor measurement technology based wide area measurement system [A]. Power System Technology [J], 2005, (02): 44- 49

[2]   Liu Xiaohu, The Design and Realization of Embedded Linux Remote Monitoring System  Journal of Guilin University of Electronic Technology [J] 2010, 30 (6): 577-580.

[3]   CaiXuejia, Li Xu, Deng Feng, A Research on Wireless Sensor Network Remote Measurement and Control System [A] Modern Electronic Technology, 2011, (16).

[4]   Chen Changshun, [A]. A Research on Senor Network-based Remote Management Platform, Computer and Digital Engineering [J], 2010, 38 (11): 76-80.

[5]   Lu Yongjian, Wang Ping, Wu Jia and etc, The Transplantation of Embedded Web Server Boa and Its Application [A]. [J] Journal of Hehai University Changzhou Campus, 2005, 19 (4): 44-47.

[6]   WEAVER A, LUO J, ZHANG X. Monitoring and control using the Internet and Java[C] / / Proceedings of the IEEE 1999 25th Annual Conference on Industrial Electronics Society. San Jose, CA: IEEE, 1999 (3): 1152- 1158

[7]   BARBOURN, SCHMIDTG. Inertial sensor technology trends[A]. IEEE Sensors Journal[J], 2001, 1 (4): 332-339.

[8]   Liu Yang, Jin Taidong, CAN Bus-based Intelligent Temperature and Humidity Data Acquisition System Measurement and Control Technology [J] based on CAN bus, 2010 (1): 126-128.

[9]   Wang Fang, Intelligent control-based Fire Alarm System Design [A]. Hunan Agricultural Machinery [J], 2010, 37 (2): 44- 45

[10]  Luo Juan, GuChuanli, Li Renfa, A Research on Role-based Wireless Sensor Network Middleware [A]. Communication Journal 2011, (1): 79-86.

[11]  Liu Zhenqiang, Wang Li, Xu Chao and etc, High-speed SCM-based Missile-borne GPS Signal Acquisition [A] Control and Detection Journal [J]  2010 (2): 42-46.

[12]  Sun Mengyu, Zhao Min, Wu Yijie and etc, ARM-based Electronic Load Network Monitoring SystemElectronic Science and Technology [J] ARM, 2010 (3): 46- 49.

# 19

# Performance evaluation of cooperative multi-layer IDMA communications with amplify-and-forward protocol

**Basma A. Mahmoud, Esam A. A. Hagras, Mohamed A. Abo El-Dhab**

Department of Electronics & Communications, Arab Academy for Science, Technology & Maritime Transport, Cairo, Egypt

**Email address:**

basmaamahmoud@yahoo.com(B. A. Mahmoud), esamhagras_2006@yahoo.com(E. A. A. Hagras), mdahab@aast.edu(M. A. A. El-Dhab)

**Abstract:** In this paper, we consider the Bit Error Rate (BER) performance analysis of a cooperative relay communication system for Multi-Layer IDMA (ML-IDMA) using Maximal-Ratio-Combining (MRC) technique. We examine the effect of layers number on the performance and derive the average bit error probability of the Amplify-and-Forward (AF) relay scheme by using the closed-form relay link Signal-to-Noise Ratio (SNR). The proposed system has been simulated to evaluate the performance of ML-IDMA with different numbers of layers and different number of relays in the ML-IDMA cooperative environment. The simulation results show that the BER performance has been improved by about 4dB in the case of double layer (K=2) IDMA system with 4-relays, also, the saving in the band width is 50%. Finally, the BER performance has been degraded with increasing in the number of layers and the proposed system has improved the band width by about 1/K.

**Keywords:** Cooperative Transmission, Multi Layers, MRC, AF

## 1. Introduction

Multiple-Inputs Multiple-Output (MIMO) technology is widely used due to its ability to offer high diversity and multiplexing gain. The Impracticality of mounting multiple antennas on a mobile device favor other techniques to be used in wireless communication and these techniques are also designed to have less impact on the size and power consumption of the devices. Recently, it has been shown that, in a cooperative system, two or more users cooperate with each other to transmit information to the destination. The cooperative users can share each other's antennas to form a virtual multiple antenna system so that a single antenna device can also benefit from the spatial diversity provided by the cooperative users. In such a way, cooperative communication allows a source node with a single antenna to share the antennas of other nodes, resulting in a form of virtual MIMO system.

User cooperation systems that utilize different cooperative signaling methods are known to improve cellular system capacity and coverage. The relay node physical limitation and allowed signaling complexity are the two criteria that limit the used cooperation system. In [1, 2] several cooperative diversity protocols were developed and analyzed; Amplify-and-Forward (AF), Decode-and-Forward (DF), Detect-and-Forward (DtF), Estimate-and-Forward (EF) and Selective-DF (S-DF).

In AF protocol, relay nodes amplify the signals received from a source node and transmit the amplified version of the signals to a destination node. In EF, the relays forward an estimate of its received signals to the destination .For DtF, the relays detect the received signals and forward the detected symbols to the destination. For DF, the relay nodes decode the information received from the source and re-encode the signal before transmitting it to the destination. For S-DF, only those nodes that can correctly decode are selected to forward the signals to the destination [3].

As a kind of non-orthogonal multiple access scheme, Interleave Division Multiple Access (IDMA) has been widely researched [4], which is regarded as a special form of Code Division Multiple Access (CDMA) by treating interleaving index sequences as multiple access codes. IDMA performance is better than the conventional CDMA regarding the power and bandwidth efficiency. IDMA has common advantages with CDMA, diversity against fading and mitigation of the user interference problem, are two important ones. A low-cost turbo-type Multi-User Detection (MUD) algorithm applicable to the system with large numbers of users, which is crucial for high-rate multiple access communication, is an important unique advantage of the IDMA that can add to its bandwidth efficiency as well as high transmission speed of data [5].

The principle of the IDMA systems is that the chip-level interleavers should be different for different users. In addition, a low-cost chip-by-chip iterative detection scheme can be utilized in the IDMA systems. Motivated by the concept of IDMA, Superposition Coded Modulation (SCM) partitions the data to multi layer, where each layer is treated by a user equivalently [6]. Multi-layer IDMA (ML-IDMA) is a special form of superposition coding scheme and it can be considered as a joint modulation/channel coding transmission scheme. Based on these backgrounds, according to the above observations, we propose a cooperative transmission scheme based on ML-IDMA.

In [5, 7-9], authors' have paid attention to the one layer IDMA cooperative system and without using any combiner technique at the destination. In this paper, we carry out the performance analysis of cooperative single user ML-IDMA scheme for equidistant relaying geometry with different number of layers and a Maximal-Ratio-Combining (MRC) technique; relay protocol that is used in this paper is Amplify-and-Forward (AF). The rest of the paper is organized as follows: in Section 2, we discuss the ML-IDMA system with AF Protocol; the experimental result is presented in Section 3; while conclusions and future works are presented in Section 4.

## 2. ML-IDMA System with AF Protocol

Amplify-and-forward is a simple cooperation scheme, in which relay does not require extra processing capability but still can achieve full diversity [10]. The only drawback in this scheme is that it amplifies the received signal along with inter-user channel noise. That means that it cannot eliminate the noise from the received signal, also termed as non-regenerative relaying protocol.

In an IDMA technique with an AF protocol, the source transmits a signal to a relay followed by amplification of the received signal that is controlled by an amplification factor as well as the power constraints. The amplification factor was shown to have an inverse relation to the received power [11].

We consider the cooperative system in Figs. 1-4. The system consists of a source (S), M relays, and a destination (D). The signal is transmitted from the source through two phases, first the source broadcasts the signal to the relays then the relays transmit the signals to the destination.

In our design, the source/user generates an input data sequence $d = [d (1), d (2) . . . , d (N)]$ which is converted by serial-to-parallel converter into $K$ subsequences layers. Then the data, in each layer, is spread, interleaved and modulated, independently. Finally, all data in $K$ layers are linearly superimposed to transmission then the source broadcasts the superimposed signal to the relays.

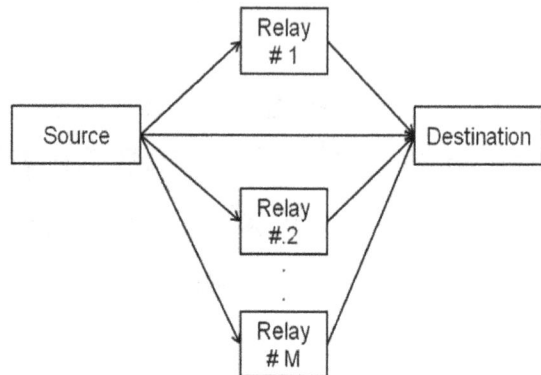

**Figure 1.** *System model of cooperative ML-IDMA with multiple relays.*

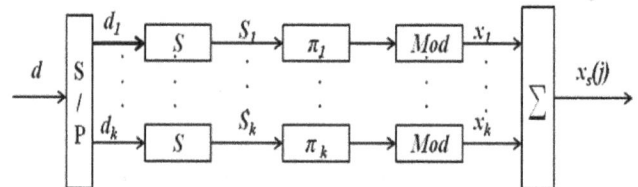

**Figure 2.** *ML-IDMA transmitter (Source)*

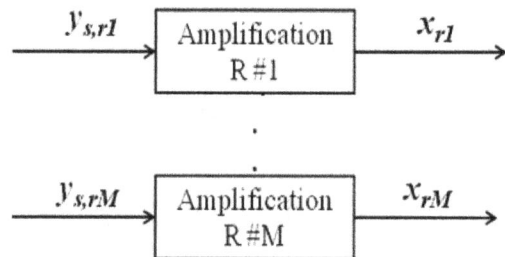

**Figure 3.** *Relays with amplify-and-forward protocol*

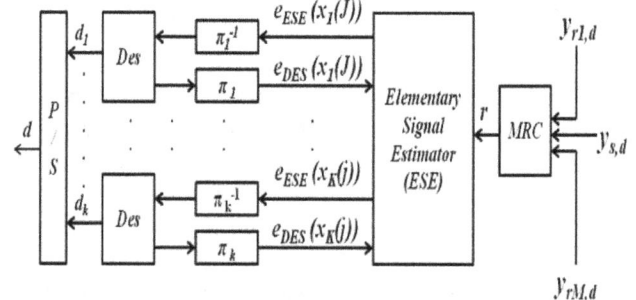

**Figure 4.** *ML-IDMA Receiver (Destination)*

For $K$-layer, the data sequence $d_k = [d_k (1), d_k (2)..., d_k (I)]$ is first spreaded, generating a spreaded sequence $s_k = [s_k (1), s_k (2)...., s_k (J)]$ then the spreaded sequence $s_k$ is interleaved by a distinct chip-level interleaver $\pi_k$ to produce a permuted sequence $s_k$. After interleaving, the randomly sequence $s_k$ is modulated to $x_k = [x_k (1), x_k (2) . . . , x_k (J)]$ by BPSK. Where $I=N / K$, $N$ is the user data length and I is the layer data length.

The transmitted signal $x_s$, after sum mission from $K$ Layers, is then given by:

$$x_s(j) = \sum_{k=1}^{K} x_k(j) \qquad j=0, 1..., J \qquad (1)$$

By assuming that each terminal has one antenna; each relay amplify the received signal and retransmit it to the destination and that the channels between the source and each relay is a quasi-static Rayleigh fading channel with Additive White Gaussian Noise (AWGN), the received signal $y_{s,r}$, at the relay, is given by:

$$y_{s,r} = \sqrt{P_t}\, h_{s,r}\, x_s + n_{s,r} \tag{2}$$

and the received power is given by:

$$E[y_{s,r}^2] = E[(\sqrt{P_t}\, h_{s,r}\, x_s + n_{s,r})^2] \tag{3a}$$

$$= E[(\sqrt{P_t}\, h_{s,r}\, x_s)^2] + E[n_{s,r}^2] \tag{3b}$$

$$= P_t |h_{s,r}|^2 + N_0 \tag{3c}$$

where $N_0 = 2\sigma^2$ is the average noise power.

The amplification coefficient is given by:

$$\beta = \sqrt{\frac{P_t}{P_t |h_{s,r}|^2 + N_0}} \tag{4}$$

By normalizing the power to $P_t = 1$, $y_{s,r}$ is given by:

$$y_{s,r} = h_{s,r}\, x_s + n_{s,r} \tag{5}$$

$$\beta = \sqrt{\frac{1}{|h_{s,r}|^2 + N_0}} \tag{6}$$

The fading amplitude and noise were considered by amplification coefficient. Some other amplification coefficients were proposed in [12]. Amplification coefficients similar to this paper were also suggested by [10, 11, 13, 14]. This kind of amplification factor considers the channel response of inter-user channel and effect of noise when added to the received signal.

The signal after amplification is given by:

$$x_r = \beta\, y_{s,r} \tag{7a}$$

$$x_r = \beta(x_s h_{s,r} + n_{s,r}) \tag{7b}$$

The relayed signal when received at destination is given by:

$$y_{r,d} = h_{r,d} x_r + n_{r,d} \tag{8a}$$

$$= h_{r,d}\left(\beta(x_s h_{s,r} + n_{s,r})\right) + n_{r,d} \tag{8b}$$

The received signal at the destination at the first phase is given by:

$$y_{s,d} = h_{s,d} x_s + n_{s,d} \tag{9}$$

The total received signal for Maximum-Ratio-Combining (MRC) receiver is given by:

$$r = h_{s,d}^* y_{s,d} + \sum_{m=1}^{M}(h_{s,r_m}^* * h_{r_m,d}^*)\, y_{r_m,d} \tag{10}$$

Using a single-path / one relay system, Eq. (10) is:

$$r = (h_{s,r}^* * h_{r,d}^*)\, y_{r,d} \tag{11a}$$

$$r = (h_{s,r}^* * h_{r,d}^*) h_{r,d} x_r + (h_{s,r}^* * h_{r,d}^*) n_{r,d} \tag{11b}$$

Where $h = (h_{s,r}^* * h_{r,d}^*) h_{r,d}$ and $n = (h_{s,r}^* * h_{r,d}^*) n_{r,d}$

$$r = h \sum_{k=1}^{K} x_k(j) + n(j) \tag{11c}$$

Where "n" represents the composite noise, $h_{s,d}$ is the channel coefficient between the source and the destination, $h_{s,r}$ is the channel coefficient between the source and the relay, $h_{r,d}$ is the channel coefficient between the relay and the destination and "n" is a sample of an AWGN process with zero mean and variance $\sigma^2$ per dimension.

We adopted an iterative sub-optimal receiver structure, as illustrated in Fig. 4 which is composed by an elementary signal estimator (ESE) and K de-spreaders (DESs). They are applied to solve inter-layer interference and the spreading constraint separately. The receiver performs the iterative processes to update the extrinsic information between ESE and DESs [15, 16].

For the detection of layer-$k$, we can rewrite Eq. (11c) as:

$$r(j) = h x_k(j) + \xi_k(j), \quad j=1, 2..., J \tag{13}$$

where, $\xi_k(j)$, represents the interlayer interference with respect to layer-k.

The ESE function is used to calculate the extrinsic log-likelihood Ratios (LLR) for estimating the transmitted signal. From the definition of the extrinsic LLR, the output of ESE function can be obtained by:

$$e_{ESE}(x_k(j)) = 2h \cdot \frac{r(j) - E(\xi_k(j))}{Var(\xi_k(j))} \tag{14}$$

where,

$$E(r(j)) = h \sum_k E(x_k(j)) \tag{15}$$

$$Var(r(j)) = |h|^2 \sum_k Var(x_k(j)) + \sigma^2 \tag{16}$$

$$E(\xi_k(j)) = E(r(j)) - h\, E(x_k(j)) \tag{17}$$

$$Var(\xi_k(j)) = Var(r(j)) - |h|^2 Var(x_k(j)) \tag{18}$$

The mean and variance of $x_k(j)$ can be calculated by the feedback from DESs, as follow:

$$E(x_k(j)) = \tanh\left[\frac{e_{DES}(x_k(j))}{2}\right] \tag{19}$$

$$Var(x_k(j)) = 1 - (E(x_k(j)))^2 \tag{20}$$

The mean and variance of the interlayer interference can be used to analyze and detect the signal of each layer. Then the updated extrinsic LLR from ESE function was proven to go through the layer-specific de-interleaver and gets into the DESs iteratively [17].

# 3. Simulation Results

In this paper, we carry out the performance analysis of cooperative single user ML-IDMA scheme for equidistant relaying geometry with different number of layers and a MRC technique. The applied relay protocol is Amplify-and-Forward (AF), which is implemented for the system that uses Binary Phase Shift Keying (BPSK). MATLAB is used to simulate the obtained results. In this paper it is assumed that all stations are arranged at the edges of a square with a length of one. That means that all channels will have the same path loss and therefore the same average Signal-to-Noise Ratio (SNR).

To simulate the cooperative ML-IDMA scheme, we assume that the channel is Rayleigh fading channel, equidistant relaying geometry, BPSK signaling is used, frame length (N) = 512, Spreading Length (SP) =32, number of relays (R) =1..., 4, number of layers (K) =2, 4 and number of iteration (it) =3.

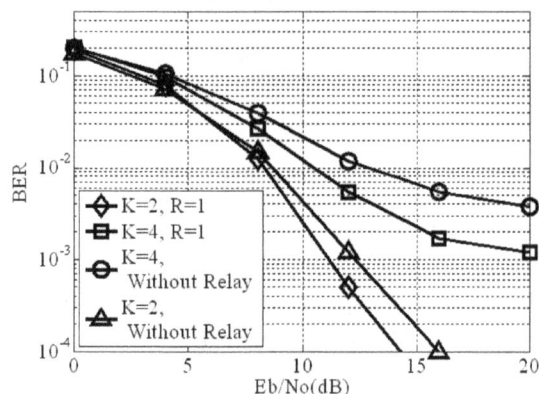

**Fig 5.** *BER of cooperative ML-IDMA in Rayleigh channels with K=2, 4, N=512, it=3, SP=32 and R = 1*

Fig. 5 shows that there is degradation in the BER performance by increasing the number of layers. In the case of without relay, K=2 and Eb/N0=15 dB, the BER performance was $1.9*10^{-4}$ while increasing K to 4, degraded BER performance to $6.6*10^{-3}$. The observed degradation in the BER performance by increasing the number of layers is due to increasing in the signal amplitude which cause signal distortion. For single relay (R=1), in the case of double layer (K=2), the improvement of the BER performance is 1.67 dB compared to without relay system. Also in the case of four layer (K=4) and Eb/N0 = 20 dB, the BER improvement was found to be about $2.6 *10^{-3}$ when compared to without relay system as shown in Fig.5.

Fig.6 shows the improvement in BER performance by increasing the number of relays to R=4 at K=2 by about 4 dB when compared to without relay system. Fig. 7 show that R= 4 is superior to the case of R=1 by about 3.5 dB at the BER of $10^{-4}$. We can conclude that the co-operative environment, which added up two signals of different links, performs better than systems designed without relay environment. We can also see that when additional relay is

deployed, the performance in the co-operative environment gets better. The analysis showed that the addition of different path powers in co-operative environment results in a lower Bit-Error-Rate (BER).

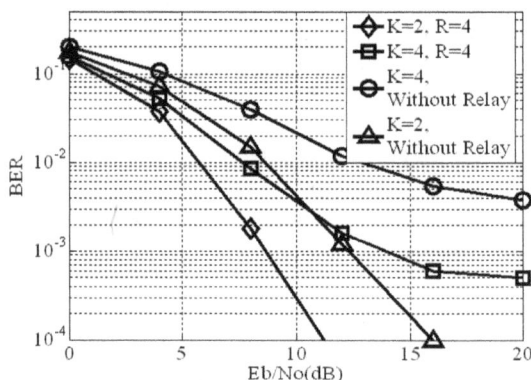

**Fig 6.** *BER of cooperative ML-IDMA in Rayleigh channels with K=2, 4, N=512, it=3, SP=32 and R = 4*

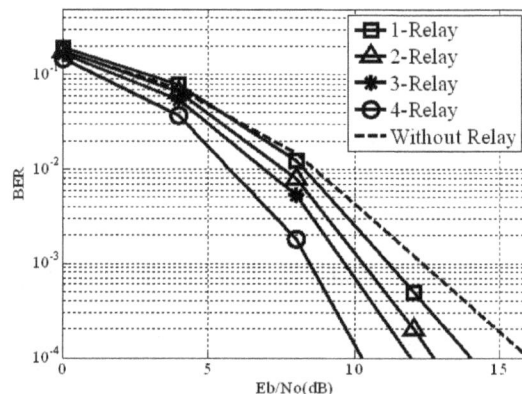

**Fig 7.** *BER of cooperative ML-IDMA system in Rayleigh channels with R= 1, 2, 3, 4, K=2, N=512, it=3 and SP=32*

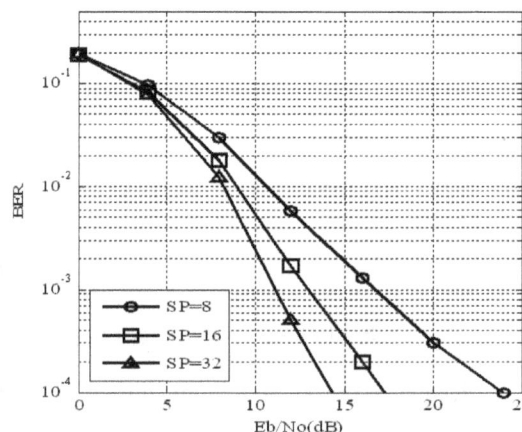

**Fig 8.** *BER of cooperative ML-IDMA system in Rayleigh channels with SP=8, 16, 32, R=1, K=2, N=512 and it=3*

Fig. 8 shows that increasing the Spreading Length (SP) can improve the performance significantly. By increasing the SP we were able to get larger spreading gain of the spread spectrum signal that improves the performance efficiency as the signal becomes larger.

# 4. Conclusion

In this paper, a cooperative communication scheme based on ML-IDMA technique is proposed for a network which has one source, one destination and multiple common relays. The proposed system is based on the chip-by-chip detection algorithm. Data reaches the destination through two different paths, i.e. a direct path from source to destination and a relayed path, where Amplify-and-Forward AF relay protocol is applied to the data that finally reaches the MRC combiner at the destination.

The proposed system has been simulated to evaluate the performance of ML-IDMA technique with different numbers of layers with Maximal-Ratio-Combining (MRC) technique in the cooperative environment. The simulation results showed that by increasing the number of layers, degradation in the performance was obtained but on the other hand the Band Width (BW) was saved by (1/K).

The simulation also showed that the cooperative environment, which added up two signals of different links, performed better than the system designed without relays. When an additional relay was deployed, the performance in the cooperative environment got better. This improvement means that the addition of different path powers in cooperative environment actually results in a lower Bit Error Rate (BER) in the Eb/N0 vs. BER curve. We also observed that, the performance of cooperative environment with two or more relays is better in all investigated layers. Future work will be done to investigate the performance of coded ML-IDMA cooperative schemes, OFDM ML-IDMA, Coded OFDM ML-IDMA.

# References

[1] J. N. Laneman, and G. W. Wornell, "Distributed Space-Time-Coded Protocols for Exploiting Cooperative Diversity in Wireless Networks," IEEE Trans. Inf. Theory, vo.49, no.10, pp.2415-2425, Oct 2004.

[2] G. Scutari, S. Barbarossa," Distributed space-time coding for regenerative relay networks," IEEE Trans. Wireless Commun., vol.4, no.5, pp.2387-2399, Sept. 2005.

[3] Jinhong Yuan, Yonghui Li, and Li Chu, "Differential Modulation and Relay Selection WithDetect-and-Forward Cooperative Relaying," IEEE Trans. ON Vehicular Technology, vol. 59, no. 1, jan. 2010

[4] L. Ping, L. Liu, K. Wu, and W. K. Leung, "Interleave-division multiple access," IEEE Trans. Wireless Commun., vol. 5, no. 4, pp. 938-947, Apr. 2006.

[5] Zhifeng Luo, Deniz Gurkan, Zhu Han, Albert Kai-sun Wong and Shuisheng Qiu," Cooperative Communication Based on IDMA," 5th International Conference on Wireless Communications, Networking and Mobile Computing, pp.1-4, Sep. 2009.

[6] J. Tong, P. Li, Z. H. Zhang, and W. K. Bhargava, "Iterative Soft Compensation for OFDM Systems with Clipping and Superposition Coded Modulation," IEEE Trans. Commun., vol. 58, no. 10, pp. 2861–2870, Oct. 2010.

[7] Xiaotian Zhou, Haixia Zhang and Dongfeng Yuan,"A multi-source cooperative scheme based on IDMA aided superposition modulation," International Conference on Information Theory and Information Security, pp.802-806, December 2010.

[8] Weitkemper, P. , Wubben, D. and Kammeyer, K.-D.," Delay-diversity in multi-user relay systems with Interleave Division Multiple Access," IEEE International Symposium on Wireless Communication Systems 2008, pp.364-368, October 2008.

[9] Chulhee Jang and Jae Hong Lee, "Novel IDM-cooperative diversity scheme and power allocation," in Proc. IEEE APWCS 2009, Seoul, Korea, pp.24-28, Aug. 2009.

[10] J. N. Laneman, D. N. C. Tse and G. W. Wornell, "Cooperative Diversity In Wireless Networks: Efficient Protocols And Outage Behavior," IEEE Trans. Info. Theory, Vol. 50, No. 12, pp. 3062-3080, November 2004.

[11] K. J. R. Liu, A. K. Sadek, W. Su and A. Kwasinksi, Cooperative Communication and Networking, New York: Cambridge University Press, 2009.

[12] D. Chen and J. N. Laneman, "Cooperative Diversity For Wireless Fading Channel Without Channel State Information," IEEE Trans. Wireless Comm. vol. 5, No. 7, pp. 1307-1312, July 2006.

[13] M. Yuksel and E. Erkip, "Diversity in Relaying Protocols with Amplify and Forward," IEEE GLOBECOM Communication Theory Symposium proceeding, San Francisco, pp. 2025-2029, December 2003.

[14] P. A. Anghel and M. Kaveh, "Exact Symbol Error Probability of a Cooperative Network in a Ryaleigh-Fading Environment," IEEE transaction on wireless communications, vol. 3, No. 5, pp. 1416-1421, September 2004.

[15] Takyu O, Ohtsuki T, Nakagawa M. Companding system based on time clustering for reducing peak power of OFDM symbol in wireless communication. IEICE Transactions on Fundamentals of Electronics Communications and Computer Sciences 2006; E89- A(7):1884–91.

[16] J. Akhtman, B. Z. Bobrovsky, and L. Hanzo, "Peak-to-average power ratio reduction for ODFM modems," in Proceedings of the 57th IEEE Semi-Annual Vehicular Technology Conference (VTC'03), vol. 2, Apr. 2003, pp. 1188–1192.

[17] Tao Peng, and Min Ye, "PAPR mitigation in superposition coded modulation systems using selective mapping" in Proceeding of the International Conference on Computer and Communication Technologies (ICCCT'2012), May 26-27, 2012,Phuket.

# 20

# A web-based remote indoor climate control system based on wireless sensor network

**Jingcheng Zhang, Allan Huynh, Patrik Huss, Qin-Zhong Ye, Shaofang Gong**

Department of Science and Technology, Linköping University, Bredgatan 33, 60174, Norrköping, Sweden

**Email address:**

Jingcheng.zhang@liu.se(J. Zhang), Allan.huynh@liu.se(A. Huynh), Patrik.huss@liu.se(P. Huss), Qin-zhong.ye@liu.se(Q. Ye), Shaofang.gong@liu.se(S. Gong)

**Abstract:** This paper presents the design and implementation of a web-based wireless indoor climate control system. The user interface of the system is implemented as a web service. People can login to the website and remotely control the indoor climate of different locations. A wireless sensor network is deployed in each location to execute control commands. A gateway is implemented to synchronize the information between the wireless sensor network and the web service. The gateway software also includes scheduling function and different control algorithms to improve the control result. Additionally, the system security and availability are highly considered in this system. The gateway software implements a warning function which sends warning messages when emergency happens. Finally, the whole wireless control system architecture is modularly designed. It is easy to add different control applications or different control algorithms into the system.

**Keywords:** Remote Control, Indoor Climate, Wireless Sensor Network

## 1. Background Introduction

Indoor climate control is very important for human comfort and health. Nowadays, almost all new buildings are installed with heating, ventilation and air conditioning (HAVC) systems which provide a good indoor air quality. However, the situation can become more complicated when providing a similar indoor climate for old buildings. Firstly, some old buildings have very high heritage preservation values so that no modification or reconstruction is allowed. Secondly, some old paintings or wooden carvings are sensitive to temperature and relative humidity change. The climate control process should be done as smooth as possible. Thirdly, some of the old buildings, e.g., churches, are not used every day. In order to save power consumption, it is only necessary to set the indoor climate into a comfort level when some activities are scheduled. Finally, from practical point of view, one person might need to maintain several old buildings. These old buildings could be hundred kilometers away from each other. It is a big waste and low efficiency to travel among different buildings to do manual control.

Many research efforts have been done in the wireless control area[1] -[9]. However, by the time when this paper is written, no scientific report has been found on using wireless sensor network to control the indoor climate of cultural heritage buildings. As described above, a system that can solve all mentioned problems should have the following features:

- A wireless control system which requires minimum reconstruction of the building during the installation.
- A wireless controller should be easy to be reused for different control purposes.
- A system which provides schedule function. The system should be able to change the indoor climate according to the user booking.
- A system that can be remotely accessed.

Such a system is researched and developed in the communication electronics research group of Linköping University, Sweden. It is a system that mainly works for remote climate monitoring and control. The system applies the ZigBee protocol[10] for wireless sensor network communication. Since the system mainly works in cultural heritage buildings, it is named as "CultureBee".

## 2. Remote Control System Architecture Description

As shown in Figure. 1, the "CultureBee" system contains three parts, the "wireless sensor network" part, the "local server" part and the "web service" part. The web service provides a website user interface. Users can login to the website to control the indoor climate of different locations. For one specific location, a local server and a wireless sensor network are deployed. The local server works as a gateway between the web service and the wireless sensor network. It communicates with the "root node" of the wireless sensor network via the USB port. It also communicates with the main server via the Internet. The sensor node of the wireless sensor network reports the indoor climate measurement to the local server periodically. The wireless controllers in the sensor network are connected with radiators or dehumidifiers according to the controller type. They listen to the command from the local server and control the indoor climate.

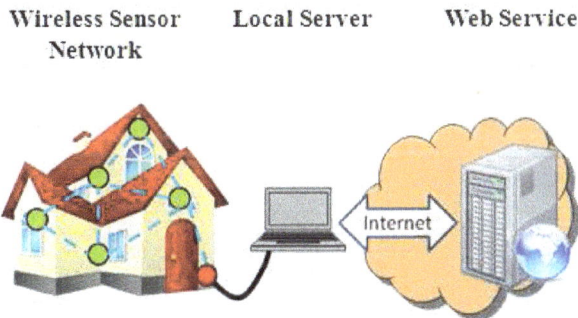

**Figure 1.** *"CultureBee"control system architecture*

Figure. 2 shows the control function message flow within the "CultureBee" system. In order to get scheduling messages from the web service, the local server sends the "schedule polling" message to the web service periodically. When a control schedule is issued, the web service stores the schedule until the polling message from the local server

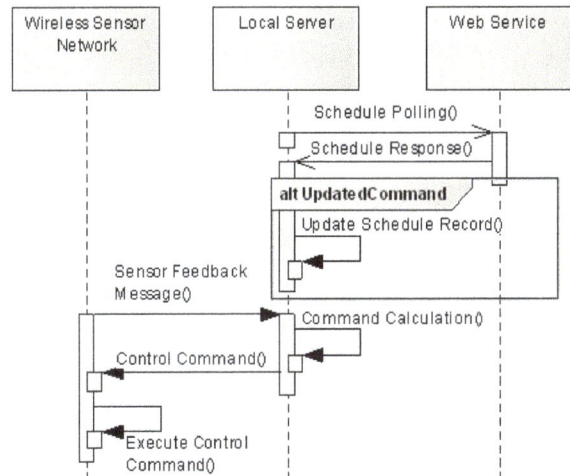

**Figure 2.** *Control function message flow*

is received. The schedule is then sent to the local server as the response of the polling message. Once the schedule is received, the local server updates the local schedule configuration accordingly. Meanwhile, if a feedback sensor message is received by the local server, the control command is calculated by the local server according to the local control schedule. The control command is forwarded from the local server to the wireless sensor network.

## 3. Wireless Sensor Network

The wireless sensor network of the "CultureBee" system mainly works for three purposes: collecting the sensor information for the local server, executing the control command sent from the local server and increasing the communication reliability of the wireless sensor network. Compared with the local server, the computational capability of the wireless device is much lower. In order to optimize the system performance, the wireless sensor network control function is designed to be as simple as possible. Most of the control logics and management functions are implemented in the local server.

### 3.1 Devices of Wireless Sensor Network

The CultureBee wireless sensor network utilizes the ZigBee protocol to transmit sensor information and climate control commands. A ZigBee wireless sensor network can contain 3 different kinds of devices: coordinator, router and end device. The coordinator is the root and creator of the network. A router works as a message relay to extend the network scale. An end device can enter sleep mode from time to time to save power. The end device is the only one that can be battery powered.

As shown in Figure. 3(a), the end device is mounted with a temperature and relative humidity sensor. It is programmed to send sensor message to the coordinator periodically. The battery lifetime of the end device is optimized so that a 1200mAh lithium battery can be used up to 10 years[11]. In order to extend the wireless sensor network scale, routers can be used to establish a multi-hop sensor network[12]. As defined by the ZigBee protocol, a router needs to be mains-powered so that the radio part can be always active. The wireless controller function is also developed base on the ZigBee router program. From the hardware point of view, as shown in Figure.3 (b), the wireless controller is a power relay which can only switch On/Off the power at the output. However, different control algorithms can be emulated by carefully selecting the switch-on interval. The coordinator is the root of the network. As shown in Figure. 3(c), it communicates with the local server via the USB port. In order to increase the radio range, all devices are equipped with a power amplifier and a low noise amplifier[13][14].

(a)Sensor module                    (b)Wireless controller

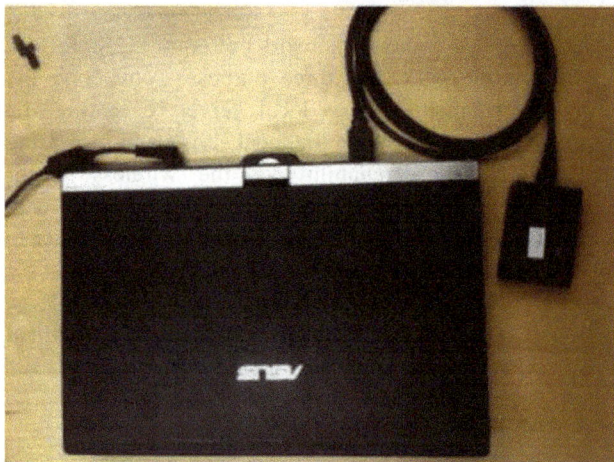

(c)Local Server

**Figure 3.** *Wireless sensor network hardware*

### 3.2. Wireless Sensor Network Message Flow

Figure. 4 shows the control function message flow of the wireless sensor network. In order to increase the communication reliability, all the wireless devices run together with the "stability enhancement" state machine[15]. From the control function point of view, the message flow can be categorized into "controller registration" cluster and "close loop control" cluster.

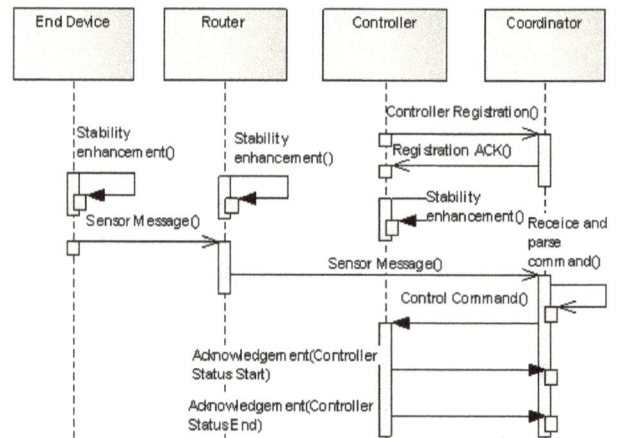

**Figure 4.** *Wireless sensor network control function message flow*

The "controller registration" message is sent by a wireless controller which newly joins the wireless sensor network. This message provides the wireless controller description to the local server. When the local server receives the message, the local server saves the registration information into its local database. Figure.5 shows the format of the "controller registration" message frame. The "Message Header" field identifies the message type which can be recognized by the local server. The "Network Address" filed contains the unique ZigBee network address[16]. The "Node ID" field contains the wireless controller device ID which is utilized by the control application. It is a unique ID all through the CultureBee system.

| Message Header | Network Address | Node ID |
|---|---|---|

**Figure 5.** *Wireless controller registration message format*

The local wireless control process is always triggered by the feedback sensor message. The end devices are programmed to periodically send temperature and relative humidity message to the local server through the coordinator. After calculation, the local server sends back the control command to the coordinator as described in Figure 6. The "Message Header" field identifies the command type. The "Network Address" field contains the address of the wireless controller. The "Interval" field contains the interval (in seconds) that the wireless controller should switch on the power relay. Once the coordinator receives the command, it forwards the message to the wireless controller whose network address is specified in the "network address" field. When the wireless controller receives the command, it turns on the power switch with the interval specified in the command.

| Message Header | Network Address | Interval |
|---|---|---|

*Figure 6. Control command message format*

For each control command, the wireless controller issue two acknowledgements to the coordinator. As shown in Figure. 4, the first acknowledgement is sent directly after the controller receives the control command. It informs the local server that the message is successfully received and executed. The second acknowledgement is sent when the command is expired. In the normal case, the time difference between the two consecutive acknowledgements should equal to the value of the "Interval" field described in Figure. 6. The frame format of the acknowledgement message is shown in Figure.7. The "Message Header" field identifies the message. The "Controller Status" field reports the On/Off status of the wireless status. If errors are detected by the wireless controller, the error code is also carried in the "Controller Status" field and sent to the local server.

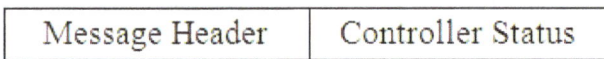

| Message Header | Controller Status |
| --- | --- |

*Figure. 7 Control command message format*

## 4. Local Server Design

The local server is the gateway between the wireless sensor network and the web service. It is a software program that runs on the Windows operating system. As shown in Figure 8, the local server communicates with the operation system mainly via three interfaces: the USB port interface which exchanges messages with the coordinator, the HTTP stack interface which synchronizes messages with the main server and the database interface which manages the local server control information. The local server also provides a local user interface so that the control system can be locally configured when the Internet is not available.

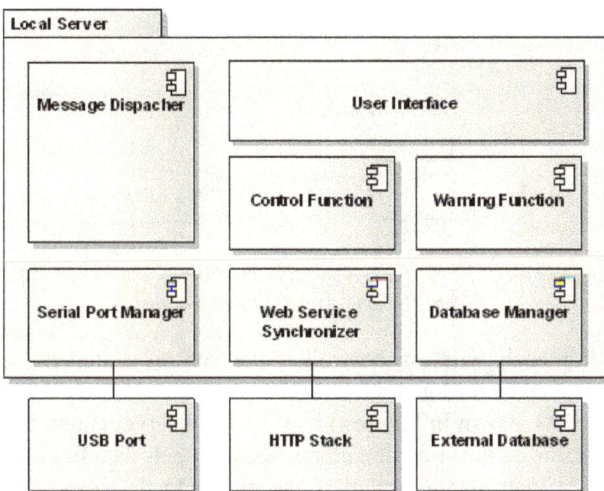

*Figure 8. Local server function block*

As shown in Figure. 8, the local server functions are implemented by different function components. The "serial port manager" component handles the communication between the local server and the coordinator. When the local server receives a serial message, the "serial port manager" forwards the message to the "message dispatcher" component. The "message dispatcher" works for all the message exchange within the local server. It parses and redirects the messages to the correct component according to the received message type. The "web service synchronizer" component is implemented to synchronize messages between the wireless sensor network and the main server. Generally, the "web service synchronizer" forwards the sensor message from the local server to the main server. It also sends polling message to the main server to ask for pending control schedules. Once the control schedule is received, the "web service synchronizer" forwards the schedule to the "message dispatcher" for further process. The "control function" component is the central software module that manages all the control operations. It contains the logic to generate the command to control the wireless sensor network with different control algorithms. The "warning function" component checks the status of the wireless device and sends out warning message when emergency happens.

### 4.1. Control Function Component Design

The goal of the "control function" component is to implement an easy to configure and high flexibility control system. A database is utilized to store the control information. An expandable design pattern[17] is applied on implementing the control algorithms. The whole local server control operations are managed by a software state machine.

#### 4.1.1. Control Function Database Design

The local server utilizes a database to store all control function related configurations. As the database structure shown in Figure. 9, the control function information are saved in 7 tables. Table "router list" and "sensor list" maintains the device information of the wireless sensor network. The table "binding relationship" contains a list that records the feedback sensor configuration for each wireless controller. In the "Router ID" filed and the "Sensor ID" field, the "binding relationship" table uses the record reference (known as the "foreign key") from the "router list" table and the "sensor list" table. In this way, the "Binding relationship" table is independent from the update of the "router list" table and the "sensor list" table. In order to increase the reliability, one controller is always configured with more than one feedback sensors. The "Priority" field tells the local server which sensor feedback should apply when there are more than one feedback sensors available for one wireless controller. The table "controller type" contains a list of all possible controller types in the system. The table "controller information" maintains the controller specific information. This table allows each controller to be configured with individual control algorithm. The "router ID" field refers to the router index in the "router list" table.

It states which router is programmed as a wireless controller. The "controller type" field refers to the index of the table "controller type". It shows the controller type that is connected with the wireless controller. The "controller status" files record the current ON/OFF status of the wireless controller which is reported from the command execution acknowledgement. The climate control schedules are saved in table "controller schedule" and "default schedule". The "controller schedule" table saves all the user issued schedules which are either synchronized from the main server or configured locally. The "default schedule" table saves the default configurations for each controller. The default schedule will be executed if there is no user specific schedule found by the local server.

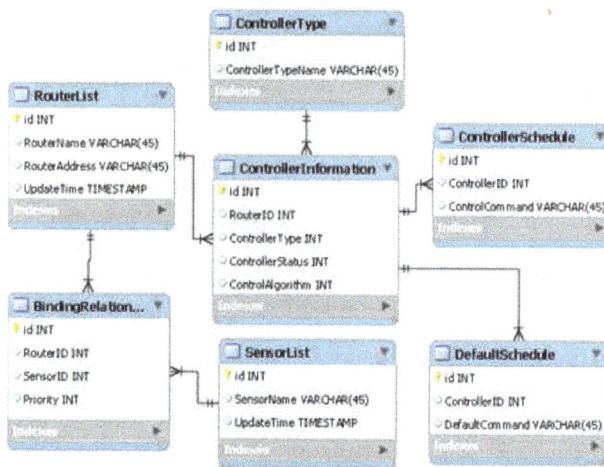

*Figure 9. Control function database design*

### 4.1.2. Control Function Algorithm Design

The "control function" component is implemented as a control algorithm container. Different control algorithms can be easily added into the component. To implement this mechanism, the "control function" component utilizes a design pattern called "abstract factory"[12]. By using this method, the control function container returns different control algorithms according to the controller configuration. As shown in Figure 10, The "Control Function Base" class is a "generalized based class"[12] which represents the most basic operations for a control algorithm. The function "Find Schedule" helps the program to find the correct schedule from the database. The function "Send Command WSN" sends the calculated command from the local server to the wireless sensor network. Three different control algorithms are implemented in the control function component, the threshold based control (TH) the proportional-integral-derivative control (PID) and mold prevention control (MP)[18]. Different control functions are implemented as the "child class" of the "Control Function Base" class. Each control algorithm child class reuses the "Find Schedule" and "Send Command" functions from the base class and implements its own "Generate Command" function. The "Generate Command" function is utilized to calculate the "switch on interval" for the controller based on

the feedback sensor message and controller information. The generated command is further executed in the wireless sensor network, as described in Fig.4.

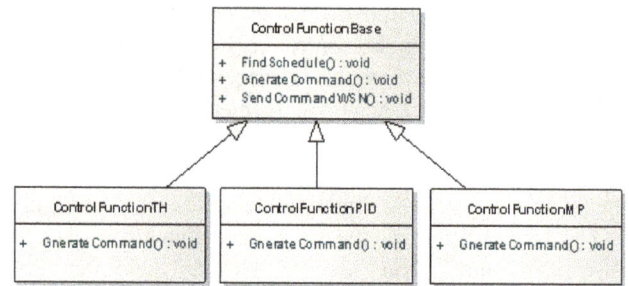

*Figure 10. Control algorithm class diagram*

### 4.1.3. Control Function State Machine

The control function operations are described by the state machines as shown in Figure.11. After start-up, the local server stays the "IDLE" state. Once a sensor message is received, the local server forwards the message to the main server using the "web service synchronizer" component. Then the local server enters the "SYNC" state. In this state, the local server checks the "Binding Relationship" table to get the bind controller list. If no controller is found, the local server jumps back to the "IDLE" state and wait for the next sensor message. Otherwise, the local server jumps to the "CONTROLLR PENDING" state. For each bound controller, the local server generates the specific control command in the "CONTROL PROCESS" state. When the control command is generated, the local server jumps back to the "CONTROL PENDING" state. If there is no more pending controller waiting for the control command, the local server jumps to the "IDLE" state and waits for the next sensor message.

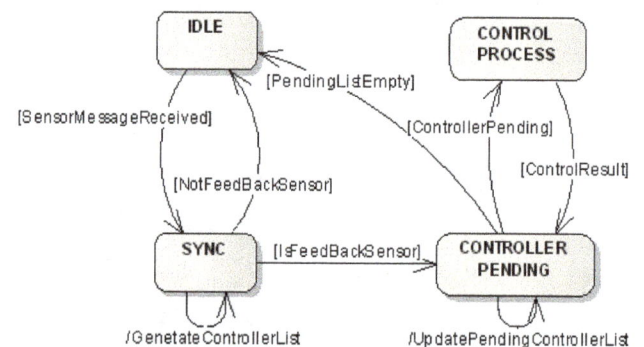

*Figure 11. Control function state machine*

For each wireless controller, the control command is generated by the local server in the "CONTROL PROCESS" state. As shown in Figure. 12, when local server enters this state, the control function component first gets data from the table "controller information" according to the device ID. The controller information contains the controller type, the controller status and the controller algorithm. Secondly, the control function component gets the correct schedule from the "controller schedule" table. It contains the current climate setting. After getting all necessary input information, the control function component sends the "get control command" request to the "controller algorithm" component. Based on the received information, the controller algorithm returns the "control command" with a "Turn On Interval" value. Once the control function component receives this value, it inserts the controller network address in the message and creates the control message as shown in Fig.6.

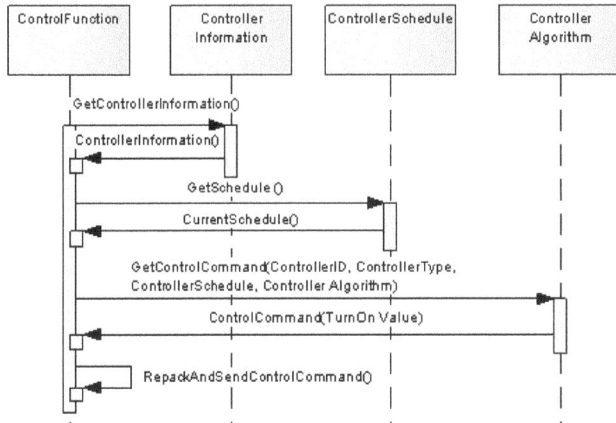

*Figure 12. Function execution in "Control Process" state*

### 4.4. Warning Function Component Design

The local server warning function is designed to notify the user when controller stops responding. Generally, a controller can stop working in two cases. One is that the wireless controller simply stops working. It can be caused by a hardware failure or loss of wireless connection. Another case is that the feedback sensor stops reporting. Normally, one wireless controller is bound with more than one feedback sensors. However, there is still a possibility that all the configured sensors lost the connection with the wireless sensor network. Moreover, the local server should also be able to configure the "sensitivity" of the controller failure. If the controller is very important, user might want to know the problem immediately. If the controllers are not so crucial, it might be considered to be offline after "keeping silence" for an hour.

A reusable wireless device status detection mechanism is designed in the local server to send warning messages. In the wireless sensor network, the sensor modules and the wireless controllers send "heart beat" messages to the local server periodically. Once the "heart beat" message is received, the local server updates the device "update time" field in the "router list" or "sensor list" table in the database according to the device ID. The warning function component is designed as shown in Figure.13. When the "warning function" starts working, it loads data from the warning configuration list. For each record in the list, the "device name" and "MAX_OFF_TIME" are saved. The device name identifies which device is enabled with the warning function in the wireless sensor network. The MAX_OFF_TIME is the threshold for the warning function component to send out warning message. The device list

from the configuration file is loaded to the local server program by calling the function "Register Device" in the "warning function" class. Another function called "check status" is invoked periodically to get "offline device list". During the function execution, the status of each registered wireless device is checked. A class "device status" is implemented to report the device status according to the device ID. During the status check, the warning function gets the device "Update Time" value from the database. The difference between the current time and "Update Time" of each device is calculated to compare with the "MAX_OFF_TIME". If the value is bigger than the "MAX_OFF_TIME", the device is considered to be "offline". When the local server has gone through the status of all registered devices, it invokes the "Send Warning Message" function to send the "offline device list" through the "web service synchronizer" component.

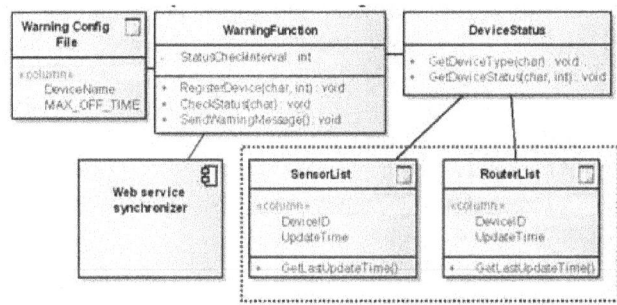

*Figure 13. Warning function class diagram*

## 5. Web Service Design

The main server is a web service which can be accessed via www.culturebee.se. For end users, the main server provides webpage user interface. Users can use the interface to issue climate setting to different locations. Figure.14 shows the user interface to control the temperature in a church. The sketch of the building is shown in the left side of the web page. The location of the controller is shown in the sketch which gives users a rough idea where the wireless controller is located. The configuration interface is shown on the right side. To configure one schedule, users need to select the start date, start time and end time of the schedule. Moreover, users can also configure the setting type as daily, weekly, monthly or yearly. The temperature setting is configured in the "set and execute" frame according to the controller type. For a radiator, the temperature is configured. For a dehumidifier, the relative humidity is configured.

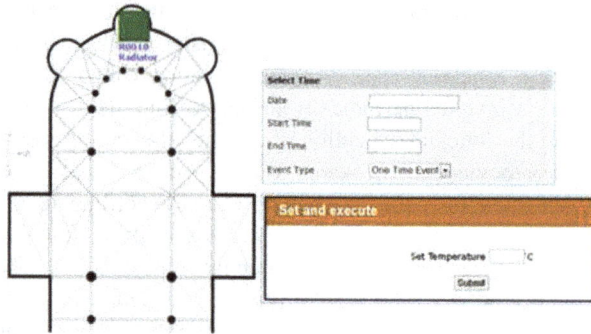

*Figure 14.* Main server schedule configuration user interface

# 6. Test Result

The "CulureBee" control system was deployed in 10 different places around Sweden for temperature control, dehumidification control and mold prevention purpose in 2012. A typical use scenario is to control the conference room temperature in VASA museum in Stockholm[19]. As shown in Figure. 15(a), a wireless sensor is placed in the center of the room. It provides the sensor feedback reading to the controller which controls the radiator. The wireless controller, as shown in Figure. 15 (b) is connected with the electric powered radiator.

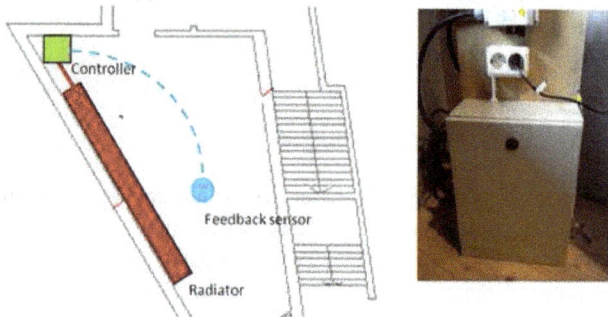

*Figure 15.* Wireless control system in VASA museum

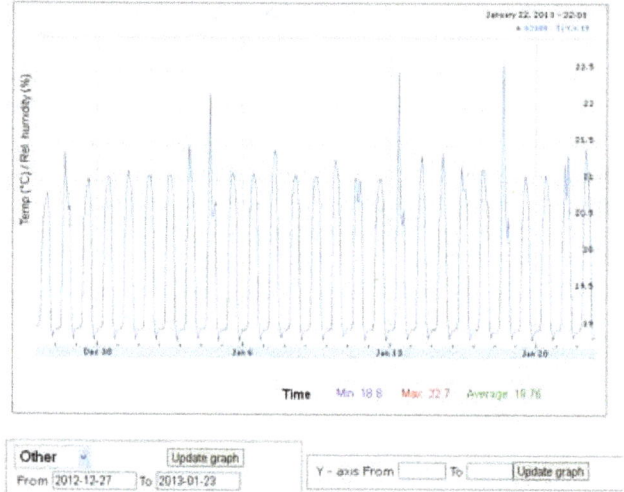

(a) Indoor temperature control result – one month

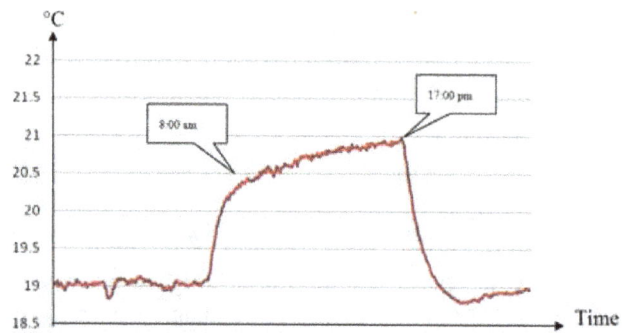

(b) Indoor temperature control result – one day

*Figure 16.* Temperature control test result

To control the indoor climate, the temperature is set to 21°C between 8:00 am to 17:00 pm and keeps 19 °C for the rest of the time every day. In order to keep the temperature stable, the PID control algorithm is utilized for the controller. Figure. 16*Figure* (a) shows the one month control result summary. As shown in the diagram, the temperature changes between 19 and 21°C every day. Figure. 16(b) shows the detailed temperature change within one day. As shown in the diagram, when temperature changes from 19 to 21°C, the temperature increases fast at the beginning. As the temperature approaching to 21°C, the temperature increases much slower until the setting is reached. In this case, no temperature vibration is introduced resulting in a much more comfortable and smoother indoor climate control. This is especially good for the heritage preservation purpose.

# 7. Discussion

Generally, the "CultureBee" system has been designed for climate control of cultural heritage buildings. The "CultureBee" system utilizes the wireless sensor network to transmit control commands. The wireless communication gives the control system a much higher flexibility compared with the wired system. Moreover, a localized climate can be established by connecting a portable radiator or dehumidifier to the place where it is necessary. In this case, the system can provide a relatively satisfactory air temperature in the human activity area, while keeping the total air temperature as stable as possible. This can be a good solution for indoor climate control of cultural heritage buildings, museums and summer houses, etc.

The local server contains all the control logic to generate the command for the wireless controller. A centralized control method is utilized in the local controller which requires all the messages to go through the local server. Alternatively, a distributed control mechanism is implemented and compared[20]. The advantage of the distributed control is that the sensors and controllers are paired in different segments. It is self-maintained by the wireless sensor network. It can also work independently from the local server. However, this also brings some problems into the system. Firstly, one feedback sensor might bind with more than one controller. In this case, the battery powered sensor module needs to send the sensor message to multiple destinations. This can reduce the battery life time of the wireless sensor device. Secondly, the wireless sensor needs to handle the control function by its own. As described in 4.1.3, it might be a too heavy task for an 8-bit microprocessor. Finally, if the local server is down, the distributed control mechanism can continue working according to the previous configuration. However, it is impossible for the user to change the configuration during the system down time. Compared with the centralized control system, no more control command will be sent out from the local server when it goes down. Thus the user knows that the control system will shut down if the local server stops working. Compared with the distributed control system, the local server failure is more scalable.

Finally, the "modular design" concept is considered in many places of the "CultureBee" control system. When the wireless controller receives the control command, it only turns on the relay with the specified interval. All the control logic is running inside the local server. In this case, no modification needs to be done when changing the controller type. In the local server, new controller types, new controllers and new feedback sensors and new warning configuration can be easily added by configuring the local server database and configuration files. A "many to many" binding relationship between feedback sensors and wireless controllers is established. It can be easily changed in the database configuration. It is also very easy to change the controller algorithm of the controller and configure the algorithm parameters. In the code level, new control algorithms can be easily added into the control algorithm container. Different control algorithms are implemented as different items inside the local server control function container.

## 8. Conclusion

A web-based remote indoor climate control system is developed and implemented for historical buildings. The main server provides a web-based user interface. Users can set schedules to control the indoor climate of different buildings. The local server contains the central logic of the control function. It is modularly designed software which provides high flexibility and re-configurability. The wireless sensor network communication reliability is enhanced. It receives and executes the control commands sent from the local server. The whole system has been proven to be robust. It is particularly suitable for the climate control of the old historical buildings where connecting wires should be avoided.

## Acknowledgements

This project is financed by the Swedish Energy Agency (Energimyndigheten) and the Churches of Sweden (svenskakyrkan).

## References

[1]   Wan-Ki Park, Intark Han, Kwang-Roh Park. "ZigBee based Dynamic Control Scheme for Multiple Legacy IR Controllable Digital Consumer Devices", IEEE Transactions on Consumer Electronics, Volume 53, Issue 1, 2007, pp 172-177

[2]   Leccese, F. "Remote-Control System of High Efficiency and Intelligent Street Lighting Using a ZigBee Network of Devices and Sensors", IEEE Transactions on Power Delivery, Volume: 28 , Issue 1, 2013, pp 21-28

[3]   Dae-Man Han; Jae-Hyun Lim. "Smart home energy management system using IEEE 802.15.4 and Zigbee", IEEE Transactions on Consumer Electronics, Volume 56, Issue 3, 2010, pp 1403-1410

[4]   Il-kyu Hwang; Jin-wook Baek. "Wireless access monitoring and control system based on digital door lock", IEEE Transactions on Consumer Electronics, Volume 53, Issue 4, 2007, pp 1724-1730

[5]   Jinsoo Han; HaeRyong Lee; Kwang-Roh Park. "Remote-controllable and energy-saving room architecture based on ZigBee communication", IEEE Transactions on Consumer Electronics, Volume 55, Issue 1, pp 264-268

[6]   Ahmad, A.W., Jan, N., Iqbal, S., Lee, C. "Implementation of ZigBee-GSM based home security monitoring and remote control system", in 2011 IEEE 54th International Midwest Symposium on Circuits and Systems (MWSCAS), Seoul, Korea (South), 07 Aug - 10 Aug 2011, pp.1-4

[7]   Jinsoo Han, Chang-Sic Choi, and Ilwoo Lee. "More Efficient Home Energy Management System Based on ZigBee Communication and Infrared Remote Controls", in 2011 IEEE International Conference on Consumer Electronics (ICCE), Las Vegas, NV, USA, 9-12 Jan. 2011, pp. 631-632

[8]   Ying-Wen Bai, Chi-Huang Hung. "Remote Power ONOFF Control and Current Measurement for Home Electric Outlets Based on a Low Power Embedded Board and ZigBee Communication", in IEEE International Symposium on Consumer Electronics, 2008. (ISCE 2008), Vilamoura,

Portugal, 14 Apr - 16 Apr 2008, pp.1-4

[9]   Cui Chengyi, Zhao Guannan and Jin Minglu, "A ZigBee based embedded remote control system", in 2010 2nd International Conference on Signal Processing Systems (ICSPS), Dalian, China, Jul 5, 2010 - Jul 7, 2010, pp. V3 373-376

[10]  ZigBee Alliance (http://www.zigbee.org/)

[11]  Jingcheng ZHANG, Allan HUYNH, Qinzhong YE and Shaofang GONG. "Remote Sensing System for Cultural Buildings Utilizing ZigBee Technology", in 8th International Conference on Computing, Communications and Control Technologies (CCCT2010), April 6th - 9th, 2010, Orlando, USA. pp.71-77

[12]  Drew Gislason, ZigBee Wireless Networking, Newnes, 2008.

[13]  Allan Huynh, Jingcheng Zhang, Qin-Zhong Ye and Shaofang Gong. "ZigBee Radio with External Low-Noise Amplifier", Sensors & Transducers Journal, Vol. 118, Issue 7, July 2010, pp.110-121

[14]  Allan Huynh, Jingcheng Zhang, Qin-Zhong Ye and Shaofang Gong. "ZigBee Radio with External Power Amplifier and Low-Noise Amplifier", Sensors & Transducers Journal, Vol. 114, Issue 8, July 2010, pp.184-191

[15]  Jingcheng Zhang, Allan Huynh, Qin-Zhong Ye and Shaofang Gong. "A Communication Reliability Enhancement Framework for Wireless Sensor Network Using the ZigBee Protocol", Sensors & Transducers Journal, Vol. 135, Issue 12, Jan 2012, pp.42-56

[16]  Jingcheng Zhang, Allan Huynh, Qin-Zhong Ye and Shaofang Gong. "Reliability and Latency Enhancements in a ZigBee Remote Sensing System", in The Fourth International Conference on Sensor Technologies and Applications (SENSORCOMM 2010), July 18 - 25, 2010 Venice/Mestre, Italy. Pp.196-202

[17]  Erich Gamma, Richard Helm, Ralph Johnson, John Vilssides. Design Pattern: Elements of reusable object oriented software. Addison-Wesley Professional, 1st edition, November 10, 1994

[18]  Willian L. Brogan. Modern Control Theory, Prentice Hall, 3rd edition, October 11, 1990

[19]  VASA museum (www.vasamuseet.se/)

[20]  Jingcheng Zhang, Allan Huynh, Qin-Zhong Ye and Shaofang Gong. "Design of the Remote Climate Control System for Cultural Buildings Utilizing ZigBee Technology", Sensors & Transducers Journal, Vol. 118, Issue 7, Jul. 2010, pp. 13-27

# 21

# Identifying and locating multiple spoofing attackers using clustering in wireless network

**AMALA GRACY[1], CHINNAPPAN JAYAKUMAR[2]**

[1]Department of Information Technology, RMK Engineering College, Anna University, Chennai, INDIA
[2]Department of Computer Science and Engineering, RMK Engineering College, Anna University, Chennai, INDIA

**Email address:**
sag.cse@rmkec.ac.in(A. GRACY), cjk.cse@rmkec.ac.in(C. JAYAKUMAR)

**Abstract:** Wireless networks are vulnerable to identity-based attacks, including spoofing attacks, significantly impact the performance of networks. Conventionally, ensuring the identity of the communicator and detecting an adversarial presence is performed via cryptographic authentication. Unfortunately, full-scale authentication is not always desirable as it requires key management, coupled with additional infrastructural overhead and more extensive computations. The proposed non cryptographic mechanism which are complementary to authenticate and can detect device spoofing with little or no dependency on cryptographic keys. This generalized Spoofing attack-detection model utilizes MD5 (Message Digest 5) algorithm to generate unique identifier for each wireless nodes and a physical property associated with each node, as the basis for (1) detecting spoofing attacks; (2) finding the number of attackers when multiple adversaries masquerading as a same node identity; and localizing multiple adversaries. Cluster-based mechanisms are developed to determine the number of attackers. The proposed model can be explored further to improve the accuracy of determining the number of attackers, by using Support Vector Machines (SVM).

**Keywords:** Wireless Network, Spoofing Attack, Identity-Based Attack, Message Digest 5, Support Vector Machines, Partitioning Around Medoids (PAM) Cluster Model

## 1 Introduction

Wireless networks are more prone towards spoofing attacks. In identity-based spoofing attacks, an attacker can forge its identity to masquerade as another device or even create multiple illegitimate identities in the networks by masquerading as an authorized wireless access point (AP) or an authorized client [1]. An attacker can launch denial-of-service (DoS) attacks, bypass access control mechanisms, or falsely advertise services to wireless clients. Therefore, identity-based attacks will have a serious impact to the normal operation of wireless and sensor networks. Spoofing attacks can further facilitate a variety of traffic injection attacks such as attacks on access control lists, rogue AP attacks, and eventually DoS[2].

Spoofing can take on many forms in the computer world, all of which involve some type fraudulent representation of information

### 1.1. IP Spoofing

Internet Protocol (IP) is the protocol used for transmitting messages over the Internet [3]; it is a network protocol operating at layer 3 of the Open Systems Interconnection (OSI) model. IP spoofing is the act of manipulated the headers in a transmitted message to mask a hackers true identity so that the message could appear as though it is from a trusted source. IP spoofing is used to gain unauthorized access to a computer. The attacker forwards packets to a computer with a source address indicating that the packet is coming from a trusted port or system.

### 1.2. ARP Spoofing

Address Resolution Protocol (ARP) is used to map IP addresses to hardware addresses [4]. A table named as ARP cache, is used to maintain a correlation between each Medium Access Control (MAC) address and its corresponding IP address. "ARP Spoofing involves constructing forged ARP request and reply packets. By sending forged ARP replies, a target computer could be convinced to send frames destined for computer A to instead go to computer B". This referred to as ARP poisoning.

### 1.3. E-Mail Spoofing

In E-Mail spoofing, an email message contains malicious objects, and appears to come from a legitimate source. But in fact, it is from an Attacker. E-mail spoofing can be used for malicious purposes such as spreading viruses / trawling for sensitive business data / other industrial espionage activities.

Normal E-mail message contains the return address at the Top left corner of the mail. This indicates where the mail is generated. But the attackers could over write any name and address in this space, which pretends to be genuine.

### 1.4. WEB Spoofing

Web or Hyperlink spoofing provides victims with false information. Web Spoofing is an attack that allows someone to view and modify all web pages sent to a victim's machine. They can observe any information that is entered into forms by the victim. This can be of particular danger due to the nature of information entered into forms, such as addresses, credit card numbers, bank account numbers, and the passwords that access these accounts.

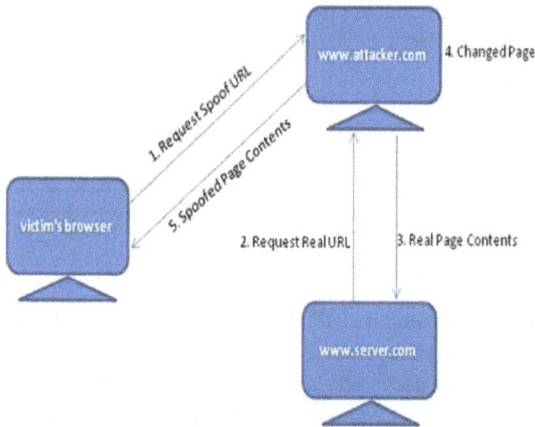

**Fig.1.** Working of WEB Spoofing

### 1.5. DNS Spoofing

A DNS spoofing attack can be defined as the successful insertion of incorrect resolution information by a host that has no authority to provide that information. It may be conducted using a number of techniques ranging from social engineering through to exploitation of vulnerabilities within the DNS server software itself. Using these techniques, an attacker may insert IP address information that will redirect a customer from a legitimate website or mail server to one under the attacker's control – thereby capturing customer information through common man-in-the-middle mechanisms.

The attacker targets the DNS service used by the customer and adds/alters the entry for www.mybank.com – changing the stored IP address from 150.10.1.21 to the attacker's fake site IP address (200.1.1.10).

The customer queries the DNS server.

The DNS responds to the customer query with "The IP address of www.bank.com is 200.1.1.10" – not the real IP address.

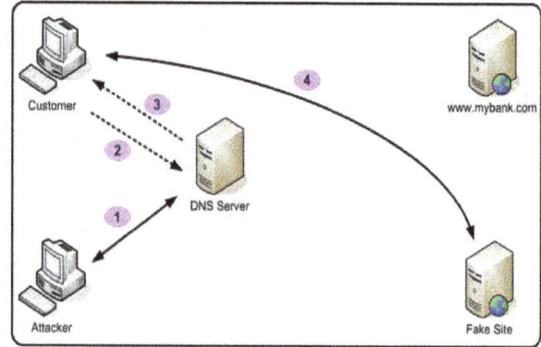

**Fig.2.** Working of DNS Spoofing

However, authentication requires additional infrastructural overhead and computational power associated with distributing and maintaining cryptographic keys.

A different approach is proposed, where in the physical property associated with each wireless node is used to assess the presence of adversaries in the wireless network. This method is hard to falsify, and not reliant on cryptography as the basis for detecting spoofing attacks. This approach enables to detect and localize multiple adversaries in the network, with high detection rate and minimal infrastructure. In a large-scale wireless network, multiple adversaries may masquerade as the same identity and collaborate to launch malicious attacks. Therefore, the problem can be divided into 2 folds such as

- Detect the presence of spoofing attacks,
- Determine the number of attackers, and localize multiple adversaries.

The identification and localization can be done in the following ways:

### Generalized Attack Detection Model

This can both detect spoofing attacks as well as determine the number of adversaries using cluster analysis methods.

### Localization of Attackers

Identify the positions of multiple adversaries even when the adversaries vary their transmission power levels.

The main contribution of the paper is organized are as follows:

- To effectively detect the presence of spoofing attack
- To count the number of attackers
- To identify the location of multiple adversaries in the network
- To provide solution to identify adversaries in the

network where in there is no additional cost or modification to the wireless devices themselves

- To avoid authentication key management
- To avoid overhead
- To develop a mechanism where in there is low false positive rate

This paper is organized as follows: section 2 deals with literature of related works in various spoofing attacks prevailing in Wireless network. Section 3 explains about proposed system architecture and how the adversaries are identified in the proposed mechanism. Section 4 presents the simulation and performance evaluation. Section 5 concludes the proposed system.

# 2. Related Works

Intrusion detection mechanisms are readily available to prevent the wireless network from exploiting vulnerabilities at IP layer or above than IP layer. Whereas, attacks that are exploiting vulnerabilities at link layer demands different set of intrusion detection mechanisms. Following works are relevant to this paper, and their summary is as follows:

Each node in a wireless network can be identified by its location information, which is hard to falsify and not reliant on cryptography. Cluster base mechanism uses this location information, to detect spoofing attacks in wireless networks. This method is capable of both detect and provide the number of adversaries in the wireless network, and spoofing the same node identity. There were experiments conducted on 2 test beds through 802.11 networks (Wi-Fi) and an 802.15.4 (ZigBee) network in two real office building environments. This method having a detection rate of above 98% and determining the number of adversaries with the hit rate of over 90%.

Sequence number-based MAC Address Spoof detection algorithm is employed to detect adversaries in the Wireless network[5]. Presence of MAC Address spoofing can be detected in real-time WLAN environments. This algorithm best work, when the casual attackers do not take advantage of false negative of the algorithm. If they exploit, then the algorithm can detect the presence of MAC spoofing, whereas, it does not detect the spoofed frames. By means of K-means cluster analysis, both detection of spoofing attacks and locating positions of adversaries in the network either Area based (and) point based localization algorithms [6]. This method can detect the presence of adversaries with high detection rate as well as low false positive rate. DEMOTE system uses Received Signal Strength (RSS) traces collected over time. Then it achieves an optimal threshold to partition the RSS traces into classes for attack detection. Temporal constraints used in this Algorithm Alignment Prediction method, to predict the best RSS alignment of partitioned RSS classes for RSS trace for reconstruction over time [7].

Supporting Quality of Service (QoS) in ad hoc is a challenging task. A lot of researches have been done on supporting QoS in the ad hoc but most of them are not

suitable .Clustering technique provides a solution to support QoS in ad hoc networks [8]. The clusters must be long-lived and stable based on the effective functioning of cluster based routing algorithms. Difficult to furnish QoS without knowing any state information. The proposed method includes the spatial and temporal stability of the nodes. The first select a node as cluster head. The node elected as the cluster head is such that it has maximum associativity as well as satisfies a minimum connectivity requirement. The cluster head collects and aggregates information in its own cluster and passes information. By rotating the cluster-head randomly, energy consumption is expected to be uniformly distributed. Effective work has been done for Cluster Reorganization. Cluster Head Re-election is periodically done so that the cluster head is centered in a cluster. many clustering proposals for increasing network lifetime are reported suggesting different strategies of cluster head selection and its role rotation in the ad hoc networks, using different parameters .Cluster management will be effective, since the cluster heads are elected based on various parameters like processing capabilities, speed of the node, number of neighbour nodes and associativity with the neighbour nodes. Though overhead is incurred initially due to cluster setup, cluster maintenance will be easier in the long run. Ultimately routing efficiency will increase due to long-lived clusters and reduced control packets.

Mobile ad hoc network (MANETs) becomes a popular research topic due to their self-configuration and self-maintenance capabilities. Security is a major concern for providing trusted communications in a potentially hostile environment. Multimodal biometric technology provides potential solutions for continuous authentication in high security mobile ad hoc networks [9]. Continuous user authentication is an important prevention-based approach to protect high security mobile ad hoc networks (MANETs).Intrusion detection systems (IDSs) are also important in MANETs to effectively identify malicious activities. A MANET is an infra structure less network for mobile devices connected by wireless link. The mobile network is often vulnerable to security attacks even though there are many traditional approaches, due to its features of open medium and dynamic changing topology.Multi-modal biometrics is deployed to work with intrusion detection systems (IDSs) to overcome the shortcomings of uni-modal biometric systems. Multimodal biometrics can be used to alleviate some drawbacks of one mode of biometrics by providing multiple verifications of the same identity.The cluster head is elected in which Dempster-Shafer theory is evaluated in order to increase the observation accuracy to maintain high security and trusted MANET. The Dumpster–Shafer evidence theory was originated by Dempster and later revised by Shafer. It's essential idea is that an observer can obtain degrees of trust about a proposition from related proposition's subjective probabilities. Since each device in the network has measurement and estimated limitations, more than one device needs to be chosen, and observations can be fused to increase observation accuracy using Dempster–Shafer theory for data fusion.

Due to shared nature of medium in wireless networks, it is simple for an adversary to launch a Wireless Denial of Service (WDoS) attack [10]. These attacks can be easily launched by using off-the-shelf equipment. For illustration, a malicious node can continually transmit a radio signal in order to block any legitimate access to the medium and/or interfere with reception. This act is called jamming and the malicious nodes are referred to as jammers. Jamming techniques vary from simple ones based on the continual transmission of interference signals, to more sophisticated attacks that aim at exploiting vulnerabilities of the particular protocol used. Various techniques were proposed to detect the presence of jammers. Finally, numerous mechanisms which attempt to protect the network from jamming attacks were discussed.

TCP SYN flooding and IP address spoofing attacks were launched on Multi-hop wireless networks. TCP SYN flooding occurs when establishing a TCP connection for data transmission. But, even after a TCP connection is established, TCP protocol is flooded by a novel connection flooding attack which aims at consuming the entire bandwidth allocated to a network. Even though numerous techniques are available to counter this attack, there is no single technique, which effectively protects the network from TCP SYN Flooding attack. It proposes a defense mechanism which involves in defending such flooding attack and also prevents IP spoofing, which is the gateway for such flooding attacks. The performance analysis is carried out and it proves the effectiveness of the proposed defense mechanism in terms of time delay and false positive rates. Exponential back-off restoration algorithm was used to counter Denial-Of-Service attack, which is considered to be one of main threats to wireless multi-hop networks [11]. This algorithm overcomes weakness of traffic burst based detection methods. The proposed method can detect anomaly and restore the network scheme to defeat low-rate DoS launched by rogue nodes. Exponential backoff restoration (EBR) algorithm is proposed to reduce performance degradation.

Trust issue in wireless sensor networks is predominant, in security schemes [12]. It is necessary to analyze how to resist attacks with a trust scheme. It categorizes various types of attacks and countermeasures related to trust schemes in WSNs. Furthermore, it provides the development of trust mechanisms, give a short summarization of classical trust methodologies and emphasize the challenges of trust scheme in WSNs. An extensive literature survey is presented by summarizing state-of-the-art trust mechanisms in two categories: secure routing and secure data. Based on the analysis of attacks and the existing research, an open field and future direction with trust mechanisms in WSNs is provided

Wireless Sensor Networks (WSNs) pose host of security issues in the area of development and secure routing

protocols [13]. Improving network hardware and software may address many of the issues, but others will require new supporting technologies. With the proliferation of usage of Wireless Sensor Networks, the security mechanisms play an important role. Recently proposed solutions address but a small subset of current sensor network attacks. Also because of the special battery requirements for such networks, normal cryptographic network solutions are irrelevant. New mechanisms need to be developed to address this type of network.

Due to vulnerability of Wireless networks, towards spoofing attacks, host of other forms of attacks can be launched on the networks [14]. A spoofing attack is the most common online attack in which an adversary or program successfully masquerades as another by falsifying data and thereby gaining an illegitimate advantage. It demands for more sophisticated defense mechanisms. Although the identity of a node can be verified through cryptographic authentication, authentication is not always possible because it requires key management and additional infrastructural overhead. Various defense mechanisms were discussed.

This work differs from the previous study mentioned earlier, by using MD5 and physical property associated with each node, to identify the presence of adversaries in the wireless network, and using SVM to identify number of such adversaries in this Network. Finally, by using NS2, to simulate and verified the results.

# 3. Proposed System Architecture

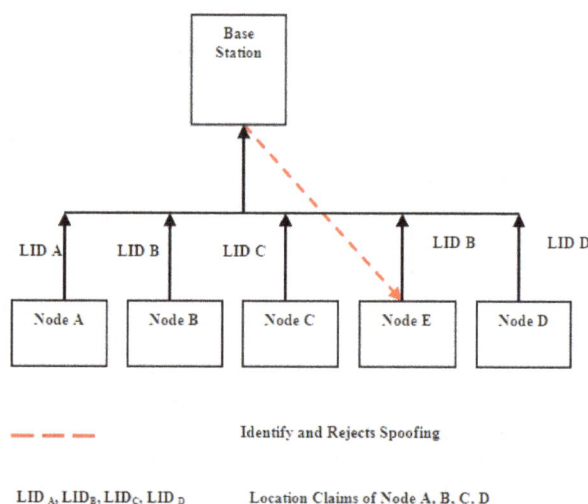

*Fig.3. Architecture of Proposed System*

A spoofing attack is a situation in which one person or program successfully masquerades as another by falsifying data and thereby gaining an illegitimate access into Wireless network. Fig.3 considers a wireless network with N nodes. Let N denote the set of all nodes in the network. Nodes are deployed in 2D platform. Each node is associated with unique location identifier using MD5. If any one of the node needs to communicate with the base station, it will check the location ID of respective node. If the base station finds that any two nodes has the same location ID (ie. Node B), then it meant that spoofing has taken place. A base station is a radio receiver / transmitter that serves as the hub of the local wireless network, and may also be the gateway between a wired network and the wireless network. It typically consists of a low-power transmitter and wireless router.

### 3.1. Message Digest 5

Fig.4 MD5(Message Digest 5) algorithm uses message of arbitrary length, as input and produces an output of 128-bit "fingerprint" or "message digest" of that input. This algorithm is intended for digital signature applications, where a large file must be "compressed" in a secure manner. MD5, with its 128bit encryption algorithm has 1,280,000,000,000,000,000 possible combinations. It mainly used for Verifies data integrity and particularly in Internet-standard message authentication.

**Figure.4.** *Working of Message Digest 5*

### 3.2. Node Deployment

Node deployment can reduce complexity in wireless networks. The nodes can be deployed in dense or in sparse manner. It depends mainly on application. The mobile nodes can change the topology of network. There are various different deployment methods or scenarios such as grid, random and square. Nodes are randomly deployed.

### 3.3. System Module

Detection and Localization of Spoofing attackers are identified from the following modules.
- Detection of Spoofing Attack
- Find the Number of Attackers
- Localization of Attackers

#### 3.3.1. Detection of Spoofing Attacks

Spoofing attack detection is performed using Cluster Analysis. As the wireless network is deployed as clusters, the attackers are identified in each and every cluster separately. Fig.5 Under the spoofing attack, the victim and the attacker are using the same ID to transmit data packets (i.e., spoofing node or victim node). Since under a spoofing attack, the data packets from the victim node and the spoofing attackers are mixed together, this observation suggests to conduct cluster analysis on location id in order to detect the presence of spoofing attackers in wireless network.

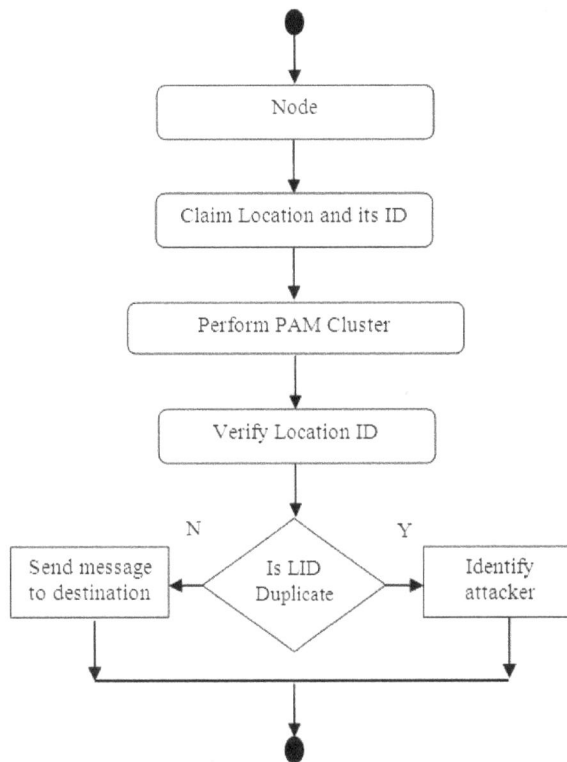

**Fig.5.** *Activity diagram of proposed system*

#### 3.3.1.1. GADE

A Generalized Attack Detection modEl that can both detect spoofing attacks as well as determine the number of adversaries using Cluster analysis. In GADE, the Partitioning Around Medoids (PAM) cluster analysis method is used to perform attack detection and then applied cluster-based methods to determine the number of attacker.

#### 3.3.1.2. PAM

The PAM Method is a popular iterative descent clustering algorithm. Compared to the popular K-means method, the PAM method is more robust in the presence of noise and outliers. Spoofing attack detection is performed using Cluster Analysis. As the wireless network is deployed as clusters, the attackers are identified in each and every cluster separately. consider the wireless nodes are composed of several clusters of Ordinary Nodes.

The PAM algorithm partitioned a dataset of 'n' objects into a number of clusters ('k'), where both the dataset and the number k is an input of the algorithm. This algorithm works with a matrix of dissimilarity, where its goal is to minimize the overall dissimilarity between the represents of each cluster and its members.

The PAM algorithm can work over two kind of input, the first is the matrix representing every entity and the values of its variables, and the second is to work with the dissimilarity matrix directly. This algorithm has following two phase.

#### Build Phase
- Choose k entities to become the Medoids, or in case

these entities were provided use them as the Medoids

- Calculate the dissimilarity matrix if it was not informed, and
- Assign every entity to its closest Medoids.

### Swap Phase

- For each cluster search if any of the entities in the cluster having value lower than the average dissimilarity coefficient, if it does select the entity that lower the most this coefficient as the medoids for this cluster;
- If at least the medoids from one cluster has changed go to (3), else end the algorithm.

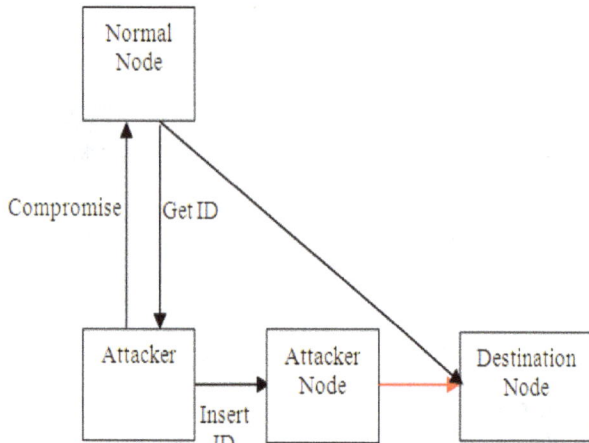

**Fig.6.** *Spoofing Attack Detection*

Fig.6 shows, how Spoofing attacks are detected by same location id. Attacker gets the ID of normal node and makes use of the same to send packets to Destination node.

### 3.3.2. Determine the Number of Attackers

SVM used to improve the accuracy of determining the number of spoofing attackers in the network. SVM is widely used in object detection & recognition. It has two types that are Linear SVM and Nonlinear SVM. SVM is used to classify the number of the spoofing attackers.

The advantage of using SVM is that it can combine the intermediate results (i.e. features) from different statistic methods to build a model based on training data acquired from cluster, to accurately predict the number of attackers. On detecting a attacker in the wireless network, SVM increment the target Value by '1', else '0'. SVM can be applied to solve classification and regression problems. Some of SVM applications are Handwriting Recognition, 3D Object Recognition, Speaker Identification, Face Detection, Text Categorization, Bio-Informatics, and Image Classification.

Fig.7 shows on detecting multiple adversaries present in a Wireless network. In Multi Spoofing attack, ID of a compromised Node is used by multiple adversaries present in the network, to send packet to Destination Node.

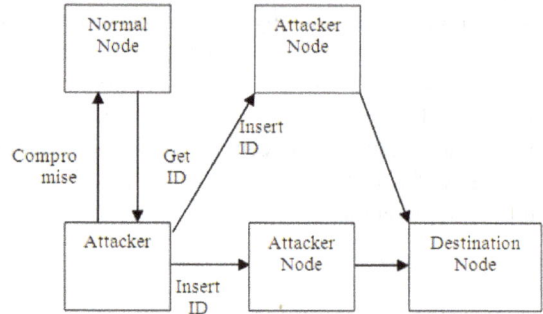

**Fig.7.** *Detection of Multi Spoofing Attack*

### 3.3.3. Localization of Attackers

The simulation is performed under Linux environment on NS2. Let us consider the number of nodes deployed in the simulation window for e.g 3000X3000. The nodes are deployed in 2D platform. Each and every position of nodes are defined, thus from the initialized value, the attackers location in the 2D area can be determined accurately. The proposed method can be extended to networks like 802.11(WiFi), 802.15.4(ZigBee) to localize the attackers, in real life environment.

## 4. Performance Evaluations

The simulation of the proposed system is done with Network Simulator 2(NS2). The results of the proposed system helps in analyzing various performance metrics such as False positive rate, detection rate, delay metric, energy level and hit rate. This graph provides an analysis for finding the overall performance of the system.

### 4.1. Setting Environment

The proposed system has executed under the NS2 environment, where in Tcl language is used to simulate. This environment is having parameters as follows:

| | |
|---|---|
| Application | : VMware |
| Operating System | : Linux - Red Hat |
| X Axis | : 1400 |
| Y Axis | : 1400 |
| Channel Type | : Wireless Channel |
| MAC Type | : 802_11 |
| Routing Protocol | : AODV |

### 4.2. Simulation Evaluation

Fig.8 compares existing and proposed method of detecting the attackers in the wireless environment.

On analyzing the relationship between false positive rate with detection rate, it is found that Detection Rate becomes Constant after a certain False positive rate. The Threshold detection rate decreases when the distance between two centroids in signal space increase. Threshold detection rate is inversely proportional to dB. If the Distance becomes more, then False Positive Rate will increase.

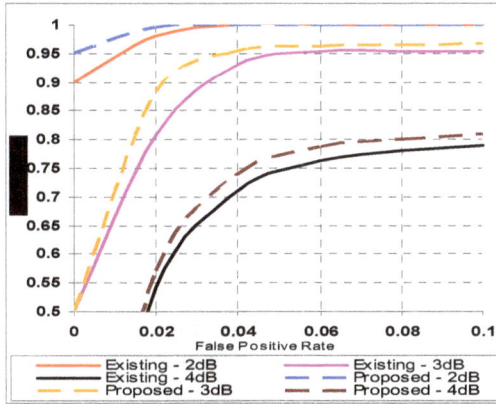

**Fig.8.** *False Positive Rate Vs Detection Rate*

### 4.2.1. Detection Rate Vs Time

Fig.9 shows about how detection rate is changing with respect to time. It is inferred that, high detection rate can be achieved after a considerable time.

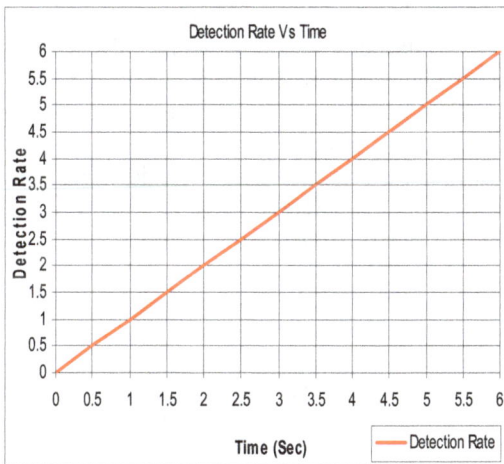

**Fig.9.** *Detection Rate Vs Time*

### 4.3. Performance Evaluation

### 4.3.1. Delay Vs Time

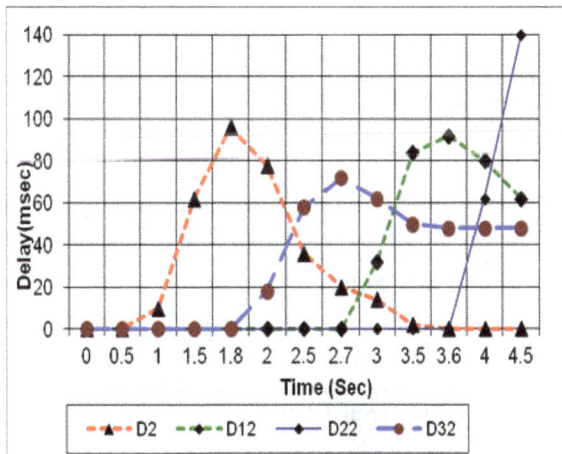

**Fig.10.** *Delay Vs Time*

Fig.10 shows the graphical representation of the delay metric, where x axis represents time and y axis represents delay. In this study, 4 nodes where chosen randomly, and measured the delay against time in seconds. The graph shows decrease in delay due to the network bandwidth and connectivity of the nodes. This mainly helps to prolong the connectivity of network and increase the performance.

### 4.3.2. Energy Vs Number of Attackers

Various Energy levels are measured against the number of attackers in the wireless environment, and the same represented as below. In this figure, Number of attackers are considered in X axis and energy levels in Joule is considered in Y axis.

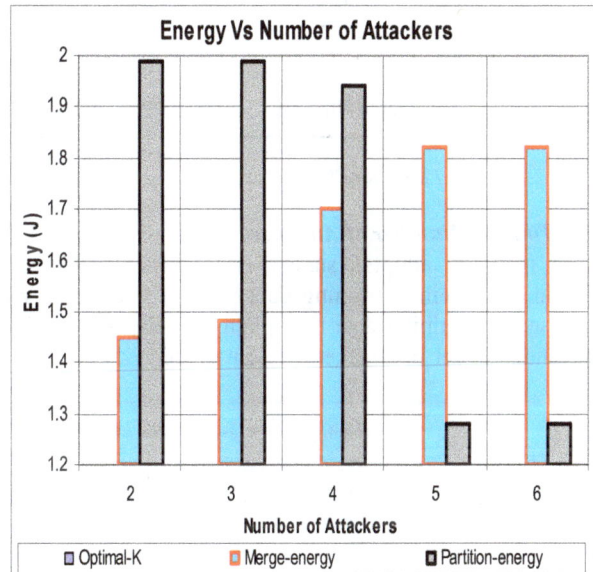

**Fig.11.** *Number of Attackers and Energy level*

From the fig.11, it is inferred that, Merge Energy level will increase as the number of attackers, increases. On the contrary, Partition energy level is decreased as the number of attackers increase. As a result, it is evident that, Merge energy and partition energy levels are inversely proportional to each other. Once the number of attackers in the environment is crossing 4, partition energy level drops drastically, and similarly there is a significant increase in the Merge energy level.

### 4.3.3. Hit Rate Vs Number of Attackers

A comparative study is conducted between the Existing and proposed method on the basis of hit rate. In X axis Number of attackers are considered and in Y axis hit rate is considered to plot this graph.

From the study, it is found that proposed method is having higher hit rate than existing method. But in both of methods, hit rate is decreased as the number of attackers increases. Proposed method is more efficient in identifying the attackers than that of existing method.

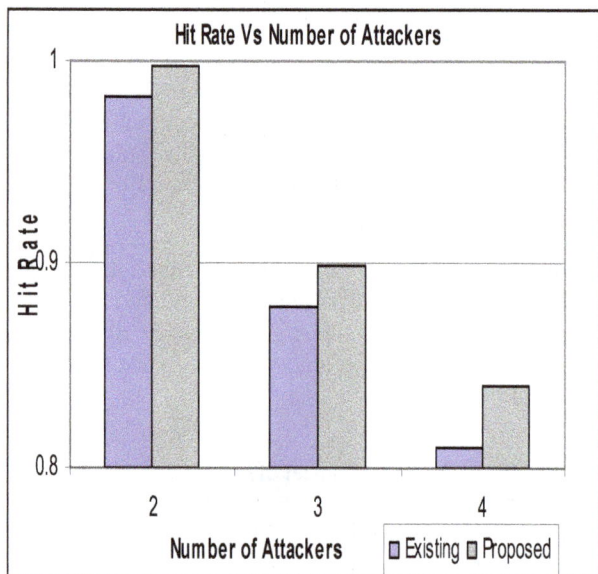

*Fig.12.* *Hit Rate Vs Number of Attackers*

### 4.3.4. Time Vs Number of Attackers
SVM method adopts supervised machine learning algorithm, wherein it training data plays a vital role. We have used the training data to explain the relation between Time and Number of attackers present in the system.

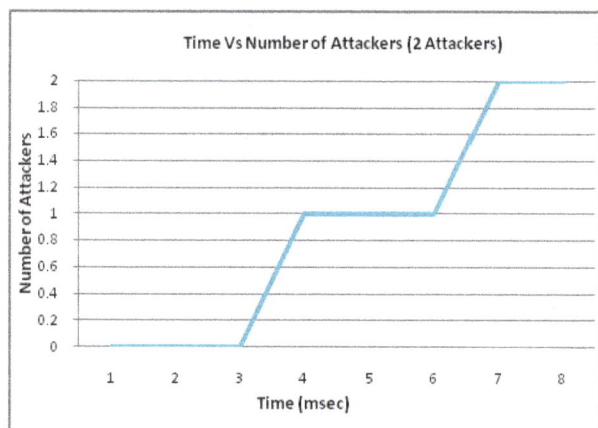

*Fig.13.* *Time Vs Number of Attackers (2 Attackers)*

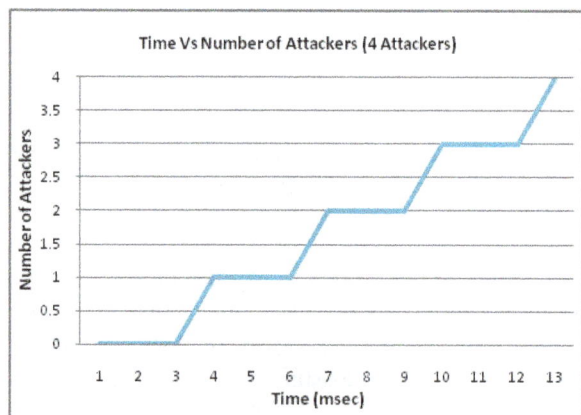

*Fig.14.* *Time Vs Number of Attackers (4 Attackers)*

# 5. Conclusion

Wireless networks are used in numerous applications. Due to its proliferation of usage, there exist threats in terms of spoofing attacks. In the proposed approach detection of the presence of attacks as well as determine the number of adversaries as same node identity. It can localize any number of attackers and eliminate them. Determine the number of adversaries in particular, is a challenging task. This mechanism that employs the minimum distance testing in addition to cluster analysis to achieve better accuracy of determining the number of attackers than other methods under study, such as Silhouette Plot and System Evolution, that use cluster analysis alone.

Further, based on the number of attackers determined by the mechanisms, our integrated detection and localization system can localize any number of adversaries even when attackers using different transmission power levels. The performance of localizing adversaries achieves similar results as those under normal conditions, thereby, providing strong evidence of the effectiveness of our approach in detecting wireless spoofing attacks, determining the number of attackers and localizing adversaries. The proposed methods can achieve over 99 percent Hit Rate and Precision when determining the number of attackers

In future, based on the outcome of this model, explore further to find ways to eliminate those identified multiple adversaries, from the wireless network. Thus way, wireless networks will be more robust and less prone to attack.

# References

[1]    Jie Yang, Yingying Chen, Wade Trappe and Jerry Cheng, "Determining the Number of Attackers and Localizing Multiple Adversaries in Wireless Spoofing Attacks", *Proceedings of 28th Annual Joint Conference of the IEEE Computer and Communications Societies*, 2009, pp. 666–674.

[2]    Wei Che, "Defending against TCP SYN Flooding Attacks under Different Types of IP Spoofing", *Proceedings of International Conference on Mobile Communication*, 2006, pp.38.

[3]    Diana Jeba Jingle, Elijah Blessing Rajsingh, "Defending IP Spoofing Attack and TCP SYN Flooding Attack in Next Generation Multi-hop Wireless Networks", *International Journal of Information & Network Security*, Volume 2, No.2, 2013, pp.160-166.

[4]    P.Ramesh Babu, S.D.Lalitha Bhaskari, CH. Satyanarayana, "A comprehensive Analysis of spoofing", *International Journal of Advanced Computer Science and Application* Volume 1, No.6, 2010.

[5]    Fanglu Guo and Tzi-cker Chiueh, "Sequence Number-Based MAC Address Spoof Detection", *Proceedings of 8th International Symposium on Recent Advances in Intrusion Detection*, 2005, pp.309-329.

[6]    Yingying Chen, Wade Trappe and Richard P. Martin, "Detecting and Localizing Wireless Spoofing Attacks",

*Proceedings of 4th Annual IEEE Communications Society Conference on Sensor, Mesh and Ad Hoc Communications and Networks,* 2007, pp.193-202.

[7] Jie Yang, Yingying Chen, Wade Trappe, "Detecting Spoofing Attacks in Mobile Wireless Environments", *Proceedings of 6th Annual IEEE Communications Society Conference on Sensor, Mesh and Ad Hoc Communications and Networks,* 2009, pp.107-189.

[8] Jayakumar C and Chellappan C, "Quality of Service in Associativity based Mobility-Adaptive K-Clustering in Mobile Ad-hoc Networks", *International Journal of The Computer, the Internet and Management,* Vol.14, No.1, 2006, pp.61-80.

[9] P. Infant Kingsly, C. Jayakumar, Mahendran Sadhasivam, S. Deepan Chakravarthy, "Smart Way for Secured Communication in Mobile Ad-hoc Networks", *International Journal Computational Intelligence and Informatics,* Vol.2 No. 1, 2012, pp.1-9.

[10] Pelechrinis, K., Krishnamurthy, S.V., "Denial of Service Attacks in Wireless Networks: The Case of Jammers", *IEEE Journal of Communications Surveys & Tutorials,* Volume 13, Issue: 2, 2011, pp.245-257.

[11] Qiang Liu, "Enhanced detection and restoration of low-rate denial-of-service in wireless multi-hop networks", *Proceedings of International Conference on Computing, Networking and Communications (ICNC),* 2013, pp. 195-199.

[12] Yanli Yu, Keqiu Li, Wanlei Zhou, Ping Li, "Trust mechanisms in wireless sensor networks: Attack analysis and countermeasures", *Journal of Network and Computer Applications,* Volume 35, Issue: 3, 2012, pp.867–880.

[13] Koffka Khan, Wayne Goodridge & Diana Ragbir, "Security in Wireless Sensor Networks", *Global Journal of Computer Science and Technology Network, Web & Security,* Volume 12, Issue: 16, Version 1.0, 2012.

[14] Divya Pal Singh, Pankaj Sharma and Ashish Kumar, "Detection of Spoofing attacks in Wireless network and their Remedies", *International Journal of Research Review*

*in Engineering Science and Technology,* Volume 1, 2012, pp.1-5.

[15] Jayakumar C and Chellappan C, "A QoS aware energy efficient routing protocol for wireless ad-hoc networks", *Asian Journal of Information Technology,* Vol.4, No.6, 2005, pp.578-582.

[16] Karmel A and C. Jayakumar, "Analysis of MANET Routing Protocols Based on Traffic Type", *IJREAT International Journal of Research in Engineering & Advanced Technology,* Vol.1, Issue 1, 2013, pp.1-4.

[17] K. Tan, Guanling Chen, D. Kotz, A Campbell, "Detecting 802.11 MAC layer spoofing using received signal strength", *Proceedings of 27th Conference on Computer Communications,* 2008, pp.1768-1776.

[18] Iyad Aldasouqi and Walid Salameh, "Detecting and Localizing Wireless Network Attacks Techniques", *International Journal of Computer Science and Security,* Volume 4, No.1, 2010, pp.82-97.

[19] Daniel B. Faria and David R. Cheriton, "Detecting identity-based attacks in wireless networks using Signal prints", *Proceedings of 5th ACM Workshop on Wireless Security,* 2006, pp.43-52.

[20] Jie Yang, Yingying Chen, Wade Trappe and Jerry Cheng, "Detection and Localization of Multiple Spoofing Attackers in Wireless Networks", *IEEE Transaction on Parallel and Distributed System,* Volume 24, No.6, 2012, pp.44-58.

[21] A. Wool, "Lightweight Key Management for IEEE 802.11 Wireless LANS with key refresh and host revocation", *Journal of ACM/Springer Wireless Networks,* Volume 11, 2005, pp.677–686.

[22] Jayakumar C and Chellappan C, "Optimized on demand routing protocol of mobile adhoc network", *Informatica,* Vol.17, No. 4, 2006, pp.481-502.

[23] M. Vijay Anand and C. Jayakumar, "Secure Routing in Manet Using Artificial Immune System Based on Danger Theory", *International Journal of Emerging Research in Management &Technology,* Vol.2, No.4, 2013, pp.73-77.

# 22

# Circularly polarized cross-slot-coupled stacked dielectric resonator antenna for wireless applications

**Rushikesh Dinkarrao Maknikar[*], Veeresh Gangappa Kasabegoudar**

Post Graduate Department, Mahatma Basaveshwar Education Society's, College of Engineering, Ambajogai, India, 431 517

**Email address:**

maknikar_rushikesh@rediffmail.com (R. D. Maknikar), veereshgk2002@rediffmail.com (V. G. Kasabegoudar)

**Abstract:** This paper presents a stacked cylindrical dielectric resonator antenna with wide circular polarization (CP) bandwidth (axial ratio < 3dB) of 16.0%. This wide CP bandwidth is achieved by stacking low permittivity dielectric (9.2) resonator on high permittivity dielectric (9.8) resonator to obtain improved impedance and axial ratio bandwidths as compared to conventional DRA. It is also shown that the asymmetrical structure used in the geometry results in a very high impedance bandwidth (more than 100%) in the frequency range of 2.1 GHz-12.0 GHz but at the expense of distorted CP operation on the off-broadside. This essentially covers the FCC band of operation (3.1GHz to10.6 GHz).

**Keywords:** Dielectric Resonator Antenna, Slot Coupling, Circular Polarization

## 1. Introduction

Dielectric resonator antennas (DRAs) were first discovered in 1983 by Long et. al., and since then they have been widely studied [1]. There have been several works reported in literature to improve various parameters (impedance bandwidth, axial ratio bandwidth, gain etc.) of DRAs [2-20]. Their inherent features of high radiation efficiency & ease of excitation, compactness, and ability to obtain radiation patterns using different excitation modes offer much to an antenna application [2]. Furthermore, the shape of a DRA can be custom, resulting in favorable operating modes or polarizations. Also, the DRAs supporting circular polarizations have been studied and can be found in the literature. Traditionally, circular polarization is achieved using quadrature couplers and elaborate feed systems which can also be done with DRAs as well [3, 4].

The design in [4] is unique in using a dual mode dielectric resonator (DR) excited by vertical strips fed in quadrature, while [3] uses an under-laid hybrid coupler. However, the bandwidth reported in [4] is only up to 8% whereas [3] offers an AR bandwidth of 33% but the geometry is not only complex but uses dual feed. Furthermore, four DRA arrays can be excited using sequential rotation to obtain wide axial ratio bandwidths [5]. In another effort [6], CP operation was accomplished by

using two crossed slots of unequal lengths to couple energy from a microstrip line to a simple cylindrical DRA. Both slots were angled at 45° with respect to the feeding microstrip line and their centers were at the same position, cantered underneath the dielectric resonator (DR). Their lengths were unequal, so that two near-degenerate orthogonal modes of equal amplitude and 90° phase difference were excited at frequencies close to that of the fundamental mode of the DRA. The resulting CP bandwidth was 3.91%.

An optimization of this topology was presented in [1] where offsetting each slot from the end of the microstrip line (thereby having the slot's intersection off from DR center) allows for AR bandwidths up to 4.7% at broadside angle (θ = 0°). This design excites the fundamental mode at 5.7GHz, where the lengths of slots and distance of slots to the end of microstrip line are tuned for the optimized performance. In [8] simulated results on a dual band single feed excited cylindrical DRA are discussed. However, this geometry offers dual band operations but they are linearly polarized.

Most of these techniques either involve laborious methods or they provide insufficient impedance and/ or axial ratio bandwidth values. Therefore, we demonstrate a geometry which offers an impedance and axial ratio bandwidth of more than 100% and 16% respectively. Although the configuration is a stacked one but uses equal radius dielectric resonators with different dielectric constant values. Hence, stacking alignment issue does not impose a problem. Also, antenna's impedance and axial ratio bandwidth values are significantly higher than the results presented in [1]. In other words, with the help of conventional stacking and suitable selection of dielectric substrate parameters, the axial ratio bandwidth of antenna reported in [1] can be increased from 4.7% to 16%, and impedance bandwidth from about 55% to more than 100% at the expense of increased height. However, this configuration is more useful where size is not the prime constraint.

Therefore, in this paper investigation results on wideband stacked DRA with wide AR and impedance bandwidth values are presented. The basic geometry of the DRA is presented in Section 2. Parametric studies and optimized geometry are explained in Section 3 followed by conclusions of this work are presented in Section 4.

## 2. DRA Geometry

The aperture coupled stacked DRA excited through cross-slots was designed and simulated using Ansoft's HFSS and is shown in Figure 1. Two slots are formed on the ground plane. Each of the slots forms an angle of 45° with respect to the microstrip line printed on backside. The slot lengths and permittivity of the substrate and dielectric resonator determine the frequency of slot resonance, while the DRA modes depend on the DR dimensions and permittivity. The ground plane having slots is sandwiched between the microstrip line and stacked dielectric resonators. The substrate used for etching the microstrip line is an FR4 with dielectric constant of 4.4, and height of 0.8mm. The two resonators having equal dimensions (discussed further in Section 3) with different dielectrics (9.2 and 9.8) are stacked and affixed on the ground plane (Figure 1). The resonant frequency of the stacked DRA structure is calculated by [7]:

$$f(GHz) = 30K_0 a/(2\pi r(cm)) \quad (1)$$

Where,

$$K_0 a = 0.8945(1 + 3.017X^{0.881} + e^{(0.962 - 1.6252X)})/\varepsilon_{reff}^{0.45} \quad (2)$$

$$X = r/2h. \quad (3)$$

The effective permittivity is calculated using,

$$\varepsilon_{reff} = \frac{h}{\varepsilon_{r1}/h_1 + \varepsilon_{r2}/h_2}. \quad (4)$$

Where

$$h = h_1 + h_2 \quad (5)$$

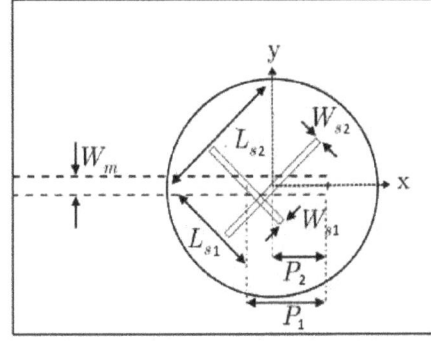

(a) Top view

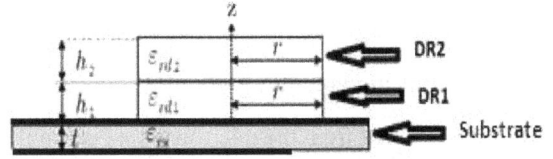

(b) Cross sectional view

**Figure 1:** *Geometry of the proposed dielectric resonator antenna.*

## 3. Parametric Studies and Discussion

Extensive parametric studies were conducted on an early variant of the final design in order to understand how best to tune for wide axial ratio (AR) bandwidth, and impedance matching. The key design parameters used to optimize the antenna geometry are stubs length ($P_1$ and $P_2$), slots lengths ($L_{s1}$ and $L_{s2}$) and their widths ($W_{s1}$ and $W_{s2}$). All these parameters have been thoroughly investigated and their effects have been presented in the following subsections.

### 3.1. Effect of Variation of Stub Lengths

The DRA is excited from the microstrip line etched on the backside through the cross-slots, where their physical dimensions play a critical role in proper DRA excitation. The slots were modified one at a time to isolate each slot's effect on the antenna performance. At first $P_1$ was swept from 4.8mm to 5.2 mm in steps of 0.1mm and optimized value was found to be 5.0mm. Similarly, $P_2$ was optimized by varying it from 4.0 mm to 4.4 mm in 0.1 mm steps.

**Table 1:** *Effect of variation of stub length $P_1$ on axial ratio bandwidth*

| Sr. No. | Stub Length $P_1$ (mm) | Stub Length $P_2$ (mm) | Axial Ratio Bandwidth (%) |
|---------|------------------------|------------------------|---------------------------|
| 1 | 4.8 | 4.2 | 4.7 |
| 2 | 4.9 | 4.2 | 14.8 |
| 3 | 5.0 | 4.2 | 16.0 |
| 4 | 5.1 | 4.2 | 10.2 |
| 5 | 5.2 | 4.2 | 7.4 |

**Table 2:** *Effect of variation of stub length P₂ on axial ratio bandwidth*

| Sr. No. | Stub Length P$_1$ (mm) | Stub Length P$_2$ (mm) | Axial Ratio Bandwidth (%) |
|---------|------------------------|------------------------|----------------------------|
| 1 | 5.0 | 4.0 | 6.6 |
| 2 | 5.0 | 4.1 | 7.4 |
| 3 | 5.0 | 4.2 | 16.0 |
| 4 | 5.0 | 4.3 | 12.0 |
| 5 | 5.0 | 4.4 | 9.75 |

From the above investigation and Table 2 it can be noted that that maximum value of axial ratio bandwidth is achieved for stub lengths of 5.0 mm and 4.2 mm.

### 3.2. Effect of Variation of Slots Length

The energy excited from the microstrip line is coupled to dielectric resonators through these slots. Hence, slot lengths and widths play important role in optimizing the AR and impedance bandwidths. In the first attempt, effect of their lengths was investigated and in the next subsection their width values will be optimized.

The effect of variation of slot 1's length ($L_{s1}$) on return loss characteristics is shown in Figure 2 (a). Initially, $L_{s1}$ is varied from 11.2 mm to 11.6 mm in steps of 0.1 mm with slot 2's length ($L_{s2}$) constant. It was observed the change in the value of $L_{s1}$ shifts the frequency band i.e., increasing $L_{s1}$ shifts the band on lower side and vice versa.

The length of slot 2 ($L_{s2}$) has a pronounced effect on both impedance bandwidth and axial ratio. As shown in Figure 2 (b), the length of $Ls2$ was varied from 8.7 mm to 9.1 mm with $L_{s1}$= 11.4 mm constant. Due to its shorter length, $L_{s2}$ dominates higher resonance frequency. In this investigation it was observed that the change in $L_{s2}$ results in the shift of operating frequency band of the CP.

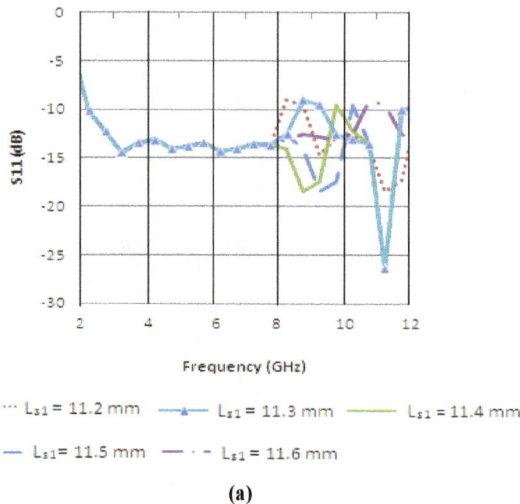

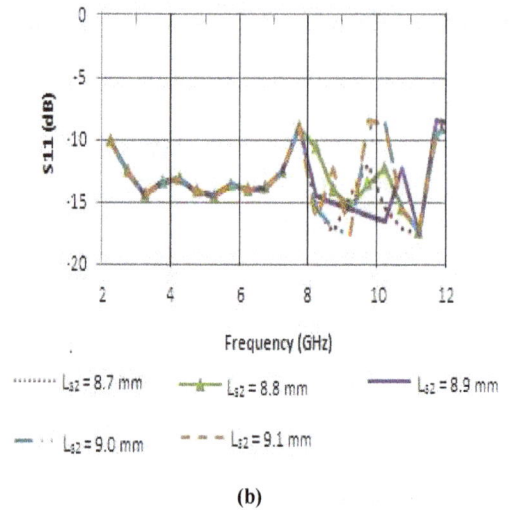

**Figure 2:** *(a) Reflection coefficient magnitude ($s_{11}$) vs. frequency for variation of length of slot 1, and (b) Reflection coefficient magnitude ($s_{11}$) vs. frequency for variation of length of slot 2.*

### 3.3. Effect of Variation of Slots Width

The effect of variation of slot width on return loss characteristics is shown in Figure 3 (a). Initially, $W_{s1}$ was varied from 0.2 mm to 0.4 mm in steps of 0.1 mm with $W_{s2}$ constant. It was observed that increase in value of $W_{s1}$ shifts the band on upper side and vice versa.

In order to observe the effect of variation of width of slot 2 on return loss characteristics, $W_{s2}$ was swept from 0.2 mm to 0.4 mm in steps of 0.1 mm while the width of slot 1 kept constant at 0.3 mm as shown in Figure 3 (b). It was observed that increase in value of $W_{s2}$ shifts the operating band on lower side.

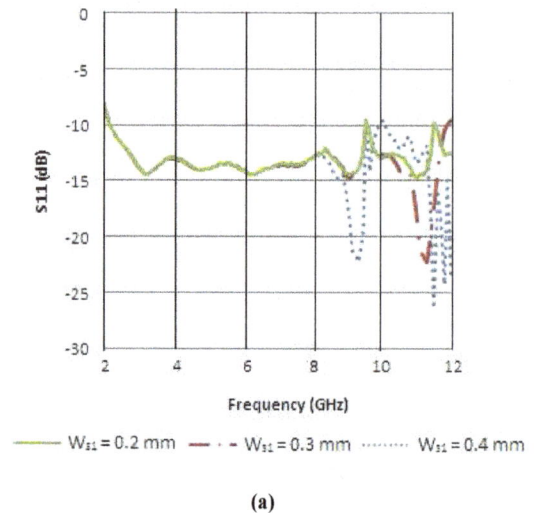

(a)

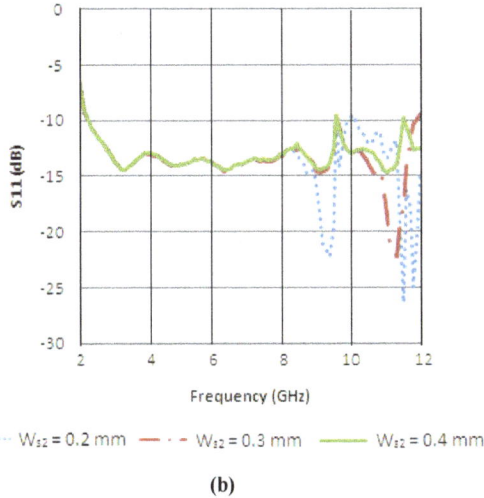

(b)

**Figure 3:** *(a) Reflection coefficient magnitude (s₁₁) vs. frequency for variation of width of slot 1, and (b) Reflection coefficient magnitude (s₁₁) vs. frequency for variation of width of slot 2.*

Based on the detailed parametric studies the optimum geometry was simulated and whose return loss characteristics are presented in Figure 4.

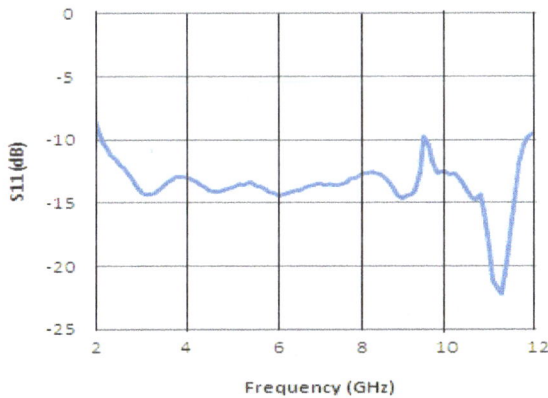

**Figure 4:** *Simulated return loss characteristics of optimized geometry shown in Figure 1.*

Also, these (optimum) dimensions are summarized in Table 3. The wide impedance bandwidth achieved here is due to the proper coupling between the slots & DRA resonances, and stacked arrangement of the DRAs. The presented antenna here achieves an impedance matching (S11< -10dB band) from 2GHz to 12GHz

**Table 3:** *Optimum values of dielectric resonator antenna geometry with $\varepsilon_{rs}$=4.4, $\varepsilon_{r1}$=9.8, and $\varepsilon_{r2}$=9.2*

| Parameter | $P_1$ | $P_2$ | $L_{s1}$ | $L_{s2}$ | $W_{s1}$ | $W_{s2}$ | $W_m$ | t | $h_1$ | $h_2$ | r |
|---|---|---|---|---|---|---|---|---|---|---|---|
| Value (mm) | 5.0 | 4.2 | 11.4 | 8.9 | 0.3 | 0.3 | 2.4 | 0.8 | 5.1 | 5.1 | 15.2 |

The axial ratio characteristics of the antenna are depicted in Figure 5. From the Figure 5 it may be noted that the CP operation (AR≤ 3dB) is in between 10.0GHz to 11.4GHz which corresponds to an axial ratio bandwidth of 16.1%.

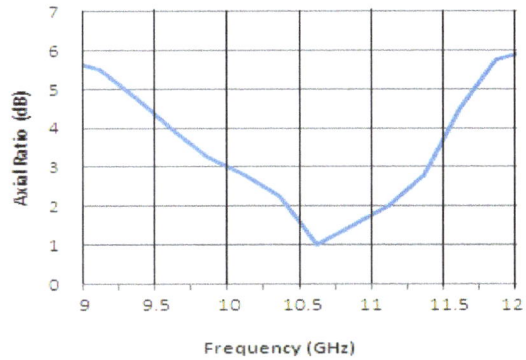

**Figure 5:** *Axial ratio vs. frequency plot.*

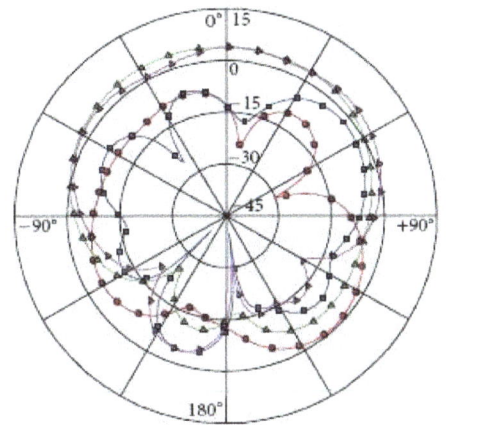

(a) LHCP pattern

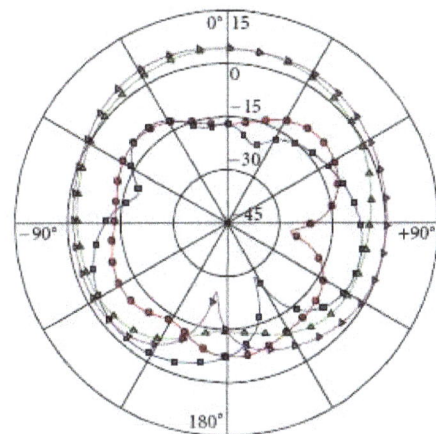

(b) RHCP patterns

**Figure 6:** *Radiation patterns (LHCP and RHCP) of DRA at resonant frequency.*

Figure 6 shows radiation patterns of the proposed antenna at 10.5GHz at which it exhibits CP operation. The circular polarization obtained was left-handed when $L_{s1}>L_{s2}$ (Figure 6 (a)). Similarly, RHCP operation was obtained by setting

$L_{s2} > L_{s1}$ and the same is depicted in Figure 6 (b). The gain vs. frequency plot in CP band is shown in Figure 7. From the Figure 7 it may be noted that the gain is positive throughout the band (LHCP) of operation and the maximum gain (about 5 dB) was observed at 10.6 GHz.

From the results presented here it may be noted that by stacking the dielectric resonators with equal diameter and height but with different dielectric constants the impedance and AR bandwidths may be significantly increased. The reason for using different dielectric constants is to make proper coupling between bottom and top resonators. This arrangement increases the axial ratio bandwidth from 4.7% [1] to 16.1% [our geometry] and impedance bandwidth from nearly 55% [1] to more than 100% in our case.

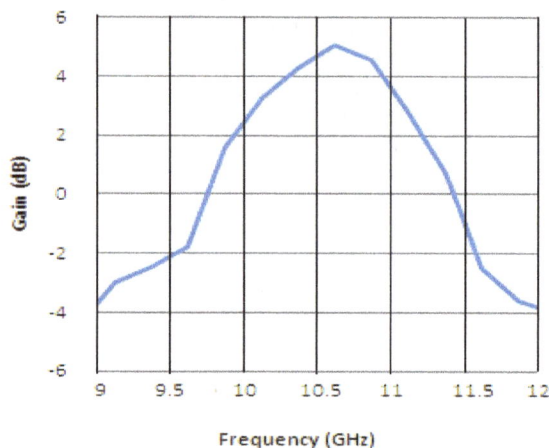

**Figure 7:** *Gain vs. frequency characteristics plot.*

## 4. Conclusions and Future Scope

The cross-slot coupled cylindrical DRA with wide impedance & AR bandwidth has been proposed. It was demonstrated that the bandwidth of cross-slot coupled DRA can be increased substantially by stacking two DRs vertically. The dielectric resonators are made up of same materials with different dielectric constants to achieve the benefit of compact size and wide bandwidth. Due to the absence of conductor loss, the proposed antenna has high gain and radiation efficiency. The future study of this work includes the fabrication and testing of the geometry proposed here. Also, all the modes and working of the geometry need to be verified analytically.

## References

[1]  G. Almpanis, C. Fumeaux, and R. Vahldieck, "Offsetcross-slot coupled dielectric resonator antenna for circular polarization," *IEEE Microwave and Wireless Components Letters*, vol. 16, no. 8, pp. 461–463, 2006.

[2]  A. Petosa, A. Ittipiboon, Y. M. M. Antar, D. Roscoe, and M. Cuhaci, "Recent advances in dielectric-resonator antenna technology," *IEEE Antennas Propag. Mag.*, vol. 40, no. 3, pp. 35–48, 1998.

[3]  E. H. Lim, K. W. Leung, and X. S. Fang, "The compact circularly-polarized hollow rectangular dielectric resonator antenna with an underlaid quadrature coupler," *IEEE Transactions on Antennas and Propagation*, vol. 59, no. 1, pp. 288–293, 2011.

[4]  K. W. Leung and K. K. So," Frequency tunable designs of linearly & circularly polarized DRA using parasitic slot," *IEEE Transaction on Antenna and Propagation*, vol. 53, no. 1, pp. 572-576, 2005.

[5]  A. Petosa and A. Ittipiboon, "Dielectric resonator antenna: a historical review and current state of art," *IEEE Antennas and Propagation Magzine*, vol. 52, no.5, 2010.

[6]  C. Y. Huang, J. Y. Wu, and K. L. Wong, "Cross-slot-coupled microstrip antenna and dielectric resonator antenna for circular polarization," *IEEE Transaction on Antennas Propagation*, vol. 47, no. 4, pp. 605–609, 1999.

[7]  K. M. Luk and K. W. Leung, *Dielectric Resonator Antennas*, Baldock, U.K.: Research Studies Press, 2003.

[8]  D. Batra, S. Sharma, and A. K. Kohli, "Dual band Dielectric Resonator Antenna for C and X band application," *International Journal of Antenna and Propagation*, Article ID 914201, pp. 1-7, vol. 2012.

[9]  K. M. Luk, K. W. Leung, and K. W. Chow, "Bandwidth and gain enhancement of a dielectric resonator antenna with the use of stacking element," *Microwave and Optical Technology Letters*, vol. 14, no. 4, pp. 215-217, 1997.

[10] S. M. Shum and K. M. Luk, "Stacked annular ring dielectric resonator antenna excited by axi-symmetric coaxial probe," *IEEE Transactions on Antennas and Propagation*, vol. 43, no. 8, pp. 889-892, 1995.

[11] G. D. Makwana, D. Ghodgaonkar, "Wideband stacked rectangular dielectric resonator antenna at 5.2 GHz," *International Journal of Electromagnetics and Applications*, vol. 2, no. 3, pp. 41-45, 2012.

[12] K. W. Khoo, Y. X. Gou, and L. C. Ong, "Wideband circularly polarized dielectric resonator antenna," *IEEE Transactions on Antennas and Propagation*, vol. 55, no. 7, pp. 1929-1932, 2007.

[13] M. B. Oliver, R. K. Mongia, and Y. M. M. Antar, "A new broadband circularly polarized dielectric resonator antenna," *IEEE International Symposium on Antennas and Propagation*, vol. 1, pp. 738-741, 1995.

[14] Y. Ge and K. P. Esselle, "A wideband probe-fed stacked dielectric resonator antennas," *Microwave and Optical Technology Letters*, vol. 48, no. 8, pp. 1630-1633, 2006.

[15] A. A. Kishk, B. Ahn, and D. Kajfez, "Broadband stacked dielectric resonator antenna," *IEE Electronics Letters*, vol. 25, no. 18, pp. 1232-1233, 1989.

[16] A. A. Kishik, "Experimental study of broadband embedded dielectric resonator antennas excited by a narrow slot," *IEEE Antennas and Wireless Propagation Letters*, vol. 4, pp. 79-81, 2005.

[17] C. Ozzaim, F. Ustuner, and N. Tarim, "Stacked conical ring dielectric resonator antenna excited by a monopole for improved ultra-wide bandwidth," *IEEE Transaction on Antennas and Propagation*, vol. 61, no. 3, pp. 1435-1438, 2013.

[18] M. R. Nikkhah, J. R. Mohassel, and A. A. Kishk, "Compact low cost phased array of dielectric resonator antenna using parasitic elements and capacitor loading," *IEEE Transaction on Antennas and Propagation*, vol. 61, no. 4, pp. 2318-2321, April 2013.

[19] M. Abedian, S. K. A. Rahim, and M. Khalily, "Two-segment compact dielectric resonator antenna for UWB application," *IEEE Antennas and Wireless Propagation Letters*, vol. 11,

pp. 1533-1536, 2012.

[20] K. S. Ryu and A. A. Kishk, "Ultra wideband dielectric resonator antenna with broadside patterns mounted on a vertical ground plane edge," *IEEE Transaction on Antennas and Propagation*, vol. 58, no. 4, pp. 1047-1053, 2010.

# 23

# An adaptive energy-aware transmission scheme for wireless sensor networks

**Abdullahi Ibrahim Abdu, Muhammed Salamah**

Computer Engineering Department, Eastern Mediterranean University , KKTC, Mersin 10, TURKEY

**Email address:**

teetex4sure@gmail.com (A. I. Abdu), muhammed.salamah@emu.edu.tr (M. Salamah)

**Abstract:** In this paper, we proposed a scheme to extend the lifetime of a wireless sensor network, in which each sensor node decides whether to transmit a message or not and with what range to transmit the message, based on its own energy reserve level and the information contained in each message. The information content in each message is determined through a system of rules describing prospective events in the sensed environment, and how important such events are. The messages deemed to be important are propagated by all sensor nodes and with different transmission ranges depending on nodes energy resource level, while messages deemed to be less important are handled by only the nodes with high energy reserves level and transmitted with different transmission ranges based on nodes energy resource level. The results show that by adapting the transmission range based on nodes energy reserve and message importance, a considerable increase in network lifetime and connectivity can be achieved.

**Keywords:** Energy-Aware, Wireless Sensor Networks, Adaptive Transmission Ranges, Priority Balancing

## 1. Introduction

A wireless sensor network (WSN) typically consists of a number of small, inexpensive, locally powered sensor nodes that communicate detected events wirelessly through multi-hop routing [1]. These networks have limited computing capability and memory [2]. Typically, a sensor node is a tiny device that includes three basic components: a sensing subsystem for data acquisition from the physical surrounding environment, a processing subsystem for local data processing and storage, and a wireless communication subsystem for data transmission. In addition, a power source supplies the energy needed by the device to perform the programmed task. This power source often consists of a battery with a limited energy budget. In addition, it could be impossible or inconvenient to recharge the battery, because nodes may be deployed in a hostile or unpractical environment [3]. WSNs are being used in a wide variety of critical applications such as traffic, defense, medical treatment, manufacturing [4], military, health-care applications [5-6], and environmental monitoring [7]. A key research area is concerned with overcoming the limited network lifetime inherent in the small, locally powered sensor nodes [1]. Many of the WSN algorithms designed to extend the network lifetime are modified routing algorithms [8]. How-

ever, a low cost and low complexity technique is the motivation of this study.

In this paper, a scheme to extend the lifetime of a wireless sensor network, named as AIRT (**A**daptive **I**nformation managed energy aware algorithm for sensor networks with **R**ule managed reporting and **T**ransmission range adjustments) is proposed. The extension in the network lifetime is achieved by adapting transmission ranges based on nodes energy reserve level and importance of messages. One main advantage of this scheme is that, nodes do not have to transmit messages with their maximum transmission ranges all the time. The node's energy reserve level and message importance are taken into consideration, and therefore the node's transmission range is adapted accordingly. AIRT also maintains high network connectivity by adjusting the minimum transmission range of a node to cover at least one other sensor node.

The rest of this paper is organized as follows: Section 2 presents the related work. Section 3 presents the proposed AIRT scheme. Section 4 gives the performance study (simulation results and discussions). Section 5 provides conclusion and future work.

## 2. Related Work

In Merret et. al. [9], IDEALS (Information manageD

Energy aware ALgorithm for Sensor networks) was introduced as a scheme that extends the lifetime of a WSN through the possible discrimination of low important messages and prioritizing important messages. The IDEALS scheme is built upon the idea of message priorities and power priorities.

Authors in [10] proposed a localized technique to extend the lifetime of a wireless sensor network; referred to as IDEALS|RMR (Information manageD Energy aware ALgorithm for Sensor networks with Rule Managed Reporting). The extension in the network lifetime is achieved at the possible sacrifice of low importance packets, as a result of a union between: Information control, quantified by a system of rules, referred to as rule managed reporting (RMR) and energy management, supplemented by energy harvesting, for example, mechanical vibrations. Furthermore, IDEALS|RMR uses a single-fixed transmission range for each sensor node regardless of whether its energy resource is high or low and this causes redundancy in energy consumption as lots of areas are covered by several sensors.

In Mingming et al. [11], the authors developed a power saving technique by combining two methods: scheduling sensors to alternate between active and sleep mode method, and adjusting sensors sensing range method. They combined both methods by dynamic management of node duty cycles in a high target density environment. In their approach, any sensor schedules its sensing ranges from 0 to its maximum range, where 0 corresponds to sleep mode.Adding the system of information control proposed in this paper could significantly save energy.

Authors in [12] try to deal with the problem of energy holes (unbalance distribution of communication loads) in a wireless sensor network. This means that, energy of nodes in a hole will be exhausted sooner than nodes in other region. As energy holes are the key factors that affects the life time of wireless sensor network, they proposed an improved corona model with levels for analyzing sensors with adjustable transmission ranges in a WSN with circular multi-hop deployment. Additionally, the authors proposed two algorithms for assigning the transmission ranges of sensors in each corona for different node distributions. These two algorithms reduce the searching complexity as well as provide results approximated as optimal solution.

Two forwarding techniques were proposed by M. Busse et al. in [13] to maximize energy efficiency. The techniques are termed single-link and multi-link energy efficient forwarding. Single-link forwarding sends a packet to only one forwarding node; multi-link forwarding exploits the broadcast characteristics of the wireless medium. This leads to efficiency because if one node does not receive a packet, another node will receive the packet and performs the forwarding process. There is however a tradeoff of delivery ratios against energy costs.

In [14], we extended the lifetime of wireless sensor network by sacrificing low importance message and also adjusting of transmission ranges based on only nodes energy

reserve. However, in our proposed AIRT technique, we improved lifetime even better by adapting transmission ranges based not only on nodes energy reserve, but also on message importance.

# 3. The Proposed AIRT Scheme

The idea of our proposed AIRT scheme was first derived from [9], where the basic principles of IDEALS (Information manageD Energy aware ALgorithm for Sensor networks) was introduced as a scheme providing an increase in the lifetime of a wireless sensor network through the discrimination of certain messages. Additionally, the work of [9] was extended by authors in [10] through coupling IDEALS with a system for determining the messages information content – RMR (Rule Managed Reporting). Furthermore, in our previous research (named as IRT) [14], we extended [10] by introducing adjustable transmission ranges based on nodes energy reserve only. Finally, in this paper, we extended our previous work [14] by

- Coupling IDEALS|RMR with transmission range adjustment (TRA), whereby the TRA is determined as a function of two parameters;
1. Nodes energy reserve level and
2. Importance of the message.
- Perform a detailed analysis through the simulation of the AIRT, IRT [14], IDEALS|RMR [9, 10], and traditional [9, 10] schemes to prove that the proposed AIRT technique is the most energy efficient.

The system diagram of the proposed AIRT scheme is shown in Fig. 1. Firstly, upon sensing data, the sensor node passes the information to the controller, which sends the sensed value (e.g. temperature) to RMR (Rule Management Reporting) unit. The purpose of RMR unit is to determine if a message worth reporting has occurred and how important that message is. The value is then received by the Rule Testing module. Its responsibility is to determine if a message worth reporting has occurred. It does that by checking the sensed data against the rules in the Rule Database module (getting history information about the previously sensed values), at the same time, it updates the history with the current information of message and sensed message. Rules may be fulfilled or not, any rules which are fulfilled are passed to the Message Priority Assignment Module to determine how important the content of the message is. It does that by assigning message priorities (MP) to each fulfilled rule. In this work, five different MPs are used (MP1-MP5). MP1 denotes the most important message which might represent drastic change in the sensed data value. Conversely, MP4 - MP5 relates to the least important messages which might represents slight or no change in the sensed value. MP2 - MP3 relates to intermediate priorities messages which might represent moderate change in the sensed value. Any number of predefined rules can be entered by the designer, and these rules describe different events that can be detected in the sensed environment. Examples of such possible rules are:

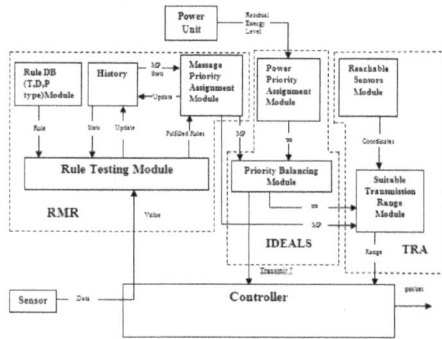

*Fig. 1. The proposed AIRT system diagram.*

1. Threshold rules (applied when the sensed value crosses a preset value).
2. Differential rules (applied when the change in the sensed value is larger or smaller than a preset value).
3. Routine rules (applied when a message of a certain importance has not been sent for a preset period) [10].

Secondly, data from RMR is then passed to IDEALS unit, its responsibility is to decide if the node should transmit a message or not, and it's done by Priority Balancing Module. The node's energy resource is characterized by Power Priority Assignment Module, which assigns a power priority (PP) for each node based on the state of its battery. A full battery life is assumed to have 100jouls of energy, whereas an empty battery life is assumed to be 0jouls. The 100jouls is divided equally into 5 levels, and priorities are assigned to each level. The highest power priority is PP5 and it is assigned to the highest level of energy for a node which is from 80 to 100j. While the lowest power priority is PP1 and it is assigned to the lowest level of energy for a node which is from 0 to 20j. When priority balancing module receives MP and PP, it compares them and if PP $\geq$ MP, then the message will be transmitted as illustrated in the priority allocation and balancing process of Figure 2.

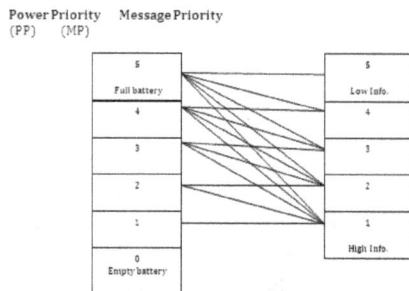

*Fig. 2. Priority Balancing.*

Finally, PP and MP are also passed to the *Transmission Range Adjustment* (TRA) unit in which the *Suitable Transmission Range* module decides with what range a sensor node shall transmit the message. The Suitable transmission range module gets PP from the power priority allocation module, MP from message priority allocation module and coordinates from *reachable sensors module*. The reachable

sensors are the entire sensors in the maximum transmission range of a sending sensor node. Now, based on the value of the PP, and MP, a suitable transmission range (TR) is determined and passed to the controller to transmit the message with the new range as illustrated in the heuristic Table 1. In this work, five different TRs are used (TR1-TR5), where TR1 represents minimum transmission range and TR5 represents maximum transmission range. For example, when a node has full battery (i.e. PP5), it will transmit MP1 messages with TR5, MP2 with TR4, MP3 with TR3, MP4 with TR2 and MP5 with TR1. However, if a node's battery decreases to its minimum PP1, it will transmit only messages with the highest message priority MP1 and with the lowest transmission range TR1. It is worth to mention that a power control mechanism is assumed to be embedded in the sensor node for the adjustment of its transmission range.

*Table 1. Heuristically determined transmission ranges.*

| MP \ PP | 5 (Highest) | 4 | 3 | 2 | 1 (Lowest) |
|---|---|---|---|---|---|
| 5 (Lowest) | 1(Min. TR) | 0 | 0 | 0 | 0 |
| 4 | 2 | | 2 | 0 | 0 |
| 3 | 3 | | 4 | 2 | 0 |
| 2 | 4 | | 4 | 3 | 2 |
| 1 (Highest) | 5(Max. TR) | | 4 | 3 | 2 |

The transmission ranges in Table 1 above were determined heuristically after several runs of simulation with different values of TR. The resultant matrix has a meaningful structure which reflects the tradeoffs between MP and PP. These near-optimal TR values for the AIRT scheme improve the network lifetime, connectivity, and maintain a considerable message loss compared to other schemes (IRT, IDEALS|RMR, and Traditional) as will be seen in the performance analysis section.

Table 1 can be seen as a matrix A with $i$ = MP, $j$ = PP, and $A[i,j]$ = TR. Then, assuming that the number of levels used for MP, PP and TR is equal to L, the transmission ranges can be derived from the following algorithm:

$$A[i,j] = \begin{cases} 0 & \text{if } i > j \\ 2 & \text{if } i = j \text{ and } j \neq L \\ j & \text{if } i < j \text{ and } j \neq L \\ L+1-j & \text{if } j = L \end{cases}$$

## 4. Performance Evaluation

Performance evaluation of the proposed AIRT scheme is done through extensive simulation. To show that the proposed scheme conserves the most amount of energy, it is compared with other related power saving schemes based on Energy Depletion Times (EDT) and Network Connectivity (NC) performance metrics. Based on these performance metrics, the simulations of Traditional, IDEALS|RMR, IRT, and AIRT schemes are conducted.

### 4.1. Simulation Setup

We simulated 20 sensor nodes in a network size of 70×70 meters. Each node has the same maximum transmission range of 20 meters. A predefined node placement topology has been considered in this study as it can be seen from Figure 4. At the beginning of the simulation, all nodes have the same initial energy of 100 Joules [10]. The following Equation [10] is used for calculating the energy required to transmit a packet:

$$E_{tx}(1,d) = E_{elec}l + E_{amp}ld^2 \qquad (1)$$

where $E_{elec}$[J] is the energy required for the circuitry to transmit or receive a single bit, $E_{amp}$[J] is the energy required for the transmit amplifier to transmit a single bit a distance of one meter, $d$ is the separation distance in meters and $l$ is the length of the packet. The simulation parameters are given in Table 2.

**Table 2**. *Simulation parameters and results.*

| Simulation area | 70×70meters |
|---|---|
| Number of nodes | 20 nodes |
| Packet length | 1000 bytes |
| Initial node energy | 100 Joules |
| Simulated Node Id | node-08 |
| Minimum transmission range | 13.04meters |
| Maximum transmission range | 20 meters |
| Simulated node Coordinates | (x = 38 , y = 37) |
| Waiting time 't' | t minutes |
| Node 08 Traditional Scheme Result, for NEDT performance Metric. | 9 Hours |
| Node 08 IDEALS/RMR Scheme Result, for NEDT Performance Metric. | 34 Hours |
| Node 08 IRT Scheme Result, for NEDT Performance Metric. | 63 Hours |
| Node 08 AIRT Scheme Result, forNEDT Performance Metric. | 82 Hours |
| Node 08 Traditional Scheme Result, for NC Performance Metric. | 9 Hours |
| Node 08 IDEALS/RMR Scheme Result, for NC Performance Metric. | MP5 MP4 MP3 MP2 MP1 4h11h 12h 22h34h |
| Node 08 IRT Scheme Result, for NC Performance Metric. | MP5 MP4 MP3 MP2 MP1 13h26h 29h 47h 62h |
| Node 08 AIRT Scheme Result, for NC Performance Metric. | MP5 MP4 MP3 MP2 MP1 23h 53h57h68h82h |

Only one sensor node is chosen for the simulation as the remaining sensors are assumed to operate in the same way. The chosen sensor node id and the distance to its closest sensor node are given as inputs. The sensor node senses data and the AIRT algorithm is performed as illustrated in Figure 3. Since the maximum transmission range (TR5) is fixed for every sensor, the other transmission ranges can be calculated by considering the minimum transmission range (TR1) as the distance to the closest sensor in the sending sensor's maximum transmission range. So the ranges between TR1 to TR5 are calculated successively by adding $\Delta TR$ = (TR5-TR1)/4. That is, TR(i) = TR(i-1) + $\Delta TR$, for i=2,3,4. For example, adding $\Delta TR$ to TR1 gives TR2, and so on. To maintain connectivity, we took TR1 as the distance to the closest sensor in the sending sensor's transmission.

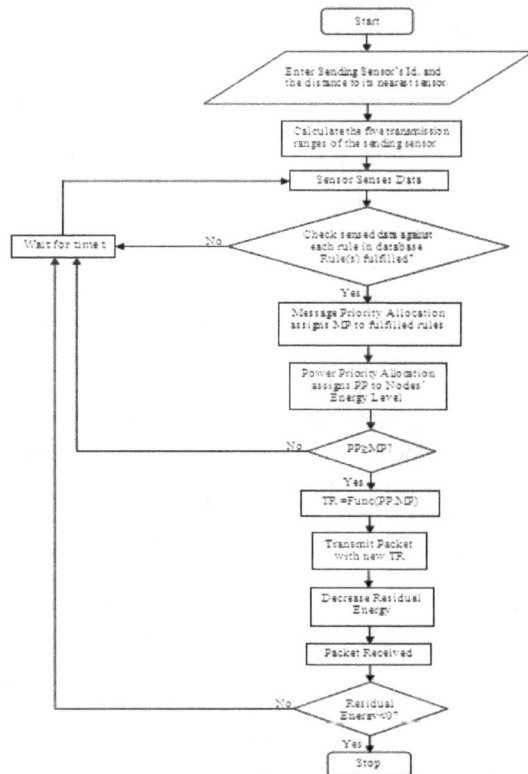

**Fig. 3.** *Simulation Flowchart of AIRT scheme.*

All nodes except the sink node (final destination of messages), performs multi-hop routing of messages by using the flooding algorithm. The simulation program is dynamic in the sense that, different coordinates from the ones used in the simulation can be entered and any node can be chosen for the simulation. Figure 4 shows a snap shot of a predefined node placement topology used in the simulation. Circles represent the maximum transmission range of sensors and lines represent possible communication link [10]. We chose node-8 as it is located in the middle. We assumed messages are transmitted every 5 minutes.

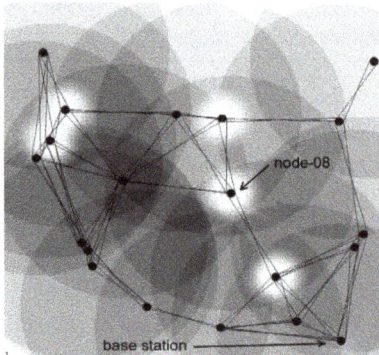

**Fig. 4.** *A snap shot of node placement used in the simulation.*

### 4.2. Simulation Results

Extensive simulation was done to show the effectiveness of the proposed AIRT scheme in extending the network life time. Rather than message generation, the simulator focused on the operation of our algorithm, whereby a data is sensed, if it is worth reporting based on user defined rules, priority is given to the related message, and the message is transmitted if PP ≥ MP. Furthermore, based on data priority (MP) and battery life (PP), a suitable transmission range is determined using Table 1.

(1)Node Energy Depletion Times (NEDT): Fig. 5 shows node-8 energy depletion times for all the four schemes. In the figure, '100' means that the node energy reserve is full, while '0' means that the energy reserve is depleted. For the traditional scheme, node-8 depleted its energy reserve after around 9h, this is because it is transmitting messages regularly every 5 minutes, and does not take into account the nodes energy reserve and message importance. For the IDEALS|RMR scheme, it can be seen that the nodes energy level drops suddenly, this is because when the node has high energy reserve, it transmits every message that comes to it. The energy level then remains constant for a while because of less importance messages which are not transmitted. For the IRT scheme, the energy level drops and then remains constant because the IDEALS|RMR scheme is embedded inside IRT. However, the network lifetime increased almost twice that of the IDEALS|RMR due to the adjustment of transmission ranges based on the nodes energy reserve level. Finally, for the AIRT scheme, the same process of sudden dropping of energy level and then remaining constant occurs. However, the network lifetime is significantly enhanced compared to all other three schemes. The achieved improvements of the AIRT scheme over traditional, IDEALS|RMR, and IRT schemes are 720%, 135%, and 33% respectively. This is due to the fact that nodes are adapting their transmission ranges based on both their energy reserve level and message importance.

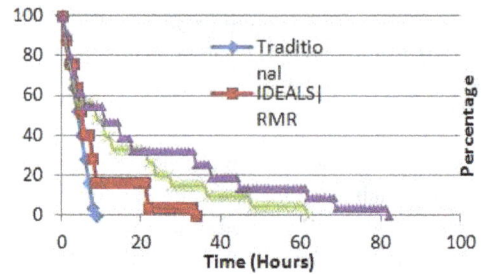

**Fig. 5.** *Node energy depletion times.*

(2) Network Connectivity (NC): The network connectivity is the measure of the ability of any node in the network to successfully transmit a message to the sink node. If the network is 100% connected, then any node in the network can successfully transmit a message to the sink node. However if the network is 50% connected, then only half of the nodes can successfully transmit a message to the sink node. Figure 6 shows that the connectivity of the Traditional scheme is completely lost after around 9 hrs. Figure 7 shows the connectivity of the IDEALS|RMR scheme. It is clear from the figure that this scheme extends the network lifetime and hence improves the network connectivity. The network lifetime for the most trivial message (MP5) is lost after around 4hrs (therefore no message of MP5 importance can reach the sink node after 4hrs) while it is connected for about 76%. However the network is still around 80% connected for the most important message (MP1) for 34 hrs. This is because the messages can not be transmitted every 5 minutes as in the traditional scheme. Figure 8 shows the performance of the IRT scheme. As it can be seen from the figure, IRT manages to extend the network lifetime almost twice that of the IDEALS/RMR scheme. This is because of the adjustment of transmission range based on nodes energy level. Additionally, network lifetime for the most trivial message is lost after around 13hrs while it is connected for about 76%. However the network is 79.6% connected for the most important message for 63hrs. Finally, Figure 9 shows the connectivity of the proposed AIRT scheme. It is obvious that AIRT scheme outperforms all other schemes as it has the longest network lifetime and maximum connectivity. It is interesting to note that AIRT extends the network lifetime twice longer that of IDEALS/RMR scheme and a little longer than the IRT scheme for MP5. Despite its lifetime finishes in 83hours, the network is 84.7% connected for the most important message MP1.

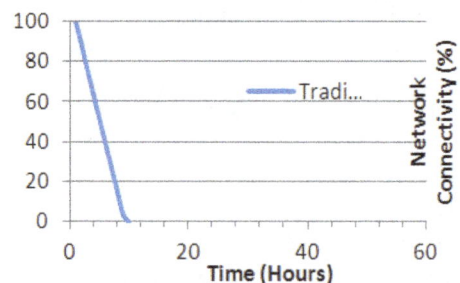

**Fig. 6.** *Connectivity of the Traditional scheme.*

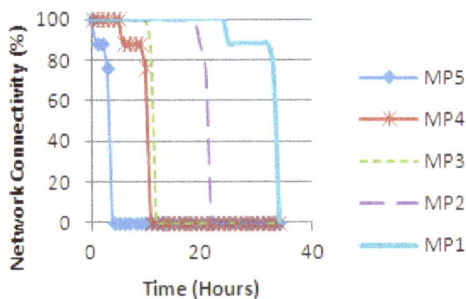

*Fig. 7.* Connectivity of the IDEALS|RMR scheme.

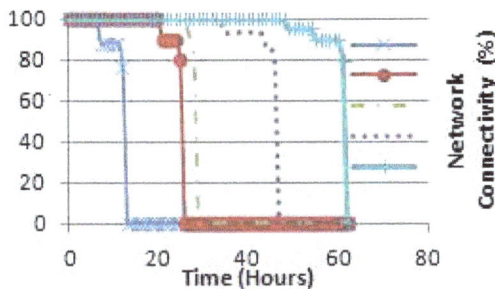

*Fig. 8.* Connectivity of the IRT scheme.

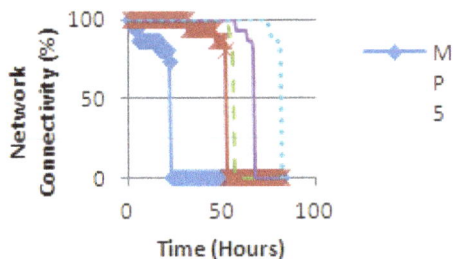

*Fig. 9.* Connectivity of the AIRT scheme.

(3) Message Loss (ML): It can be observed in fig. 5 that for all the schemes, the battery life drops drastically then it start dropping slowly, this is because of constant transmission of high priority messages. Consequently, low priority messages are ignored, leading to their loss.

As a summary, Table 3 shows the percentage improvement of the proposed AIRT scheme over traditional, IDEALS/RMR and IRT schemes in terms of network connectivity for the different types of message priorities. It is important to note that for the most important message (MP1), the AIRT scheme achieved (82-9)/9*100= 811%, (82-34)/34*100=141%, and (82–62)/62=32% improvement over Traditional, IDEALS/RMR and IRT respectively. The same way is used to calculate the percentage improvement for the remaining message priorities.

*Table 3.* Percentage improvement of AIRT Scheme over all other Schemes in terms of Network Connectivity.

|  | MP1 | MP2 | MP3 | MP4 | MP5 |
|---|---|---|---|---|---|
| Traditional | 811 | 811 | 811 | 811 | 811 |
| IDEALS|PMR | 141 | 209 | 375 | 381 | 450 |
| IRT | 32 | 45 | 97 | 69 | 104 |

## 5. Conclusion

In this paper, we introduced the AIRT scheme, which operates on the combination of information control (deciding if a message is worth processing based on the user defined rules, and how important the message is), energy management (balancing of the energy resource level and message importance) and TR adaptation (deciding on the transmission range based on nodes energy resource level and message importance) which we believe have never been considered before. Though scalability is not the primary concern of the work, but it is suitable for harsh environments. Simulation was performed using C programming language where a single node was simulated to show the operation of our scheme. The proposed scheme is preferred over others in situations when an intelligent system is used to adjust the TR of nodes. The results obtained show that the proposed AIRT scheme outperforms the other schemes (IRT, IDEALS|RMR, and Traditional) in terms of lifetime and connectivity, however, considerable amount of messages are lost due to TR adaptation.

We are currently working on coupling AIRT scheme with fuzzy logic, in the sense that, decisions on the MP, PP and TR values can be based on fuzzy logic, with the anticipation that message loss will be decreased.

## References

[1]  Akyildiz, I.F., Su, W., Sankarasubramaniam, Y., Cayirci, E.: Wireless sensor networks: a survey, Computer Networks. 38, 393-422 (2002).

[2]  Raquel, A.F.M, Antonio, A.F.L, Badri, N.: The distinctive design characteristic of a wireless sensor network: the energy map. Computer Communications 27, 935–945 (2004).

[3]  Anastasi, G., Conti, M., Francesco, M.D., Passarella, A.: Energy conservation in wireless sensor networks: A survey, Ad Hoc Networks. 7, 537–568 (2009).

[4]  Jason, L. H., David, E.C.: MICA: a wireless platform for deeply embedded networks. IEEE Micro. 22, 12-24 (2002).

[5]  Adel Gaafar, A.E., Hussein, A.E., Salwa, E.R., Magdy, M.I.: An Energy Aware WSN Geographic Routing Protocol, Universal Journal of Computer Science and Engineering Technology 1 (2), 105-111 (2010).

[6]  Lo, B.P.L., Yang G-Z., Key technical challenges and current implementations of body sensor networks, in: Proceedings of the Second International Workshop Wearable and implantable Body Sensor Networks (BSN'2005), London, UK, April 2005.

[7]  Werner-Allen, G., Lorincz, K., Ruiz, M. et al.: Deploying a wireless sensor network on an active volcano, Internet Computing, IEEE 10 18-25 (2006).

[8]  Y. Liu, W.K.G. Seah, A priority-based multi-path routing protocol for sensor networks, in: Proceedings of the 15th IEEE international Symposium on Personal, Indoor and mobile Radio Communications (PIMPC'04), September 2004, pp.216-220.

[9]   Merrett, G.V., B.M. A`l-Hashimi, N.M., White, N.R. Harris, Information managed wireless sensor networks with energy aware nodes, in: Proceedings of the 2005 NSTI Nanotechnology Conference and Trade Show (Nano Tech'05), Anaheim, CA, 2005, pp. 367-370.

[10]  Merrett, G.V., A`l-Hashimi, N.M., White, N.R.: Energy managed reporting for wireless sensor networks, Sensors and Actuators A 142, 379-389 (2008).

[11]  Mingming, L., Jie, W., Mihaela, C., Minglu, L., Energy-Efficient Connected Coverage of Discrete Targets in Wireless Sensor Networks, International Journal of Ad Hoc and Ubiquitous Computing4, 137 - 147 (2009).

[12]  Song, C., Liu, M., Cao, J., Zheng, Y., Gong, H., Chen, G.: Maximizing network lifetime based on transmission range adjustment in wireless sensor networks, Computer Communications 32, 1316–1325 (2009).

[13]  Busse, M., Haenselmann, T., Effelsberg, W.: Energy-efficient forwarding in wireless sensor Networks, Pervasive and Mobile Computing 4, 3–32 (2008).

[14]  Abdullahi, A.I., Muhammed, S.: Energy-Aware Transmission Scheme for Wireless Sensor Networks: The Third International Conference on Wireless & Mobile Networks. Ankara,           Turkey,           June           2011. http://www.purplemath.com/modules/distform.htm.

# 24

# Performance assessment of a downlink two-layer spreading encoded COMP MIMO OFDM system

**Md. Mainul Islam Mamun[1], Joarder Jafor Sadique[2, *], Shaikh Enayet Ullah[1]**

[1]Dept. of Applied Physics and Electronic Engineering, Rajshahi University, Rajshahi, Bangladesh
[2]Dept. of Electrical and Electronic Engineering (EEE), University of Information Technology and Sciences (UITS), Dhaka, Bangladesh

**Email address:**
mainul_apee@yahoo.com (M. I. Mamun), joarderjafor@yahoo.com (J. J. Sadique), enayet67@yahoo.com (S. E. Ullah)

**Abstract:** In this paper, a comprehensive BER performance simulative study has been made on data transmission in a downlink coordinated multipoint wireless communication System. The COMP MIMO OFDM system under investigation implements Turbo and LDPC channel coding, Spatially multiplexing and Space-time block coding (STBC), Doubly Spreading Minimum Mean Square Error (MMSE) and Zero Forcing (ZF) signal detection (Equalizers) schemes under 16PSK and 16QAM digital modulations based on the analysis it is remarkable that the simulated system is highly effective to combat inherent interferences under Rayleigh fading channel and provides robust performance in 16QAM, MMSE channel equalization and spatial multiplexing schemes.

**Keywords:** Coordinated Multipoint (CoMP) Transmission, LDPC and Turbo Coding, Two-Layer Spreading,
Signal Detection Schemes, Bit Error Rate (BER)

## 1. Introduction

With development of physical layer techniques , the data rates of mobile communication services have increased by about 100 times every 6–7 years  and it is predicted  that in 2020, the required data rate will be as large as 100–1000 times  the currently served data rate. The wireless transmission and networking technologies are the essential components of the mobile communication systems. Due to the recent breakthrough in transmission technologies with consideration of constraints of traditional cellular systems in terms of transmit power, complicacy in frequency of handover in high speed mobile environment (350km/h) and cell edge effect for transmission frequencies higher than 2 GHz, cellular communications have entered the era of cooperative communications. In Cooperative communication system, various types of cooperative schemes such as relay, DAS, multicellular coordination, Group Cell, Coordinated Multiple Point transmission and reception (CoMP) are used [1].

Cooperative communications have recently been migrated to one of state-of-the-art features of the 3GPP LTE-Advanced (LTE-A) system. In LTE-Advanced system, single carrier frequency division multiple access (SC-FDMA) has been adopted in the uplink communication and orthogonal frequency division multiple access (OFDMA) has been adopt in the downlink communication. In such system, base-station (BS) cooperative transmission under CoMP cooperative transmission scheme has been widely recognized as a promising technique to enhance throughput by avoiding intercell interference (ICI), particularly for cell-edge users. In January 2009, CoMP and Cooperative Relay based trial networks have been deployed in the campus of Beijing University of Posts and Telecommunication. [2, 3]

Orthogonal Frequency-Division Multiplexing (OFDM) has emerged as a successful air-interface technique. OFDM techniques are also known as Discrete Multi-Tone (DMT)  transmissions and are employed in the American National Standards Institute's (ANSI's) Asymmetric Digital Subscriber Line (ADSL), High-bit-rate Digital Subscriber Line (HDSL)  and Very-high-speed Digital Subscriber Line (VDSL)  standards as well as in the European Telecommunication Standard Institute's (ETSI's) VDSL applications. In wireless scenarios, OFDM has been advocated by many European standards such as Digital Audio Broadcasting (DAB), Digital Video Broadcasting for Terrestrial television (DVB-T), Digital Video Broadcasting for Handheld terminals (DVB-H), Wireless Local Area Networks (WLANs) and Broadband Radio Access Networks (BRANs). MIMO OFDM is the key technology for various cellular communications such as 3GPP-LTE, Mobile WiMAX and IMT-Advanced. The quality of a wireless link viz. transmission rate, transmission range and transmission reliability can be improved using MIMO aided OFDM technology [4, 5]. In this present study, MIMO-aided OFDM radio interface technology in CoMP transmission based wireless network. In Figure 1, a scenario of joint processing (JP) CoMP transmission based MIMO wireless network has been shown where it is seen that a single mobile user (user equipment, UE) is receiving identical spatially demultiplexed complex signals transmitted simultaneously from seven different geographically located multi antenna supported macro base stations (eNBs).

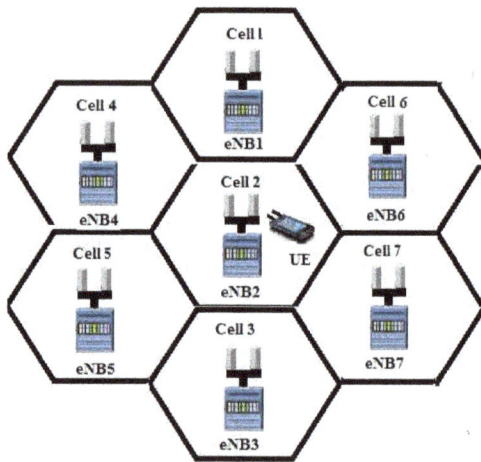

*Figure 1. Downlink Coordinated Multiple-Point Transmission (CoMP) Scenario for homogeneous macro network. User unit (UE) in Cell 2 receives data from seven multi antenna supported macro base stations (eNBs.)*

## 2. Signal Processing Schemes

In this paper, various signal processing schemes have been implemented. A brief overview of such schemes is given below in concise form.

### 2.1. LDPC Coding and Decoding

An low-density parity-check (LDPC) code invented by Gallacher, is a (n,k) linear block code of rate R= k/n. It can be defined in terms of (n-k) × n parity-check matrix [H]=[ $h_1$, $h_2$, $h_3$ ....$h_n$ ]. Each entry $h_{ij}$ of matrix [H] is an element of finite field of GF(2) viz. 0 or 1. The parity-check matrix contains only a few 1's in comparison to 0's (i.e., sparse matrix). In this present study, a ½.-rated irregular LDPC code with code length 1024 has been used. Its parity-check matrix [H] has a dimension of 512 × 1024 and it is formed from a concatenation of two matrices [A] and [P]([H]=[A]|[P]), each has a dimension of 512 × 512. The columns of the parity-check matrix [H]is rearranged/reordered to produce a

new parity-check matrix [$H_{new}$]. With reordered matrix elements, the matrix [A] becomes non-singular and it is further processed to undergo LU decomposition. The redundant or parity bits sequence [p]has been considered to be obtained from the frame based input binary data sequence [u]=[$u_1u_2u_3u_4$.......$u_{512}$]$^T$ and three matrices [P](of [$H_{new}$]),[L] and [U]using the following Matlab notation :

$$p = mod(U\backslash(L\backslash z), 2); where, z = mod(P*u , 2);$$

The LDPC encoded 1024× 1 sized frame based binary data sequence [c] is formulated from concatenation of parity check bit p and information bit u as:

$$[c]=[p;u] \qquad (1)$$

The first 512 bits of the codeword matrix [c] are the parity bits and the last 512 bits are the information bits. In Log Domain Sum-Product LDPC decoding Algorithm, data retrieval is made with adaptation of an iterative approach. The binary data sequence c is converted into another data sequence x with its each sampled value=+1(-1) when element of c=0 (1).

The received signal vector r (=x+n) is contaminated with the independent white Gaussian noise vector n of variance, $\sigma^2$

Initially, four key parametric values are set as follows:
(i)    A parametric matrix [$Lci$] of dimension 1024×1 is set as:

$$[Lci] = -4r / \sigma^2 \qquad (2)$$

In row wise iterative decoding, it is assumed that B is a 512× 1024 sized matrix with its each row containing all elements of matrix [LCi], the element wise product of matrix B and modified parity-check matrix [$H_{new}$] is given by:
[LQIJ]= [$H_{new}$][B] with its element $lq_{ij}$(i=1,2,3.....512; j=1,2,3 .....1024).

(ii)   $\alpha_{ij} \triangleq$ sign($lq_{ij}$);  (iii)  $\beta_{ij} \triangleq$ |$lq_{ij}$|  and
(iv)  $\pi\beta_{ij}$=log$_e$[ (exp($\beta_{ij}$+1))/(exp($\beta_{ij}$-1))]

If $kk_i$(i=1,2,3....512) is the number of non zero element in matrix [$H_{new}$] at its different row i.

The [SMIJ] is a 512×1 matrix containing its elements obtained from row wise summation of all non zero elements in matrix [LQIJ]. For each row wise non zero elements in matrix [LQIJ], a new value is computed as:
NWSM$_{im}$=SMIJ$_i$-LQIJ$_{im}$;  i=1, 2,....512, m is the identification number(=1,2,....kk) viz, first, second, third..... non zero element in each row of matrix [LQIJ].

To avoid division by zero/very small value , its threshold value is set at $1\times10^{-20}$ which implies that if the computed value of NWSM$_{im}$ is found to be below this level, it would be threshold value. From values of NWSM$_{im}$,another logarithm term is computed as:

LGSM$_{im}$=log$_e$ [(exp (NWSM$_{im}$+1))/(exp(NWSM$_{im}$−1))] (3)

IfALIJ$_{im}$ containsnon zero elements in each row of matrix [$\alpha_{ij}$], the previously considered matrix [LRJI] =$0_{512X1024}$ would be updated through replacing 0's at the desired locations with

LRJI$_{im}$=ALIJ$_{im}$×LGSM$_{im}$ (i=1,2,....512), m is the identification number (=1,2,....kk).

For a matrix [SUMJI] with dimension 1024× 1 with its each element computed from summation of all non-zero elements column wise in matrix [[LRJI], the updated matrix is

$$[LQi]=[Lci]+[SUMJI] \qquad (4)$$

If element of matrix [LQi] < 0, it indicates 1 and if the element of matrix [LQi] > 0, it indicates 0, the decoded bit sequence b contains merely 1024 binary bits(0/1). Its first 512 bits are the parity and the rest 512 are the retrieved bits [6, 7].

## 2.2. Turbo Coding and Decoding

Turbo codes are formed by concatenating in parallel two recursive systematic convolutional (RSC) codes separated by an interleaver. The conventional convolutional code is constructed in a feed-forward fashion; that is, the encoder consists of no feedback. In contrast, its RSC equivalence involves feedback in the encoding process. Apparently, the turbo code is a systematic code. Its coding rate is $\frac{1}{3}$, that is, for every input bit, the Turbo encoder produces three code bits. In maximum a posteriori (MAP) turbo decoding, the transmitted message bits can be retrieved iteratively through computation of their log likelihood ratio (LLR).

Let $\bar{c} = c_0, c_1, c_2, c_3 \ldots \ldots \ldots c_{N-1}$ be a coded sequence produced by the rate 1/2 RSC encoder and $\bar{r} = r_0, r_1, r_2, r_3 \ldots \ldots \ldots r_{N-1}$ be the noisy received sequence where the code word is $c_k = \left( c_k^{(1)} \; c_k^{(2)} \right)$ with the first bit being the message bit and the second bit being the parity bit. The corresponding received word is-

$$r_k = \left( r_k^{(1)} \; r_k^{(2)} \right) \qquad (5)$$

The coded bit 0/1 is converted to a value +1/-1. The maximum a posteriori (MAP) decoding is carried out as:

$$c_k^{(1)} = \begin{cases} +1, if & P(c_k^{(1)} = +1|\bar{r}) \geq P(c_k^{(1)} = -1|\bar{r}) \\ -1, if & P(c_k^{(1)} = +1|\bar{r}) < P(c_k^{(1)} = -1|\bar{r}) \end{cases} \qquad (6)$$
$$(i = 0,1,2...N-1)$$

A posteriori log likelihood ratio (LLR) of $c_k^{(1)}$ is given by

$$L\left( c_k^{(1)} \right) \triangleq \ln\left[ \frac{P(c_k^{(1)} = +1|\bar{r})}{P(c_k^{(1)} = -1|\bar{r})} \right] \qquad (7)$$

The MAP decoding rule in Equation (6) can be presented alternatively as:

$$c_k^{(1)} = sign\left[ L\left( c_k^{(1)} \; |\bar{r} \right) \right] \qquad (8)$$

The magnitude of LLR, $| L(c_k^{(1)}|\bar{r}) |$ measures the likelihood of $c_k^{(1)} = +1$ or $c_k^{(1)} = -1$. The LLR can be expressed as a function of the probability $P(c_k^{(1)} = +1|\bar{r})$

$$L\left( c_k^{(1)} \right) = \ln\left[ \frac{P(c_k^{(1)} = +1|\bar{r})}{P(c_k^{(1)} = -1|\bar{r})} \right]$$
$$= \ln\left[ \frac{P(c_k^{(1)} = +1|\bar{r})}{1 - P(c_k^{(1)} = +1|\bar{r})} \right] \qquad (9)$$

### 2.3. Signal Detection Schemes

In Spatially multiplexed MIMO (SM-MIMO) wireless communication system, the transmitted signal Xs, Received signal Y, channel coefficient matrix H and addititive white Gaussian noise term N can be written as the following signal model:

$$Y=HXs+N \qquad (10)$$

With implementation of Signal Detection/ channel equalization techniques, the transmitted signal is recovered/detected using the following relation:

$$\tilde{X}s_{det\,ected} = WY \qquad (11)$$

Where, W is the assigned weight matrix for different channel equalization schemes.

In Minimum mean square error (MMSE) channel equalization scheme, the MMSE weight matrix in terms of equivalent channel matrix H and noise variance $\sigma_n^2$ is given by-

$$W_{MMSE} = (H^HH + \sigma_n^2I)^{-1}H^H \qquad (12)$$

In Zero-Forcing (ZF) scheme, the ZF weight matrix is given by [5]-

$$W_{ZF} = (H^HH)^{-1}H^H \qquad (13)$$

## 3. System Model

A simulated single user 2 x 2 spatially multiplexed and FEC encoded Two-Layer Spreading aided COMP MIMO OFDM wireless communication system is depicted in Figure 2. This is merely a comprehensive block diagram where signal is transmitted from a single cell to a mobile user unit although the present study has taken into consideration of downlink simultaneous data transmission from seven cells. However, in such a communication system, a single user is receiving synthetically generated binary bit stream from the Base station. In transmitting section, we consider that a binary data sequence D of length K are channel encoded using $\frac{1}{3}$-rated Turbo coding / ½-rated LDPC. The channel encoded binary data $D_{FEC}$ is interleaved and mapped into digitally modulated symbols with its size L depending upon the order of modulation considered and copied. The number of digitally modulated symbols is increased sixty four times in copying section (as the processing gain of the Walsh Hadamard codes is sixty four) and multiplied with Walsh Hadamard (W-H) spreading codes. The spreaded data symbol vector X are spatially demultiplexed/space time block encoded with implementation of Alamouti scheme to produce two complex data streams $X_1$ and $X_2$. The data of each stream are rearranged into $M(=32*L/(N_c))$ number of blocks. In each block, $N_c$ number of modulated source symbols, $S_{m,q}(n)$, n=0,1,2,3..................$N_c$-1 are processed with serial to parallel converter (S/P). The time domain signal OFDM signal using inverse Fast Fourier Transform (IFFT) can be written as

$$x_{m,q}(t) = \frac{1}{N_c} \sum_{n=0}^{N_c-1} S_{m,q}(n) e^{j(2\pi f_n t)} \tag{14}$$

$$0 \le t \le T_s$$

where, m and q are the transmitting antenna and block identifiers ; m=1,2 and q=1,2,3.....M, $T_s(=N_cT_d)$ is the OFDM symbol duration, $T_d$ is the source symbol duration and $N_c$ is the total number of sub-carriers(=1024). In each of the two data stream, the complex data symbols are again multiplied with Walsh Hadamard (W-H) spreading codes after OFDM modulation. However, in each OFDM block, each sub-carrier is used for modulating each source symbol. The subcarrier spacing is assigned to a value of $1/T_s$ and $N_c$ sub-carrier frequencies are located at

$$f_n = \frac{n}{T_s},$$
$$n = 0,1,2,3 ............ \quad ...... \quad N_c - 1 \tag{15}$$

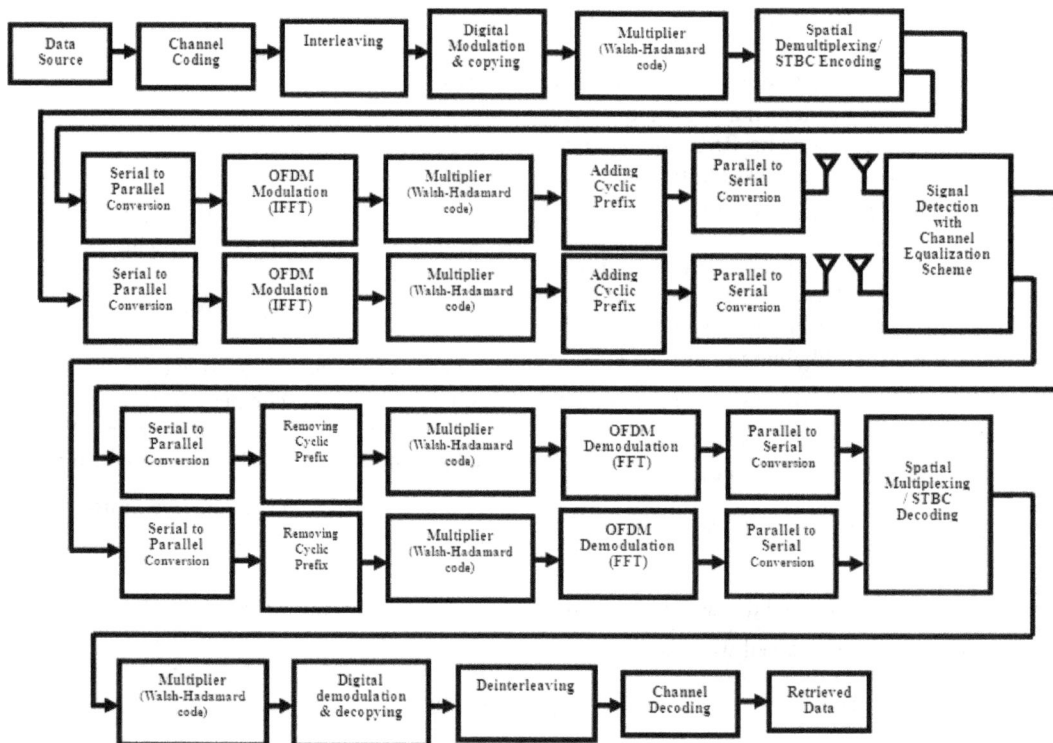

**Figure 2.** *Block diagram of a Downlink Two-Layer Spreading Encoded COMP MIMO OFOM wireless communication system*

The sampled sequence of the complex envelope $x_m(t)$ of an OFDM symbol presented in Equation (14) with a rate $1/T_d$ can be written as

$$x_{m,q}(v) = \frac{1}{N_c} \sum_{n=0}^{N_c-1} S_{m,q}(n) e^{j2\pi n v/N_c}, \tag{16}$$
$$v = 0,1,2,3 .......N_c - 1$$

We assume that the OFDM symbol duration $T_s$ is large as compared to the duration of the impulse response of the channel $\tau_{max}$ and ISI is reduced significantly. In order to avoid completely the effects of ISI and to maintain the

orthogonally between the signals on the sub-carriers for avoiding ICI, a guard interval of duration $T_g \ge \tau_{max}$ is inserted between adjacent OFDM symbols. The guard interval is a cyclic extension of each OFDM symbol, which is obtained by extending the duration of an OFDM symbol to

$$T_s' = T_g + T_s \tag{17}$$

The discrete length of the guard interval is-

$$L_g \geq \left\lceil \frac{\tau_{max} N_c}{T_s} \right\rceil \qquad (18)$$

The sampled sequence with cyclic extended guard interval results in the following expression [9].

$$x_{m,q}(v) = \frac{1}{N_c} \sum_{n=0}^{N_c-1} S_{m,q}(n) \; e^{j2\pi nv/N_c} \qquad (19)$$

$$v = -L_g \ldots\ldots\ldots\ldots N_c - 1$$

Considering applicability of signal model presented in Equation (19) for all blocks of signals transmitted from all antennas, we can write the transmitted signal vector Xs in terms of its two signal vector components $Xs_1$ and $Xs_2$ as:

$$Xs = \begin{bmatrix} Xs_1 \\ Xs_2 \end{bmatrix}$$
$$= \begin{bmatrix} x_{1,1}(v) & x_{1,2}(v) & \cdots & x_{1,M-1}(v) & x_{1,M}(v) \\ x_{2,1}(v) & x_{2,2}(v) & \cdots & x_{2,M-1}(v) & x_{2,M}(v) \end{bmatrix} \qquad (20)$$

If $H_1, H_2 \ldots\ldots H_7$ are considered to be the 2×2 channel matrices for the base stations to the user unit and $n_1, n_2 \ldots\ldots\ldots\ldots n_7$ are the corresponding zero mean circularly symmetric complex Gaussian noises, Equation (5) can be written under such special case as

$$Y = (H_1 + H_2 + H_3 + H_4 + H_5 + H_6 + H_7)Xs$$
$$+ (n_1 + n_2 + n_3 + n_4 + n_5 + n_6 + n_7) \qquad (21)$$

Equation (21) can be written as in terms of equivalent channel matrix H and Equivalent noise N as signal model presented in Equation (10).

With implementation of channel equalization techniques, the transmitted signals are recovered/detected using signal models presented in Equation (11) through Equation (13)

The detected signals can be represented in matrix form as

$$\widetilde{Xs}_{detected} = \begin{bmatrix} \widetilde{Xs}_1 \\ \widetilde{Xs}_2 \end{bmatrix} = \begin{bmatrix} \widetilde{x}_{1,1}(v) & \widetilde{x}_{1,2}(v) & \cdots & \widetilde{x}_{1,M-1}(v) & \widetilde{x}_{1,M}(v) \\ \widetilde{x}_{2,1}(v) & \widetilde{x}_{2,2}(v) & \cdots & \widetilde{x}_{2,M-1}(v) & \widetilde{x}_{2,M}(v) \end{bmatrix} \qquad (22)$$

In Equation (22), first $L_g$ samples of each element of the two rows are induced with ISI and these $L_g$ samples are removed from each cyclically extended OFDM block and Walsh Hadamard (W-H) spreading codes prior to multi-carrier demodulation with exploitation of Fast Fourier Transform (FFT). The FFT operated OFDM blocks are undergone parallel to serial conversion and fed into spatial multiplexer/STBC decoder. Its output is multiplied with Walsh-Hadamard codes. The de-spreaded complex symbols are decopied, demodulated, de-interleaved and turbo/LDPC decoded to recover the transmitted data [10,11]

## 4. Results and Discussion

In this section, we present a series of simulation results to illustrate the significant impact of system performance in terms of BER in Coordinated Multiple Point transmission and reception. The simulation study has been made using MATLAB 2012a based on the parameters given in Table 1. It is assumed that the channel state information (CSI) is available at the receiver and the fading process is approximately constant during whole transmission time from each macro base station to user unit. The graphical illustrations presented in Figure 3 through Figure 6.

*Table 1. Summary of the Simulated Model Parameters*

| | |
|---|---|
| No. of synthetically generated binary data used | 8192 and 8188 |
| Communication System Type | Cooperative based on Coordinated Multiple Point transmission and reception (CoMP) |
| No. of cells in a group | 7 |
| Channel Coding | 1/3-rated Turbo and ½-rated LDPC |
| Digital modulation | 16QAM and 16PSK |
| No. of subcarriers (FFT Size) | 2048 |
| CP length | 205 symbols |
| Decision method adopted in LDPC decoding | Soft decision with log-likelihood ratios (LLR) computation |
| Method used in Turbo decoding | A-posteriori probability (APP) |
| No of iterations considered in LDPC and Turbo decoding | 10 |
| Antenna Configuration (User Equipment and Base station) | (2,2) |
| Signal Detection Scheme | Minimum Mean square error(MMSE) and Zero Forcing (ZF), |
| Channel | AWGN and Rayleigh fading |
| Signal to noise ratio (SNR) | -5 to 5 dB |

In Figure 3, it is quite evident that the spatially multiplexed system is incapable of showing its performance acceptability under LDPC channel coding, 16PSK digital modulation and ZF channel equalization schemes. At a quite hostile environment (-3dB SNR) where noise power is greater that signal power by 3dB, the system shows a performance enhancement of 9.04 dB in case of MMSE receiver as compared to ZF receiver with both 16PSK digital modulation. In Figure 3, it is also observable that at a typically target 1% ($10^{-2}$)BER, the MMSE linear equalizer and ZF linear equalizer with both 16QAM digital modulation require approximately 1.6 dB and 2.0 dB higher SNR respectively as compared to MMSE linear equalizer with 16PSK. In Figure 4, the BER results are shown for representing STBC encoded system performance. With MMSE linear equalizer receiver, it is noticeable that the noise enhancement is significant and the COMP scheme is incapable of providing satisfactory performance with LDPC and 16PSK schemes. With ZF linear equalizer receiver, the system performance is well defined and quite satisfactory. At a typically assumed SNR value of -3dB, the estimated BER values are 0.1377 and 0.2802 in case of ZF and MMSE linear equalizer receiver with both 16PSK digital modulations which is indicative of performance improvement by3.09 dB. In Figure 5, Turbo encoded system performances are well defined. At low SNR regime with 16QAM digital modulation, both MMSE and ZF linear equalizer receiver shows almost identical system performance. At SNR value of -3dB, the simulated system is found to have performance improvement of 7.51dB in case of MMSE linear equalizer receiver with 16QAM as compared to ZF linear equalizer receiver with 16PSK. In Figure 5, it is also observable that at a typically target 1%($10^{-2}$)BER, the MMSE and ZF linear equalizer with both 16PSK digital modulation require approximately 0.6 dB and 2.9 dB  higher SNR respectively as compared to MMSE linear equalizer  with 16QAM. On crucial examination of the simulation results presented in Figure 3 through Figure 6, it is observable that the COMP aided simulated system shows comparatively better performance in Turbo channel coding as compared to LDPC. In Figure 6, the turbo encoded simulated system shows robust performance with MMSE linear equalizer receiver and 16QAM and comparatively worst performance with ZF linear equalizer receiver and 16PSK. At -3 dB SNR value, the estimated BER values are 0.0250 and 0.1020 in case of MMSE with 16QAM and ZF with 16PSK which implies a system performance improvement of 6.11 dB. In Figure 6, it is also remarkable that at a typically target 1%BER, the  ZF linear equalizer with 16QAM,   MMSE  linear equalizer with 16PSK and ZF linear equalizer with 16PSK digital modulation require approximately 0.5 dB, 1.1 dB  and  1.4 dB higher SNR respectively as compared to MMSE linear equalizer with 16QAM.

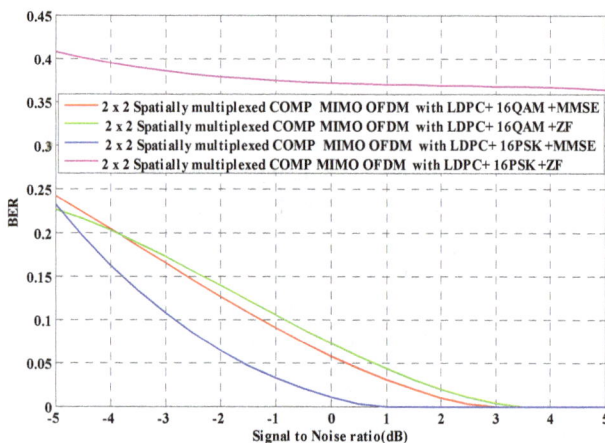

**Figure 3.** BER performance of the system under deployment of LDPC channel coding, Spatial multiplexing various channel equalization and digital modulation schemes.

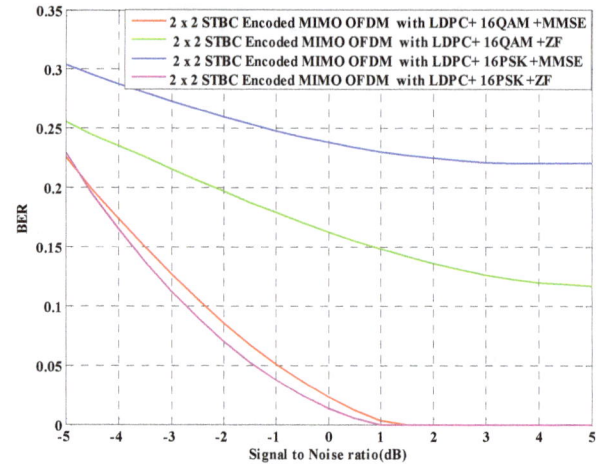

**Figure 4.** BER performance of the system under deployment of LDPC channel coding, Space-Time Block Coding, various channel equalization and digital modulation schemes

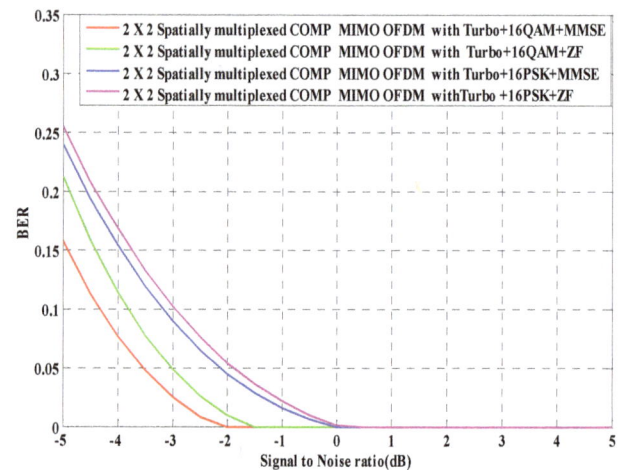

**Figure 5.** BER performance of the system under deployment of Turbo channel coding, Space-Time Block coding, various channel equalization and digital modulation schemes.

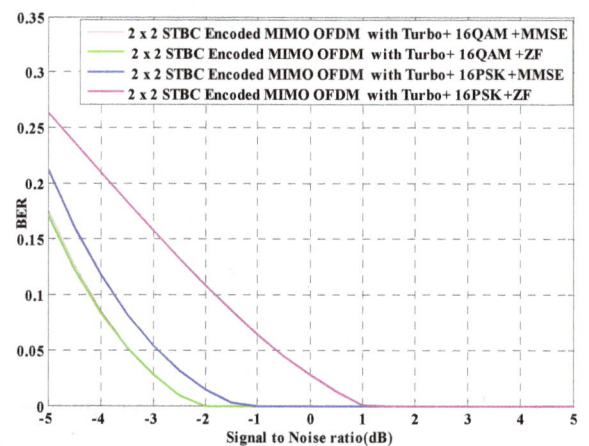

**Figure 6.** BER performance of the system under deployment of Turbo channel coding, Spatial multiplexing, various channel equalization and digital modulation schemes.

# 5. Conclusion

In this paper, we have presented simulation results concerning the adaptation of various signals detection and two-layer spreading schemes in a FEC encoded COMP MIMO OFDM wireless communication system. A range of system performance results under the regime of low SNR highlights the impact of Coordinated Multiple Point transmission and reception scheme on data transmission. In the context of system performance, it can be concluded that the spatially multiplexed and Turbo encoded COMP MIMO OFDM wireless communication system with Minimum Mean Square Error (MMSE) signal detection and 16QAM digital modulation schemes provides robust system performance.

# References

[1] XiaofengTao ,Qimei Cui XiaodongXu and Ping Zhang: "Group Cell Architecture for Cooperative Communications", Springer Publisher, New York, 2012

[2] Xiaofeng Tao, XiaodongXu, and Qimei Cui: "An Overview of Cooperative Communications", IEEE Communications Magazine, pp.65-71, 2012

[3] Guillaume de la Roche, Andr´esAlay´on Glazunov and Ben Allen: "LTE-advanced and next generation wireless networks channel modelling and propagation", John Wiley and Sons Ltd, United Kingdom, 2013.

[4] LajosHanzo, Yosef (Jos) Akhtman , Li Wang and Ming Jiang:: "MIMO-OFDM for LTE, Wi-Fi and WiMAX," John Wiley and Sons Ltd, United Kingdom, 2011.

[5] Yong Soo Cho, Jaekwon Kim, Won Young Yang, Chung G. Kang: "MIMO-OFDM Wireless Communications with MATLAB", John Wiley and Sons (Asia) Pte Limited, Singapore, 2010.

[6] Christian B. Schlegel and Lance C. Perez,: "Trellis and turbo coding", John Wiley and Sons, Inc., publication, Canada, 2004.

[7] BagawanSewuNugroho, https://sites.google.com/site/bsnugroho/ldpc

[8] Yuan Jiang: "A Practical Guide to Error-Control Coding Using MATLAB", Jiang Artech House, Boston, USA, 2010.

[9] K. Fazel and S. Kaiser: "Multi-Carrier and Spread Spectrum Systems From OFDM and MC-CDMA to LTE and WiMAX", John Wiley and Sons, Publication Ltd, United Kingdom, 2008.

[10] Goldsmith, Andrea: "Wireless Communications", First Edition, Cambridge University Press, United Kingdom, 2005.

[11] L. J. Cimini, Jr: "Analysis and simulation of a digital mobile channel using orthogonal frequency division multiplexing", IEEE Transaction on Communication, vol. COM-33, pp. 665–675, 1985.

# 25

# The RCS of a resistive rectangular patch antenna in a substrate-superstrate geometry

**Amel Boufrioua**

Electronics Department, Technological Sciences Faculty, University Constantine 1, Ain El Bey Road, 25000, Constantine, Algeria

**Email address:**

boufrioua_amel@yahoo.fr

**Abstract:** The scattering radar cross section (RCS) and the resonant frequency problem of a superstrate loaded resistive rectangular microstrip patch which is printed on isotropic or uniaxial anisotropic substrate are investigated, where an accurate design based on the moment method technique is developed. The choice of two types of basis functions (entire domain and roof top functions) is illustrated to develop the unknown currents on the patch. The accuracy of the computed technique is presented and compared with other computed results.

**Keywords:** Superstrate, Anisotropy, Resistive, Radiation, Patch, Antenna

## 1. Introduction

Microstripantennas are now extensively used in various communication systems due to their compactness, economical efficiency, light weight, low profile and conformability to any structure. However, microstrip patch antenna is limited by its inherent narrow bandwidth. Therefore, this problem has been addressed by researchers and many configurations have been proposed for bandwidth enhancement [1-6]. The study of the superstrate layer is of interest, it can affect the performance of printed circuits and antennas and may prove beneficial or detrimental to the radiation characteristics, depending on the thicknesses of the substrate and superstrate layer, as well as relative dielectric constants [1]. In this paper we extend our study [2, 3] to the case of a superstrate-loaded resistive rectangular microstrip structure, where the superstrate layer loaded on the microstrip structure is often used to protect printed circuit antennas from environmental hazards, or may be naturally formed during flight or severe weather conditions [1, 4]. The full-wave moment method has been applied extensively and is now a standard approach for analysis of microstrip geometry [1-4], [7].

In this paper the integral equation includes a superstrate resistive boundary condition on the surface of the patch and the effects of anisotropic substrate are developed. It is worth noting that the effects of non-zero surface resistance and the uniaxial anisotropy on the scattering properties of a superstrate loaded rectangular microstrip structure has not

yet been treated. A novel proposed structure pertaining to this case will be presented in this paper.

## 2. Theory

The geometry for the superstrate-loaded resistive rectangular patch antenna is shown in Figure 1. The resistive patch with length $a$ and width $b$ printed on the grounded substrate, which has a uniform thickness of $h$ and having a relative permittivity $\varepsilon_{r1}$ (region 1). The superstrate of thickness $d$ with relative permittivity $\varepsilon_{r2}$ is obtained by depositing a dielectric layer on the top of the substrate (region 2).

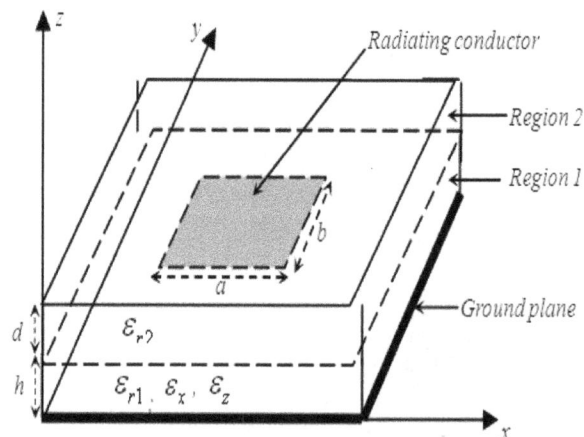

**Figure 1.** Resistive rectangular patch insubstrate-superstrate geometry

In the case of a uniaxially anisotropic substrate $\varepsilon_{r1}$ can be represented by a tensor or dyadic of this form [2]:

$$\overline{\varepsilon}_{r1} = \varepsilon_0 . \mathrm{diag}[\varepsilon_x, \varepsilon_x, \varepsilon_z] \qquad (1)$$

$\varepsilon_0$ is the free-space permittivity; all the dielectric materials are assumed to be nonmagnetic with permeability $\mu_0$.

$\varepsilon_z$ is the relative permittivity in the direction of the optical axis.

$\varepsilon_x$ is the relative permittivity in the direction perpendicular to the optical axis.

The study is performed by using a full wave analysis and Galerkin's moment method to examine the scattering properties of a superstrate loaded rectangular patch antenna with a surface resistance, in which we extend our study [2, 3] to the case of this proposed geometry.

The principal modifications are done especially at the Green's functions and at the resistance surface. We have included the effect of the superstrate in the Green's function formulation as [4] which are efficiently determined by the (TM, TE) representation [3].

$$\overline{G}(\mathbf{k}_s) = \begin{bmatrix} G^{TM} & 0 \\ 0 & G^{TE} \end{bmatrix}$$

$$= -\frac{i}{\omega\varepsilon_0} \begin{bmatrix} \dfrac{k_1 D_m}{T_m} & 0 \\ 0 & \dfrac{k_0^2 D_e}{T_e} \end{bmatrix} \cdot \sin k_1 h \qquad (2)$$

Where

$$T_m = \cos k_2 d [\varepsilon_1 k_3 \cos k_1 h - ik_1 \sin k_1 h]$$
$$+ i \sin k_2 d \left[ \frac{\varepsilon_1}{\varepsilon_2} k_2 \cos k_1 h + i \frac{\varepsilon_2 k_1 k_3}{k_2} \sin k_1 h \right] \qquad (3)$$

$$T_e = \cos k_2 d [k_1 \cos k_1 h - ik_3 \sin k_1 h]$$
$$+ i \sin k_2 d \left[ \frac{k_1 k_3}{k_2} \cos k_1 h + ik_2 \sin k_1 h \right] \qquad (4)$$

$$D_m = k_3 \cos k_2 d + i \frac{k_2}{\varepsilon_2} \sin k_2 d \qquad (5)$$

$$D_e = \cos k_2 d + i \frac{k_3}{k_2} \sin k_2 d \qquad (6)$$

$$k_i^2 = \varepsilon_i k_0^2 - k_s^2, \qquad i = 1, 2, 3, \varepsilon_3 = 1.0 \qquad (7\text{-a})$$

$$k_s^2 = k_x^2 + k_y^2, \qquad k_0^2 = \omega^2 \mu_0 \varepsilon_0 \qquad (7\text{-b})$$

$\omega$ The angular frequency;

$\mathbf{k}_s$ The transverse wave vector; $\mathbf{k}_s = \overline{\mathbf{x}} k_x + \overline{\mathbf{y}} k_y$ and

$k_s = |\mathbf{k}_s|$ ;

$\overline{\mathbf{x}}, \overline{\mathbf{y}}$ Unit vectors in x and y direction, respectively;

$j = \sqrt{-1}$

An integral equation can be formulated by using this Green's function on a thick dielectric substrate to determine the electric field at any point. The details of the solution of the transformed integral equation are presented in [1-3].

Entire domain sinusoid basis functions and roof top subdomain basis functions are introduced to expand the unknown current on the metal patches of this proposed antenna[2, 3].

Since that we have included the effect of uniaxial anisotropic substrate and the effect of superstrate in the Green's functions, also the effect of the non-zero surface resistance at the resistance matrix and consequently at the impedance matrix the different antenna characteristics can be easily obtained similar to [1-4], [7, 8].

## 3. Numerical Results

The moment method technique with entire domain and roof top basis functions has developed to examine the resonant frequency and the scattering properties of a rectangular patch antenna.

For all our computations the mode that we will be studying is the TM01 mode with the dominant component of the current in the y direction. To simplify the analysis, the antenna feed will not be considered.

To ensure that the computer programs are correct, comparisons are shown in Figure 2 for a perfectly conducting patches of different sizes without dielectric substrates (air) and with no superstrate, the substrate has a thickness h = 0.317 cm. It is important to note that the normalization is with respect to $f_0$ of the magnetic wall cavity. The calculated results for the two sets of basis functions shown in Figure 2 agree very well with experimental results obtained by other authors [7], we found an excellent agreement with a slight shift in the resonant frequency when we use entire domain basis functions compared to the measured results given by [7], on the other hand computations show that the roof top subdomain basis functions provides a significant improvement in the computations time with less iterations in the evaluation of the resonant frequency of a microstrip patch compared to the entire domain sinusoid basis functions [3, 8]. It should be noted that the convergence of the solution was investigated by varying the number of subsections.

As mentioned previously, good convergent solutions are reached by using entire domain basis functions, for this reason, the use of these basis functions has developed to examine the resonant frequency and the scattering properties for the following results.

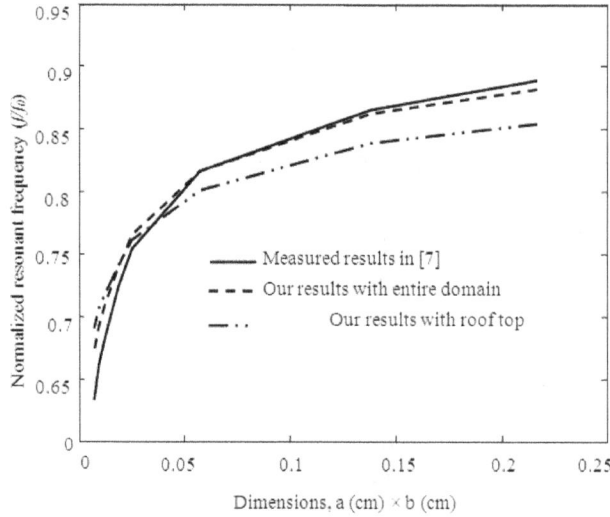

**Figure 2.** *Measured and calculated resonant frequencies versus the dimensions of the patch for a rectangular microstrip antennas with no superstrate.*

The substrate has a relative permittivity of $\varepsilon_{r1}$ =2.35 with a uniform thickness of h=0.1cm and the patch dimensions is 6.0cm x5.0cm.

In Figures 3 and 4, the resonant frequency (the real part of the complex resonant frequency) and the band width versus the superstrate thickness for different dielectric constants of the superstrate are shown. The obtained results show that when the superstrate thickness as well as the superstrate permittivity is increased, the resonant frequency decreases.

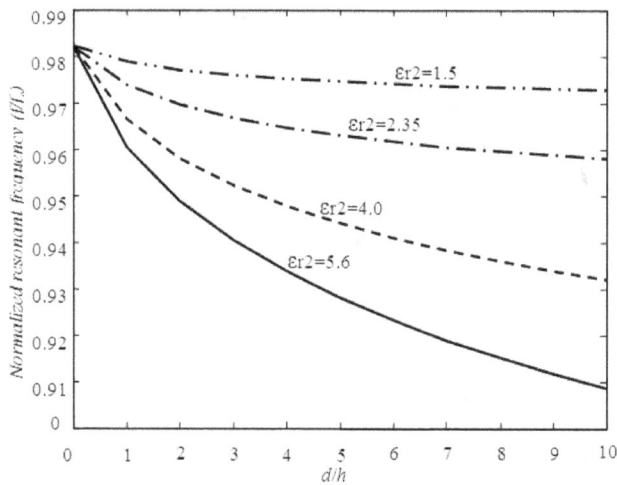

**Figure 3.** *Normalized resonant frequency of a superstrate-loaded rectangular patch versus the superstrate thickness d.*

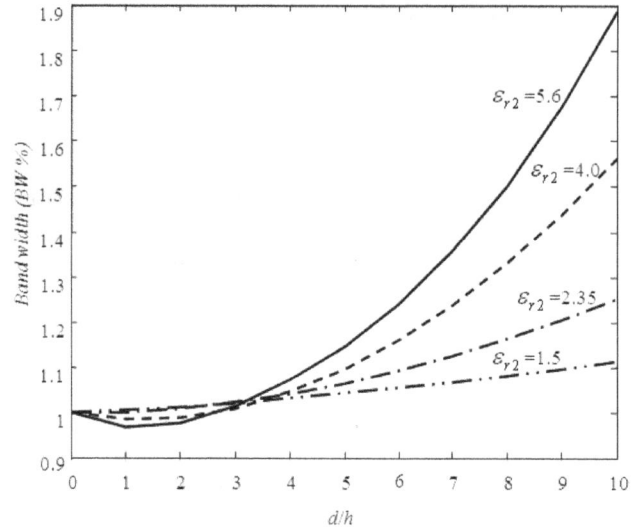

**Figure 4.** *Normalized band width of a superstrate-loaded rectangular patch versus the superstrate thickness d.*

The variation of the band width is very small for d less than about 4h. As the superstrate thickness increases (4h<d) the variation becomes significant for high superstrate permittivities. The obtained results show that the resonant frequency and the band width vary more significantly when the superstrate permittivity is greater than that of the substrate. These behaviors agree very well with those obtained by [4] with slight shifts in frequency and band width between our results and those of [4] are noted. It is worth noting that for these two figures the normalization is with respect to that of the perfectly patch (Rs=0 Ohm) with no superstrate (d=0).

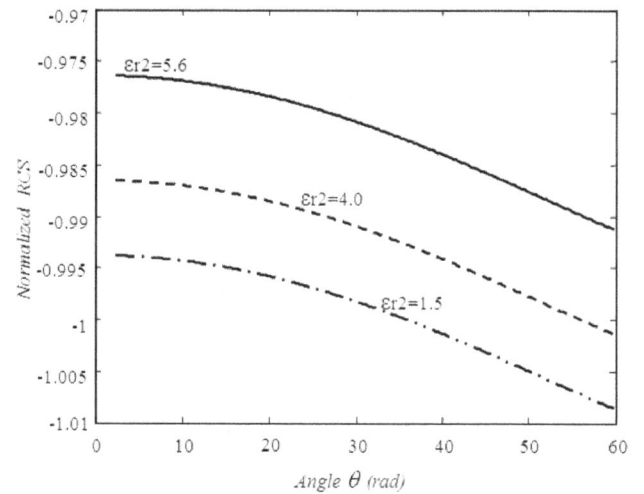

**Figure 5.** *Normalized radar cross section versus angle $\theta$ for different dielectric constants of the superstrate loaded rectangular patch antenna, d=0.1cm at $\phi = 0$ .*

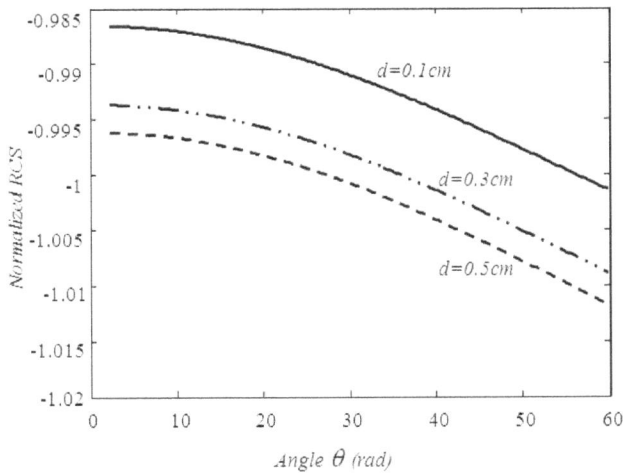

**Figure 6.** *RCS versus angle $\theta$ for various values of thicknesses of the superstrate loaded rectangular patch antenna, $\varepsilon_{r2}$ =4.0.*

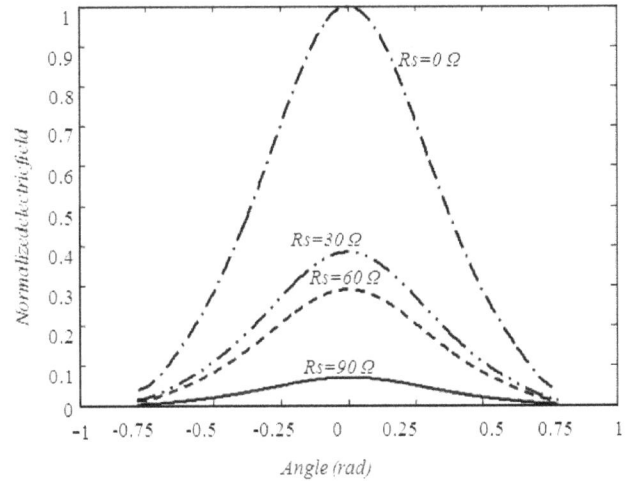

**Figure 8.** *The effect of the surface resistance on the electric field $E_\theta$, at $\phi = 0$.*

Figures 5 and 6 show the normalized scattering radar cross section RCS of a superstrate-loaded rectangular microstrip versus the angle $\theta$ for different dielectric constants and different thicknesses of the superstarate loaded rectangular patch antenna. Results showing RCS reduction are presented for high superstrate thicknesses and low superstrate permettivities.

Figure 7 shows the normalized scattering radar cross section RCS of a superstrate-loaded rectangular microstrip versus the angle $\theta$ for different surface resistance Rs. We observe that when the surface resistance is increased, the level of the radar cross section decreases. Consequently the addition of a resistance on the surface of a microstrip patch antenna has been shown to decrease the scattered energy from the antenna.

For Figures 5, 6 and 7 the normalization is with respect to that of the perfectly patch (Rs=0 Ohm) with no superstrate (d=0).

Figure 8 shows the scattering properties for the $E_\theta$ component of the electric field at $\phi = 0°$ plane displayed as a function of the angle $\theta$ and as a function of surface resistance. It is clear that when the surface resistance on the patch is increased, the level of the components $E_\theta$ decreases consequently. However, it is important to note that our results for the $E_\phi$ component do not change with the surface resistance at $\phi = 0°$.

According to the simulation it is worth noting that the component $E_\phi$ is dominant than the component $E_\theta$ and the total field take the shape of the component $E_\phi$.

Table 1 shows the scattering radar cross section RCS for an imperfectly conducting patch with the surface resistance Rs=30 $\Omega$ compared to a perfectly one and printed on a substrate of thickness h=0.2 cm, where isotropic, positive and negative uniaxial anisotropic substrates are considered. The patch dimensions are: a= 1.5cm, b=1.0 cm. It can be seen clearly that the permittivity $\varepsilon_z$ has a stronger effect on the scattering radar cross section than the permittivity $\varepsilon_x$ for both cases. Also it is noted that the study in this table is done with no superstrate.

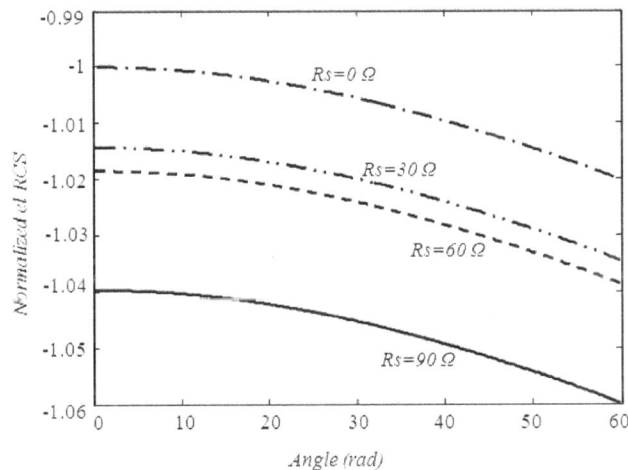

**Figure 7.** *Radar cross section versus angle $\theta$ for different surface resistance Rs of a superstrate loaded rectangular microstrip patch; h=0.2cm, d=0.159cm $\varepsilon_{r1} = \varepsilon_{r2}$ =2.35 at $\phi = 0$.*

**Table 1.** *Effects of the surface resistance on the radar cross section for isotropic, negative and positive uniaxial substrates, $\theta$ =60°, $\phi$ =0°, a=1.5cm, b=1.0cm, h =0.2cm.*

| $\varepsilon_x$ | $\varepsilon_z$ | AR | RCS(dBsm) | |
|---|---|---|---|---|
| | | | Rs($\Omega$)=30 | Rs($\Omega$)=0 |
| 2.32 | 2.32 | 1 | -29.27 | -28.57 |
| 4.64 | 2.32 | 2 | -29.39 | -28.82 |
| 2.32 | 1.16 | 2 | -29.12 | -28.05 |
| 1.16 | 2.32 | 0.5 | -29.25 | -28.68 |
| 2.32 | 4.64 | 0.5 | -29.64 | -29.50 |

# 4. Conclusion

The moment method technique has been developed to examine the complex resonant frequency, the half band width, the radar cross section (RCS) and the radiation of a superstrate loaded resistive rectangular microstrip patch which is printed on isotropic or uniaxial anisotropic substrate with the optical axis normal to the patch. The formulation is carried out in the spectral domain. The choice of roof top subdomain basis functions and the entire domain were illustrated to develop the unknown currents on the patch. The accuracy of the computed technique was presented and compared with other computed results.

# References

[1]  N. G.Alexopoulos, D. R.Jackson, "Fundamental superstrate (cover) effects on printed circuit antennas," IEEE Trans. Antennas Propagat,vol. 32, 1984,pp. 807–816.

[2]  A.Boufrioua, A.Benghalia, "Effects of the resistive patch and the uniaxial anisotropic substrate on the resonant frequency and the scattering radar cross section of a rectangular microstrip antenna" Elsevier, AST, Aerospace Science and Technology, vol. 10, 2006,pp. 217-221.

[3]  A.Boufrioua, "Resistive rectangular patch antenna with uniaxial substrate", In: Antennas: Parameters, Models and Applications, Editor. Albert I. Ferrero: Nova Publishers. New York. 2009, pp. 163-190.

[4]  J-S.Row,andK. L.Wong, "Resonance in a superstrate-loaded rectangular microstrip structure," IEEE Microwave Theory and Techniques,vol. 41, 1993, pp. 1349–1355.

[5]  B-L. Ooi, S.Qin,andM-S.Leong, "Novel design of broadband stacked patch antenna," IEEE Trans. Antennas Propagat, vol. 50, 2002,pp. 1391-1395.

[6]  A. A.Deshmukh, G.Kumar, "Formulation of resonant frequency for compact rectangular microstrip antennas," Microwave and Optical Technology Letters, vol. 49, 2007,pp. 498-501.

[7]  W. C.Chew, Q.Liu, "Resonance frequency of a rectangular microstrip patch," IEEE Trans. Antennas Propagat, vol. 36, 1988,pp. 1045–1056.

[8]  A.Boufrioua, A.Benghalia, "Radiation and resonant frequency of a resistive patch and uniaxial anisotropic substrate with entire domain and roof top functions," Elsevier, EABE Engineering Analysis with Boundary Elements, vol. 32, 2008,pp. 591-596.

# 26

# Analysis of arrays of printed strip dipole antennas and broadband antennas of six series fed strip dipoles

**Arianit Maraj[1], Ruzhdi Sefa[2]**

[1]Faculty of Computer Science, Public University of Prizren, Prizren, Republic of Kosova
[2]Faculty of Electrical and Computer Engineering, University of Prishtina, Prishtina, Republic of Kosova

**Email address:**
arianit.maraj@uni-prizren.com (A. Maraj), ruzhdi.sefa@uni-pr.edu (R. Sefa)

**Abstract:** Dipole antennas are suitable in wireless multi-band technology, because they are simple to design and have excellent radiation properties. Also, it is well known that analysis of series-fed strip dipoles antennas are very hard to do because of the presence of dielectric inhomogeneity. In addition, the strip dipoles are closely spaced and connected through transmission lines that are integral part of the antenna. Therefore, an analysis method that takes into account the presence of dielectric substrate and the effect of connecting lines is presented to accurately design these antennas. In this paper, arrays of printed strip dipoles antennas are analyzed based on moment method in the spectral domain. The double-sided configuration has been selected because it offers several practical advantages. In this paper, further we will complicate the analysis by taking into account broadband antennas comprised of six series fed strip dipoles.

**Keywords:** Dipole Antennas, Moment Method, Stripline, Double-Sided Configuration

## 1. Introduction

Printed dipole antennas are commonly used as radiating elements in wireless communications as they are easy to fabricate, have low cost, and high performance. The antennas are comprised of a number of closely spaced narrow conductive strips printed on electrically thin dielectric substrates. Although the antennas are easy to make and inexpensive, they posses tremendous performance capabilities including high gain, dual polarization, broadband or multi-band operation [1-4]. In addition when these antennas are fed in series they generally produce endfire radiation making them suitable for use with plane or shaped reflectors, a configuration commonly employed in base stations of wireless communications systems [5].

Printed dipole radiators have been popular candidates for phased array antennas that contain many elements because of their suitability for integration with microwave integrated circuit modules [6-8]. These antennas have also been considered for application in far-infrared and millimeter-wave imaging systems [9, 10].

Although printed strip antennas are intrinsically simple, the presence of the dielectric inhomogeneity presents a significant challenge in finding computationally reasonable CAD tool for designing these antennas.

Single strip dipoles printed on electrically thin, low permittivity substrates behave similarly to ordinary dipoles, and they have often been designed based on experimental trial and error [11, 12]. However, when dipoles of different lengths are closely spaced and/or connected with each other with transmission lines, the radiation characteristics depend on a large number of design parameters. An analysis method that can handle three-dimensional structures containing both dielectric and conducting materials should be used to correctly analyze these antennas. However such an analysis is computationally intensive and it provides no easy way to select optimal design parameters. A more efficient full wave analysis would be to consider dipoles printed on an infinite substrate, and apply a moment method in the spectral domain. Such an approach is used here to analyze strip dipoles printed on a thin ungrounded dielectric substrate. Similar techniques have been employed for the analysis of strip dipoles printed on a grounded dielectric substrate [13] and on a dielectric half-space [10].

The procedure for analyzing arrays of printed strip dipoles antennas based a moment method in the spectral domain is presented. The dipoles can be connected to each other.

## 2. Antenna Configuration

Fig. 1 is a schematic drawing of an array of series-fed printed strip dipoles. The antenna consists of several printed strip dipoles of different lengths, with the arms printed on opposite sides of an electrically thin dielectric substrate that are connected through parallel striplines.

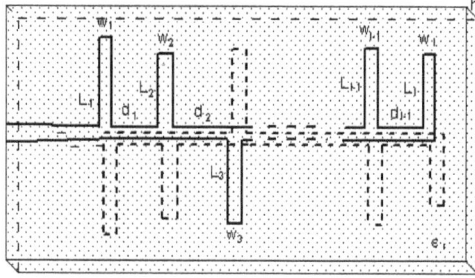

**Figure1.** *Drawing of a series-fed printed strip dipole antenna*

The parallel stripline consists of two broadside-coupled strips; each section may have different widths, and is characterized its characteristic impedance $Z_0$ and effective relative permittivity $\varepsilon_{eff}$. The antenna is fed from and the other end is open circuited. In general, some strip dipoles may be parasitically placed and several dipoles can be fed simultaneously. Usually a microstrip-to-parallel stripline transition is used to feed the antenna from a conventional coaxial connector [14]. The double-sided configuration has been selected because it offers several practical advantages such as: when required the line polarity between the strip dipoles can be easily reversed as shown in Fig. 1, parallel striplines offer low as well as large values of line impedances with practically reasonable conductor widths, and parallel striplines do not radiate since they are closely spaced.

## 3. Analysis Model

Analyses of series-fed printed strip dipole antennas are complicated by the presence of the dielectric inhomogeneity. In addition, the strip dipoles are closely spaced and connected through transmission lines that are integral part of the antenna. Since they are printed on finite dielectric substrates, they represent three-dimensional composite structures consisting of conducting and dielectric parts. A space-domain moment method analysis would require solving for the conducting currents on the strips and the polarization currents in the dielectric volume, resulting in a large number of unknowns [15]. To simplify the analysis, it is assumed that the conductive strips are printed on an infinite dielectric substrate and the effect of the connecting line is accounted for separately by using a standard transmission line theory. The approach used to analyze the antennas is described in the following.

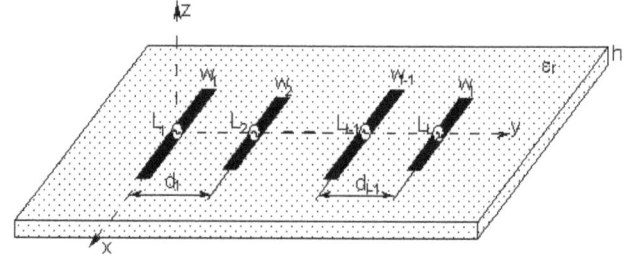

**Figure2.** *Geometry of a group of l parallel printed strip dipoles.*

Fig. 2 shows the geometry of $l$ strip dipoles printed on a dielectric substrate of height $h$ and relative permittivity $\varepsilon_r$. The strips have dimensions $L_1$, $w_1$, $L_2$, $w_2$, . . ., $L_l$, $w_l$ and are separated by distances $d_1$, $d_2$, ..., $d_l$. They are assumed to be thin ( $w_i << L_i$, $i = 1,2,...,l$ ) so only the $x$ directed currents are considered. All the strip dipoles that are connected with transmission lines are assumed to be fed by delta gap generators. The procedure is similar to that used for analysis of microstrip dipoles [13]. The $x$ component of the electric field on the strip side of the dielectric substrate due to strip currents can be expressed as:

$$E_x(x,y,0) = \frac{1}{4\pi^2} \int\limits_{-\infty}^{\infty} \int\limits_{-\infty}^{\infty} G_{xx}(k_x,k_y) J_x(k_x,k_y) e^{j(k_x x + k_y y)} dk_x dk_y \quad (1)$$

where $J_x(k_x,k_y)$ is the Fourier transform of the surface current density on the strips, $j_x(x,y)$, and $G_{xx}(k_x,k_y)$ is the spectral domain Green's function given as:

$$G_{xx}(k_x,k_y) = -\frac{Z_a}{k_0\beta^2} \left[ \frac{k_x^2 k_1 k_2}{T_m}(\varepsilon_r k_2 \cos k_1 h + jk_1 \sin k_1 h) + \frac{k_0^2 k_y^2}{T_e}(k_1 \cos k_1 h + jk_2 \sin k_1 h) \right] \quad (2)$$

$$T_m = 2(\varepsilon_r k_2 \cos k_1 h/2 + jk_1 \sin k_1 h/2)(k_1 \cos k_1 h/2 + j\varepsilon_r k_2 \sin k_1 h/2) \quad (3)$$

$$T_e = 2(k_2 \cos k_1 h/2 + jk_1 \sin k_1 h/2)(k_1 \cos k_1 h/2 + jk_2 \sin k_1 h/2) \quad (4)$$

Where $k_1^2 = \varepsilon_r k_0^2 - \beta^2$, $\text{Im}(k_1) < 0$, $k_2^2 = k_0^2 - \beta^2$, $\text{Im}(k_2) < 0$, $\beta^2 = k_x^2 + k_y^2$, $k_0 = \omega\sqrt{\varepsilon_0\mu_0}$, and $Z_a = \sqrt{\mu_0/\varepsilon_0}$.

An electric field integral equation is formed by requiring that the total electric field, the scattered electric field of Eq. (1) plus the incident field be zero on the strip surfaces. In order to solve this equation, first the unknown current density on the strips is expanded in a set of $N = N_1 + N_2 + ... + N_l$ basic functions with unknown coefficients $I_n$ as:

$$j_x(x,y) = \sum_{n=1}^{N} I_n f_n(x,y) \quad (5)$$

$$f_n(x,y) = \frac{1}{\pi\sqrt{(w_i/2)^2 + (y-y_n)^2}} \frac{\sin k_e(l_n - |x-x_n|)}{\sin k_e h},$$

$$|x - x_n| < l_n, \quad |y - y_n| < w_i/2 \qquad (6)$$

Where $l_n$ is the half-length of the expansion mode, $x_n$ and $y_n$ are the coordinates of the center of mode $n$, $w_i$ is the width of the strip to which the mode belongs, and $k_e$ is chosen as $k_e = k_0\sqrt{(\varepsilon_r + 1)/2}$. Eq. (1) is then solved using a Galerkin moment method solution, leading to the following equation:

$$[Z][I] = [V] \qquad (7)$$

where $[Z]$ is the impedance matrix of the antenna with elements given by:

$$Z_{mn} = -\frac{1}{4\pi^2} \int\limits_{-\infty}^{\infty} \int\limits_{-\infty}^{\infty} G_{xx}(k_x, k_y) F_m^*(k_x, k_y) F_n(k_x, k_y) dk_x dk_y, \qquad (8)$$

$F_n(k_x, k_y)$ is the Fourier transform of the expansion mode $n$

$$F_n(k_x, k_y) = \int\limits_{x_n-l_n}^{x_n+l_n} \int\limits_{y_n-w/2}^{y_n+w/2} f_n(x,y) e^{-jk_x x} e^{-jk_y y} dx dy, \qquad (9)$$

and * denotes the complex conjugate. The elements of the excitation voltage matrix $[V]$ are all zero except the elements corresponding to the generator locations.

The impedance matrix elements are obtained through numerical integration. The residues at the Green's function poles are computed with pole singularities extracted before the integration. These poles correspond to the zeros of Eqs. (2) and (3), that correspond to the characteristic equations for the *TM* and *TE* surface-wave modes of the ungrounded dielectric substrate [16]. There is always an amount of power coupled into these surface-waves because both the first *TM* and *TE* modes have zero cutoff frequencies. As the substrate thickness increases, the power coupled to the first *TE* surface-wave mode increases rapidly [9], so strip dipole antennas are efficient only if printed on electrically thin dielectric substrates. Nevertheless, the amount of power coupled to surface-waves can be determined using the described method.

For planar antennas analyzed using the spectral domain approach, a stationary phase method is usually used to obtain the radiation patterns [10,13]. This method gives convenient and useful results except for points close to the dielectric substrate. In fact, patterns calculated in this way appear to have nulls at the air-dielectric interface since the substrate is assumed to be infinite. In practice, the H-plane patterns of strips printed on electrically thin ungrounded dielectric substrates of finite size show no nulls there. If strips are not printed close to the dielectric edge, the strip currents obtained via Eq. (7) would be accurate even though the dielectric is of infinite size, and accurate radiation patterns are obtained by assuming the strips radiate into free space. In this way the contribution of the polarization currents is not taken into account, but they are

negligible for electrically thin dielectric substrates [17].

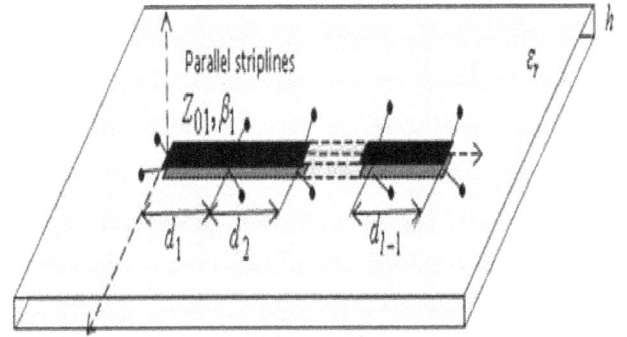

**Figure3.** *Geometry of a group of $(l-1)$ connecting printed stripline.*

To include the effect of the connecting line, the complete antenna system is considered as a parallel connection of two $l$-terminal networks, which can be represented by their admittance matrixes. One $l$-terminal network, $[Y_A]$, represents radiating elements with self and mutual admittances of Fig. 2, which are obtained by using the presented approach. The other $l$-terminal network, $[Y_L]$, represents the connecting lines with terminals at places where strip dipoles are connected as shown in Fig. 3. The elements of the $[Y_L]$ matrix are determined using standard transmission line equations (see next equation):

$$[Y_L] = \begin{cases} \begin{matrix} -Y_{01}\cot\beta_1 d_1 & \pm jY_{01}\csc\beta_1 d_1 & 0 \\ \pm jY\csc\beta_1 d_1 & -jY_{01}\cot\beta_1 d_1 - jY_{02}\cot\beta_2 d_2 & \pm jY_{02}\csc\beta_2 d_2 \\ 0 & -jY_{02}\csc\beta_2 d_2 & -jY_{02}\cot\beta_2 d_2 - jY_{03}\cot\beta_3 d_3 & \cdots \\ \cdots & \cdots & \cdots & \cdots \\ 0 & 0 & 0 & \cdots \\ 0 & 0 & 0 & \cdots \end{matrix} \\ \\ \begin{matrix} \cdots & 0 & & 0 \\ \cdots & 0 & & 0 \\ \cdots & 0 & & 0 \\ \cdots & \cdots & & \cdots \\ \cdots & -jY_{0(l-2)}\beta_{(l-2)}d_{(l-2)} - jY_{0(l-1)}\beta_{(l-1)}d_{(l-1)} & \pm jY_{0(l-1)}\csc\beta_{l-1}d_{(l-1)} \\ \cdots & \pm jY_{0(l-1)}\csc\beta_{(l-1)}d_{(l-1)} & -jY_{0(l-1)}\cot\beta_{(l-1)}d_{(l-1)} \end{matrix} \end{cases}$$

$Z_{0i} = 1/Y_{0i}$, $\beta_i$, and $d_i$, $i = 1,2,...l$ are the characteristic impedance, the propagation constant, and the length, respectively, of the connecting line sections between strip dipoles. If the line polarity between the strip dipoles is not-reversed (reversed), the plus (minus) sign must be used for off-diagonal elements. The $l$-port networks, $[Y_A]$ and $[Y_L]$, are connected in parallel as shown in Fig. 4. The overall $l$-port network as shown in Fig. 1 has $(l-1)$ ports open circuited, so that the input impedance at the first port can be readily obtained, and hence the performance of the antenna. The equivalent admittance matrix is also used to find the complex voltage of delta gap generators that are used to derive radiation characteristic of the antenna. In general not all printed dipoles have to be connected with transmission lines. Some of them can be made parasitic to form for example Yagi-Uda arrays of printed dipoles.

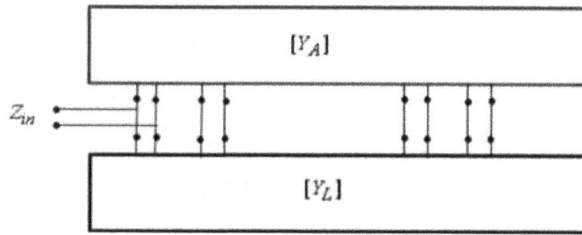

**Figure4.** *Representation of the array of printed strip dipoles and the connecting lines with admittance matrices.*

## 4. Analysis Results of Arrays of Printed Strip Dipole Antennas

In this section, analysis results for a series-fed array of two printed strip dipoles that operates as a dual-frequency antenna are presented. First, the effect of dielectric substrate on the input impedance of this antenna is shown in Fig.5 It can be seen that the calculated results agree well with experiments when the dielectric substrate is taken into account.

Antenna dimensions are like below:

Strip width: $w_1 = w_2 = 6mm$

Dipole lengths: $L_1 = 134\ mm$, $L_2 = 78\ mm$

Distance between dipoles: $d = 50mm$

Dielectric substrate thickness: $h = 1.6mm$

Dielectric substrate permittivity: $\varepsilon_r = 3.2$

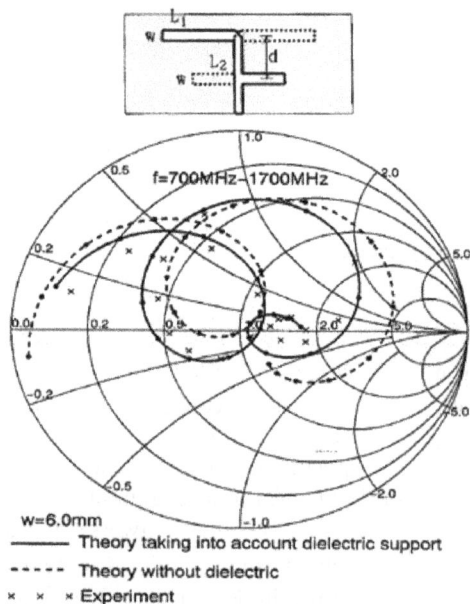

**Figure5.** *Calculated and measured input impedance of the series-fed array of two printed strip dipoles*

The dual-frequency operation characteristics of this antenna can be explained by observing the current distribution on each strip dipole at the two operating frequencies. At the first frequency of operation, (0.9GHz), the first dipole mainly contributes to radiation (see Fig 6-a), whereas in the second frequency of operation (1.55 GHz),

the second dipole mainly contributes to radiation (see Fig 6-b).

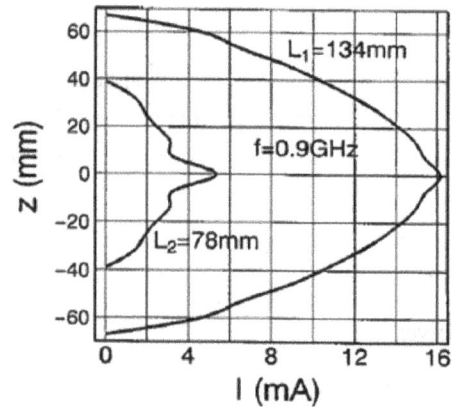

a)      Distribution of each strip dipole at 0.9 GHz

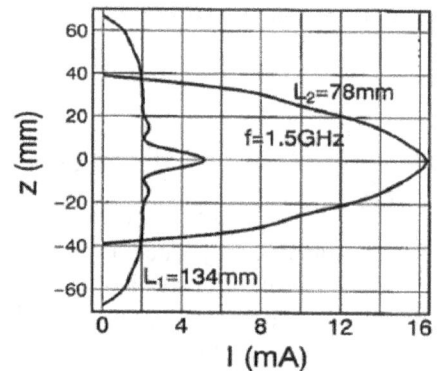

b)      Distribution of each strip dipole at 1.55 GHz

**Figure6.** *Calculated current amplitudes on the two strips of dual-frequency antenna.*

We also analyzed the effect of the dielectric substrate on the radiation characteristics of series-fed strip dipoles and present a series fed array antenna of six-printed dipoles featuring broadband operation. This simple, easily manufactured antenna is well suited for the base station antennas of public mobile communication systems. Figure 7 shows a schematic diagram of the array antenna and input return loss of this antenna.

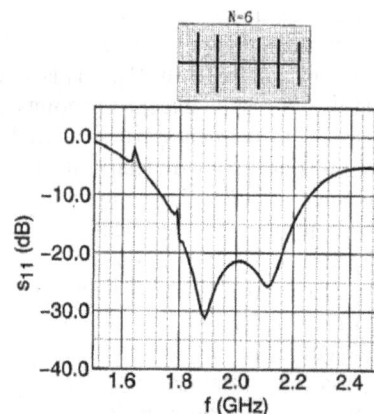

**Figure7.** *Return loss of the series-fed array of six printed strip dipoles*

The end-fire characteristics of this antenna are shown in Fig. 8, where the relative front and back lobe levels and antenna gain against frequency are plotted.

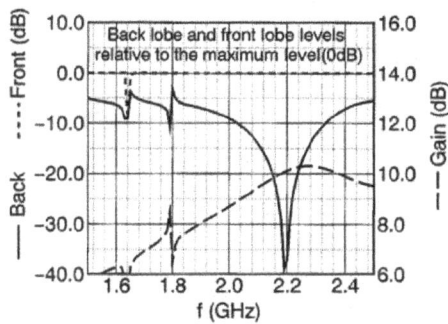

*Figure8.* *Relative front and back lobe levels and gain of the series-fed array of six printed dipoles.*

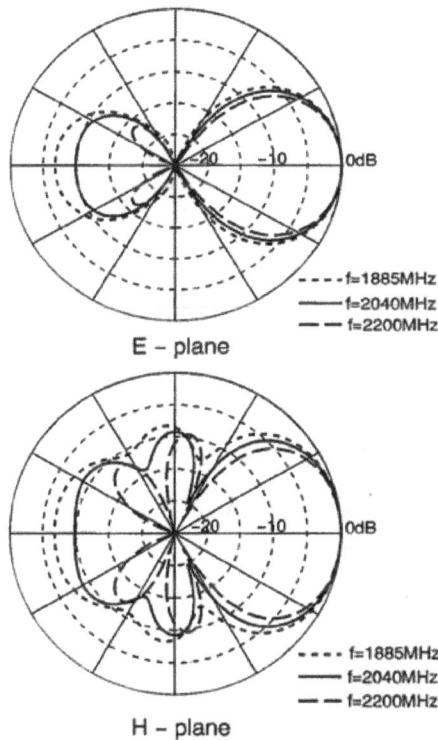

*Figure9.* *E- and H-plane radiation pattern of the series-fed array of six printed strip dipoles.*

The calculated radiation patterns at two principal planes at frequencies 1885 MHz, 2040 MHz and 2200 MHz are plotted in Fig. 9. Small variations on beamwidth of both E- and H-plans can be observed, showing broadband radiation characteristics of these antennas.

# 5. Analysis and Results of Broadband Antenna of Six Series Fed Strip Dipoles

Wireless technology in general is developing more and more. Increasing demand for high data rate transmission through wireless technology has urged implementation of

broadband antennas for fulfilling these requirements with low cost. These wireless technologies are operating in multiple frequency bands. So, the idea is to use only one antenna that radiates in multiple frequencies. The benefits of using only one antenna are multifold, but here we will mention the cost reduction and the space.

In this section, first, we have analyzed a case of series-fed arrays of six printed strip dipoles in front of plane reflector. The dimensions are: Distance between dipoles D=160 mm and dielectric substrate thickness h=30mm.

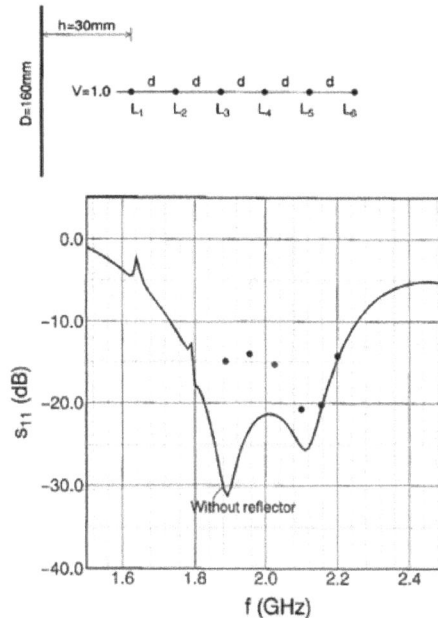

a)     Calculated return loss

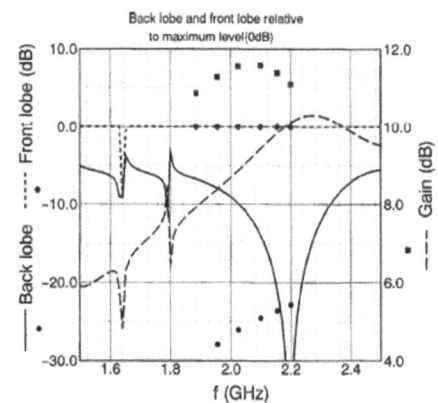

b)     Calculated radiation characteristics

*Figure10.* *a) Calculated return loss and b) Calculated radiation characteristics*

In this case we have calculated only six frequencies. In this case it is used reflector calculation which is very intensive calculation. In the figure 10, we have presented calculated results of return loss (S11 parameters), for different frequencies.

We also analyzed the effect of the dielectric substrate on the radiation characteristics of series-fed arrays of six printed strip dipoles in front of plane reflector [18]. This simple, easily manufactured antenna is well suited for the base station antennas of public mobile communication systems. Figure 7 shows a schematic diagram of the array antenna and input return loss of this antenna.

Whereas, H plane radiations pattern are shown in the figure below:

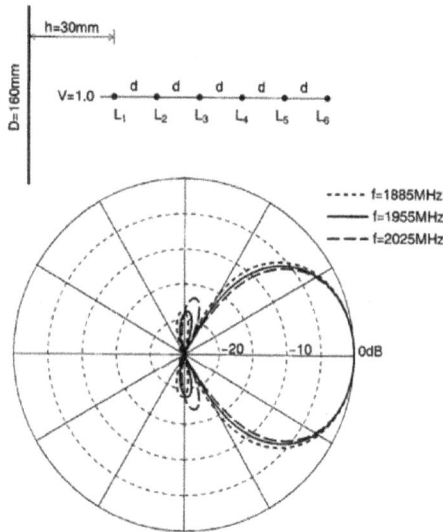

a)      H-plane radiation pattern f= 1885Mhz, f=1955Mhz, and f=2025Mhz

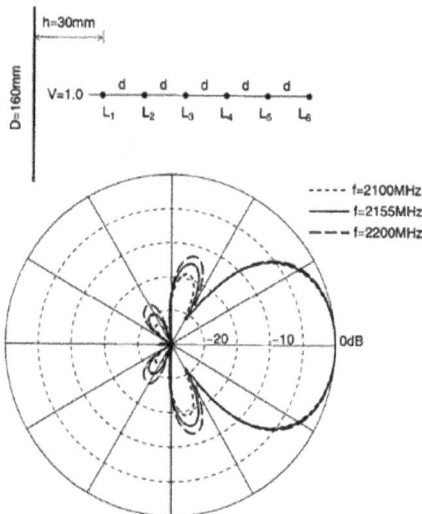

b)      H plane radiation pattern for f= 2100Mhz, f=2155Mhz, and f=2200Mhz

**Figure11.** *H plane radiation pattern for different frequencies*

From figure 11, small variations on beamwidth of H-plans can be observed, showing broadband radiation characteristics of these antennas.

Second case scenario: Two parallel arrays of series-fed printed strip dipole in front of a plane reflector. Only six frequencies calculated with reflector. Lines are for the same case without reflector for comparison

The end-fire characteristics of this antenna are shown in Fig. 12, where the relative front and back lobe levels and antenna gain against frequency are plotted.

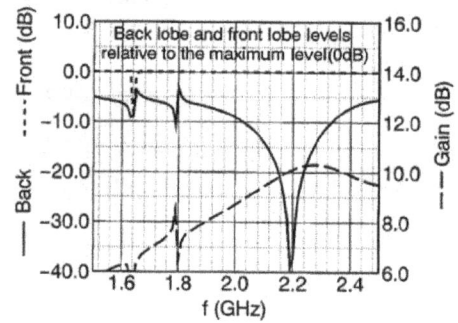

**Figure12.** *Relative front and back lobe levels and gain of the two parallel arrays of series-fed strip dipoles*

Figure 13 shows a schematic diagram of the two parallel of series – fed printed strip dipoles as well as a input return loss of this antenna's system.

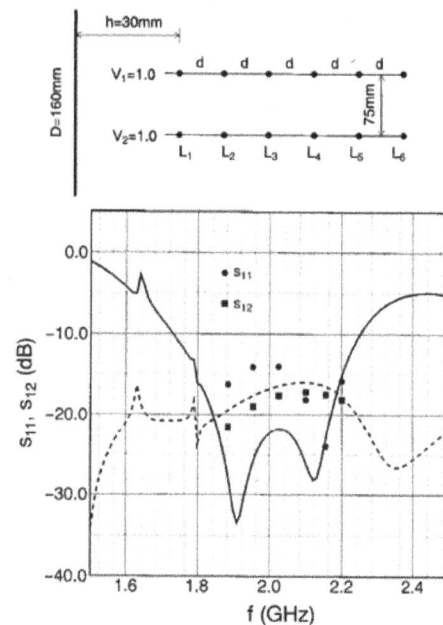

a)      Calculated return loss

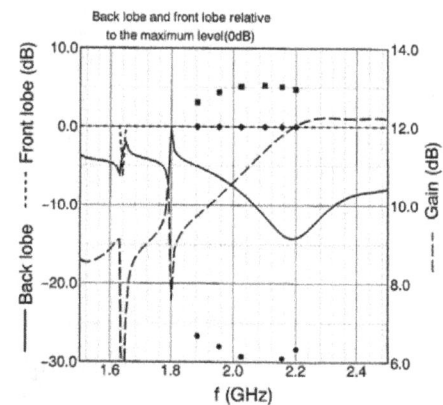

b)      Calculated radiation characteristics

**Figure13.** *a) Calculated return loss and b) Calculated radiation characteristics*

The calculated radiation patterns at frequencies 1885 MHz, f=1955 MHz, and f=2025 Mhz are plotted in Fig. 14.a. Whereas, the calculated radiation patterns at frequencies f= 2100 MHz, f=2155 MHz, and f=2200 MHz are plotted in Fig. 14.b.

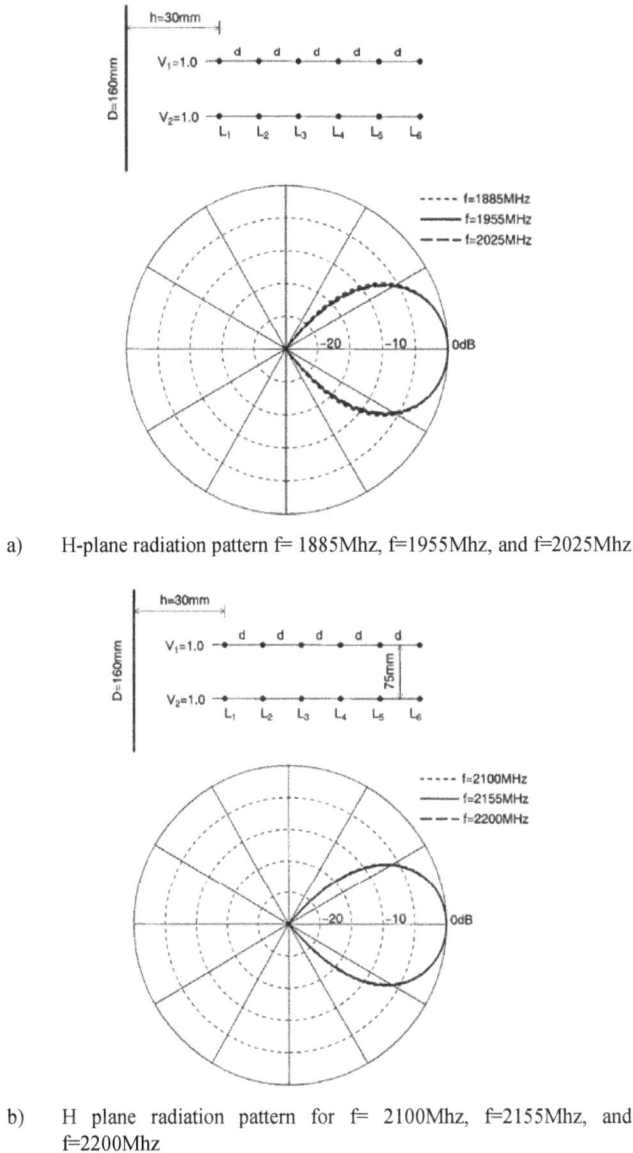

a) H-plane radiation pattern f= 1885Mhz, f=1955Mhz, and f=2025Mhz

b) H plane radiation pattern for f= 2100Mhz, f=2155Mhz, and f=2200Mhz

**Figure14.** *H plane radiations pattern for a) f= 1885 MHz, f=1955Mhz, and f=2025Mhz and b) f= 2100Mhz, f=2155Mhz, and f=2200Mhz*

In figure 14, it can be observed variations on beamwidth of H-plans, showing broadband radiation characteristics of these antennas. As we can see, in this figure x4.a, there is a very small differentiation on beamwidth for different frequencies, whereas in fig x.4b, it is hard to see any difference for different frequencies.

## 6. Summary

In this paper have been analyzed different scenarios of printed strip dipole antennas. These analyses are done based on moment method in spectral domain. First, we have analyzed a case of series-fed arrays of six printed strip dipoles in front of plane reflector. In this case we have calculated only six frequencies. In this case it is used reflector calculation which is very intensive calculation.

Also we have analyzed two parallel arrays of series-fed printed strip dipole in front of a plane reflector. Only six frequencies calculated with reflector. Lines are for the same case without reflector for comparison. For this case, we have presented radiation patterns at frequencies 1885 MHz, f=1955 MHz, f=2025 MHz, f= 2100 MHz, f=2155 MHz, and f=2200 MHz. From this case, we can see observed variations on beamwidth of H-plans, showing broadband radiation characteristics of these antennas.

In this paper have been analyzed arrays of printed strip dipole antennas. As in [18], here also it is used moment method in the spectral domain. The double-sided configuration was selected because it offers several practical advantages such as: when required the line polarity between the strip dipoles can be easily reversed. To simplify the analysis, it was assumed that the conductive strips are printed on an infinite dielectric substrate and the effect of the connecting line is accounted for separately by using a standard transmission line theory. In this paper we also analyzed the effect of the dielectric substrate on the radiation characteristics of series-fed strip dipoles and present a series fed array antenna of six-printed dipoles featuring broadband operation. We have also analyzed different scenarios of printed strip dipole antennas. First, we have analyzed a case of series-fed arrays of six printed strip dipoles in front of plane reflector. In this case we have calculated only six frequencies. In this case it is used reflector calculation which is very intensive calculation.

Also we have analyzed two parallel arrays of series-fed printed strip dipole in front of a plane reflector. Only six frequencies calculated with reflector. Lines are for the same case without reflector for comparison. For this case, we have presented radiation patterns at frequencies 1885 MHz, f=1955 MHz, f=2025 MHz, f= 2100 MHz, f=2155 MHz, and f=2200 MHz. From this case, we can see observed variations on beamwidth of H-plans, showing broadband radiation characteristics of these antennas.

## References

[1] K. Fujimoto and J. R. James, *Mobile Antenna System Handbook*, Artech House, Boston, 1994, ch. 3.

[2] F. Tefiku and C. A. Grimes, "Design of broadband and dual-band antennas comprised of series-fed printed-strip dipole pairs," IEEE *Trans. Antennas Propagat.*, vol. 48, pp. 895-900, Jun. 2000.

[3] F. Tefiku and E. Yamashita, "Double-sided printed strip antenna for dual-frequency operation," IEEE AP-S *Int. Symp. Dig.*, pp. 50-53, July 1996.

[4] F. Tefiku, "A broadband antenna of double-sided printed strip dipoles," *Int. Symp. on Antennas and Propagat., Japan* (ISAP'96), pp. 361-364, Sep. 1996.

[5]   F. Tefiku, "Broadband sector zone base station antennas," IEEE AP-S *Conf. on Antennas and Propagat. for Wireless Communications*, pp. 109-112, Nov. 1998.

[6]   A. J. Parfitt, D. W. Griffin, and P. H. Cole, "Analysis of infinite arrays of substrate-supported metal strip antennas," IEEE *Trans. Antennas Propagat.*, vol. 41, pp. 191-199, Feb. 1993.

[7]   J. R. Bayard, M. E. Cooley, and D.H. Schaubert, "Analysis of infinite arrays of printed dipoles on dielectric sheet perpendicular to a ground plane," IEEE *Trans. Antennas Propagat.*, vol. 39, pp. 1722-1732, Dec. 1991.

[8]   B. Edward and D. Rees, "A broadband printed dipole with integrated balun," *Microwave J.*, pp. 339-344, May 1987.

[9]   D. B. Rutledge, D. P. Neikirk, and D. P. Kasilingam, "Integrated –circuit antennas," in *Infrared and Millimeter Waves*, vol.10, pp. 1-91, K. J. Button, Ed., New York: Academic, 1983.

[10]  M. Kominami, D. M. Pozar, and D. H. Schaubert, "Dipole and slot elements and arrays on semi-infinite substrates," IEEE *Trans. Antennas Propagat.*, vol. 33, pp. 600-607, June 1995.

[11]  A. K. Agrawal and W. E. Powell, "A printed circuit cylindrical array antenna," IEEE *Trans. Antennas Propagat.*, vol. 34, pp. 1288-1293, Nov.1986.

[12]  E. Levine, S. Shtrikman, and D. Treves, "Double-sided printed arrays with large bandwidth," IEE *Proc. H, Microwave, Opt. & Antennas*, vol. 135, pp. 54-59, Feb 1988.

[13]  D. M. Pozar, "Analysis of finite phased arrays of printed dipoles," IEEE *Trans. Antennas Propagat.*, vol. 33, pp. 1045-1053, Oct.1985.

[14]  R. Sefa, A. Maraj, "Design of microstrip to balanced striplines transitions" The 10th WSEAS International Conference on TELECOMMUNICATIONS and INFORMATICS (TELE-INFO '11), Included in ISI/SCI Web of Science and Web of Knowledge.

[15]  I. Tekin and E. H. Newman, "Space-domain method of moments solution for a strip on a dielectric slab," IEEE *Trans. Antennas Propagat.*, vol. 46, pp. 1346-1348, Sep.1998.

[16]  R. F. Harrington, *Time Harmonic Electromagnetic Fields*, McGraw-Hill, New York, 1961, chap. 4.

[17]  R. Janaswamy, "An accurate moment method model for the tapered slot antenna," IEEE *Trans. Antennas Propagat.*, vol. 37, pp. 1523-1528, Dec. 1989.

[18]  R. Sefa, F. Tefiku, A. Maraj, "Analysis of arrays of printed strip dipole antennas" Software, Telecommunications and Computer Networks (SoftCOM), 2011 19th International Conference, Date of Conference: 15-17 Sept, IEEE Conference, 2011

# 27

# The details of virtual contention window concept for 802.11 IBSS wireless local area network mathematic modelling

**Anton V. Lazebnyy[1], Volodymyr S. Lazebnyy[2]**

[1]Scientific and Production Enterprise "Ukrservisbud" Ltd., Kyiv, Ukraine, system administrator, MA
[2]National Technical University of Ukraine "Kyiv Polytechnic Institute", Kyiv, Ukraine, Associate Professor, Ph.D. NTUU "KPI", Kyiv, Ukraine

**Email address:**
smartvs@mail.ru (V. S. Lazebnyy)

**Abstract:** The concept of a virtual contention window for assessment of temporal and probabilistic characteristics of the processes occurring in the wireless LAN 802.11 is considered. The relations for determining the transmission time delay of the data package, the uneven of transmission time, throughput of wireless channel, the probability of packet loss for networks with saturated load are proposed in this paper.

**Keywords:** Collision Probability, Contention Window, Jitter, Saturated Payload, Throughput of Wireless Channel, Time Delay, Wireless Network

## 1. Introduction

The wireless local area networks (WLAN) of 802.11 [1] standard are widely use for privet nets on the one hand and corporative nets with large number of users on the other hand.

If we shall analyze the development of 802.11 standard during a time, we shall see different changes on physical and channel levels (802.11 a, b, g, n), which was done in order to improve some characteristics of local wireless telecommunication systems. In this paper we shall speak only about physical and channel levels of 802.11 standard.

Accordingly with 802.11 standard it may be two types of wireless networks: BSS - Basic Service Set and IBSS Independent Basic Service Set.

In BSS mode all subscribers not connected directly but through AP (Access Point). Such a wireless network controlled by PCF (Point Coordination Function).

In IBSS mode all subscribers connected one to another directly by using CSMA/CA (Carrier Sense Multiple Access/Collision Avoidance) competitive access method and DCF (Distributed Coordination Function) will be used to control the communications.

If we use IBSS mode, we can get a most throughput in comparison with BSS mode but reliability of connections in IBSS network will be less, because the competitive access method on physical channel will be used.

There are a lot of investigations and scientific works corresponded with analysis of processes and parameters of IBSS networks. Some books [2-12] contain fundamental principles of wireless 802.11 standard networks functioning analysis. There we can see the most of models proposed for wireless network description. It possible to estimate the accuracy of different models and it's convenience on practice. Some of these models [2] give us a general description of probable events on 802.11 wireless networks by using of integral expressions. Such expressions are not convenient on practice because they don't give a direct dependence between network parameters and it's performance. Other models [3,4] based on the concept of saturated network load proposed by Giuseppe Bianchi. This models give ability to estimate the influence of system parameters on network performance but only in general and calculation may be done only for limited range of parameters. If we attentively look on one of such expression [3, formula (36)]:

$$\tau = \frac{2(1-2p_c)(1-p_c^{R+1})}{(1-2p_c)(1-p_c^{R+1})+W(1-p_c)(1-(2p_c)^{R+1})}, (1)$$

we can understand some features. Fist of all we see that

the meaning of collision probability $p_c$ must be less then 0,5, because in the case of $p_c = 0,5$ it will be «zero» in the numerator of shown expression. The second singularity of considered expression corresponded with $p_c = 0$. In this case the probability of successful transmitting of data pocket $\tau$ will depends only by meaning of contention window W. Because W may be equal to 8, 16 or 32 corresponding with different specifications of 802.11 standard, the calculated meaning of successful transmitting will be significantly less then «one». Such result takes place because formula (1) was proposed for estimation of successful transmitting probability for process, which has begun during some elementary time slot. That's why it is difficult to use such a formula in order to execute an accurate analysis of processes, which take place in wireless networks.

Due to problems which are considered and some other problems many scientists make their efforts in order to improve approaches for wireless networks performance estimation.

## 2. The Concept of Virtual Contention Window

It's well-known that in order to get access to wireless channel in Wi-Fi LAN every station use the parameter which named Contention Window ($CW$). This interval form by using the $CW$, the value of which is chosen randomly from the set of numbers $\{0,1,2, ..., CW\}$. $CW$ defines the number of elementary timeslots which form the transmitting delay.

We propose to use the concept of saturated network load proposed by Giuseppe Bianchi [2] and consider the process of packets transmitting as a quasistationary process.

The main idea of the Virtual Contention Window concept is that in average every station of network will transmit its packets by equal time intervals, which we called the Virtual Contention Window.

Virtual Contention Window as a parameter $VCW$ is a stochastic parameter of unplanned wireless network standard 802.11 ($BSS$), which is numerically equal to the averaged number of elementary time intervals (time slots) during which the backoff counter provides a delay interval after transmitting the previous frame before transmitting the next data frame of one station.

Duration of time interval corresponding to the virtual contention window depends on the number of active stations in the network N, the minimum value of the competitive window $CW_{min}$, the value of payload $PL$, contained in each frame of data, such as a network protocol used to transfer data blocks, the probability of collisions $p_c$ and from the impact of external environment on signal propagation, resulting in the loss of some number of frames due to the effects of noise and interference.

To develop a model of the network using a virtual contention window, the following statements and assumptions was taken into account:

each network station shall transmit the packet of infor-

mation after the backoff counter will decrease to zero. The time delay for every station defined by meaning of $CW_i$. Because some times it may be collision, $CW_i$ will defined with accordance to general rules of 802.11 standard;

if the number of network active stations is N, the danger of collision for chosen station generates ($N-1$) active stations with equal probability;

number of attempts to transmit the current data frame, which is the first in the queue of station, limited by the number $R = (m + 1)$, where $m$ is a parameter of network;

average length of the transmitted data frame is the same for each station provided a stationary regime of the network (the condition is not mandatory, but allows to simplify the final description of model).

In order to determine the value of virtual contention window for wireless network under ideal conditions (excluding environmental impact on quality of radio transmission) first of all it is necessary to determine the value of collision probability.

In the case of collision station will transmit the packet one's more. So, we say that next attempt take place. The data frame transmitting during the next attempt is characterized by the collision probability ($p_{ci}$) and the probability of successful transfer which is $p_{sc\,i} = (1 - p_{ci})$. The collision probability is important parameter of every model of IBSS wireless network. Almost all previous analytical models, the parameter $RSI$ considered as a constant or known value based on an intuitive understanding of the processes occurring in the wireless channel. Some authors use $p_{ci}$ but don't give some expression, which give us ability to calculate this probability directly [2-4]. For the correct assessment of $VCW$ value it necessary to investigate additionally the accuracy of statements on a constant value of the probability of collisions.

## 3. Evaluation of Collisions Probability in Unplanned Wireless Network

Let we consider an IBSS network, which contains only two active stations.

During the first attempt both stations will load the backoff counter by some number from the set $\{0, 1, 2, ..., CW_{min}\}$. Under these conditions, the probability that the second station will load the same meaning as contained in the backoff counter of the first station will equal to $p_{c1} = 1/CW_1 = 1/(CW_{min}+1)$ (the first attempt, $i = 1$, $CW_1 = CW_{min}+1$). With accordance to 802.11 protocol if collision took place on the first attempt the next meaning of CW will change with accordance to binary law, in particular $CW_2 = [2(CW_{min}+1) - 1]$. It means that collision probability on second attempt will be equal to $p_{c2} = 1/CW_2$. The number of attempts is limited to the number ($m + 1$), where $m = log_2 (CW_{max}/ CW_{min})$. $CW_{max}$ and $CW_{min}$ is the largest and smallest value of parameter $CW$ defined in 802.11 specification. In the case of several consecutive collisions the value of $CW_i$ will vary accordingly to binary exponential law, whereas values of

$p_{ci} = 1/CW_i$. The value of the probability of collisions on the network with two active stations and during $i$-attempt, provided that $CW_{min} = 31$, $CW_{max} = 1023$, $m = 5$ are presented in Table 1.

**Table 1.** *The probability of collisions during the first attempt of data frame transmitting*

| $i$ | 1 | 2 | 3 | 4 | 5 | 6 |
|---|---|---|---|---|---|---|
| $p_{ci}$ | 0,031 | 0,016 | 0,008 | 0,004 | 0,002 | 0,001 |

From the values in Table 1 implies that if only two stations are present in the network the probability of collisions during each subsequent attempt is not constant and decreases by half.

Consider now the network, which contain only three active stations. We will consider only the collisions that occur between two stations, because the probability of collisions between the three stations is very small (more than an order of magnitude less). We consider the probability of collisions for the selected station, which due to previous collisions will make $(m +1)$ attempts in order to convey data frame.

During the first attempt the collision may be with second station or with third station of network with the same probabilities equal to $1/CW_1$. The probability that the selected station will avoid conflict with the second station is equal to $(1-1/CW_1)$, then the probability of collision with a third station will be $(1-1/CW_1) \cdot (1/CW_1)$. Under these conditions the full probability of collision for selected station in the first attempt will be equal to

$$p_{c1} = \frac{1}{CW_1} + \frac{1}{CW_1} \cdot (1 - \frac{1}{CW_1}). \quad (2)$$

On the second attempt in the system will be two stations that took part in the collision during the first attempt and use some random number to load their backoff counters from the range defined by contention window $CW_2$. The third station that has not experienced conflict during the first attempt will use the number from range defined by $CW_1$. So, the total probability of collision at second attempt for chosen station will be equal to

$$p_{c2} = \frac{1}{CW_1} + \frac{1}{CW_2} \cdot (1 - \frac{1}{CW_1}). \quad (3)$$

If the third attempt ($i = 3$) take place, selected station will use a contention window $CW_3 = [2^{i-1} \cdot (CW_{min}+1) - 1]$, the other two stations – depending from their history will use the contention windows which defined by $CW_1$ for one of them and $CW_2$ for other or $CW_1$ and $CW_3$ respectively. Under these conditions the total collision probability may become one of the following:

$$p_{c3}^{(I)} = \frac{1}{CW_1} + \frac{1}{CW_2} \cdot (1 - \frac{1}{CW_1}),$$

$$p_{c3}^{(II)} = \frac{1}{CW_1} + \frac{1}{CW_3} \cdot (1 - \frac{1}{CW_1}). \quad (4)$$

Based on the result for the probability of collisions on the third attempt we can conclude that during the $m$-th attempt probability of collision will be determined by one of the ratios:

$$p_{cm}^{(I)} = \frac{1}{CW_1} + \frac{1}{CW_2} \cdot (1 - \frac{1}{CW_1}),$$

$$p_{cm}^{(II)} = \frac{1}{CW_1} + \frac{1}{CW_3} \cdot (1 - \frac{1}{CW_1}), \dots,$$

$$p_{c(m+1)}^{(M)} = \frac{1}{CW_1} + \frac{1}{CW_m} \cdot (1 - \frac{1}{CW_1}). \quad (5)$$

Superscript notation of collision probability characterizes one of the possible values that can take the probability of collisions during the $i$-th attempt. Numerically $M = m$. Results of calculations for the probability of collisions for one station in a network with three active stations are in Table 2.

**Table 2.** *The probability of collisions on the network with three active stations*

| $i$ | 1 | 2 | 3 | 4 | 5 | 6 |
|---|---|---|---|---|---|---|
| $p_{ci}^{(I)}$ | 0,063 | 0,048 | 0,048 | 0,048 | 0,048 | 0,048 |
| $p_{ci}^{(II)}$ | | | 0,040 | 0,040 | 0,040 | 0,040 |
| $p_{ci}^{(III)}$ | | | | 0,036 | 0,036 | 0,036 |
| $p_{ci}^{(IV)}$ | | | | | 0,034 | 0,034 |
| $p_{ci}^{(V)}$ | | | | | | 0,033 |

Based on the analysis of the process of collisions, the ratios of (2) – (5), we can conclude that with the increase in the number of active stations in the network divergence values of collision probability during the various attempts to transmit a frame of information will decrease.

From the analysis of values of collisions probability in a network with saturated payload of stations during different attempts to transfer data frame we can do the following conclusions:

on networks which contain two or three active stations probability of collisions in the first attempt is considerably different from the probabilities of collisions which take place during next attempts. As a result, the application of universal analytical models for the analysis of such networks, which used a constant probability of collision, there are significant errors and discrepancies with the results of full-scale tests [4]. Consequently, to calculate the parameters and characteristics of such networks should apply some special mathematical models for two active stations and three active stations;

probability of collisions in wireless networks, where only two or three stations  compete in order to get  the channel, is  small ($p_c \ll 1$);

in order to develop a universal analytical model of the processes occurring in the channel of wireless network 802.11 with a number of stations $N \geq 4$ it possible to  consider that the probability of collisions is constant for all attempts to transmit information frame and this probability is determined by the minimum value of contention windows and the number of stations in the network.

Based on the findings, it's possible to propose the equations for the probability of collisions during first attempt of data frame transfer in unplanned network with the number of active stations, more then four.

Let's find the probability of collisions as the total probability, provided that collision involved each time only two stations. We shall find the probability of collisions that may happen during the first attempt. We must note that received so probability of conflict will be slightly inflated. The probability of collision between chosen station and one other station is $p_{1c}$, with the second station $(1-p_{1c}) \cdot p_{2c}$. Parameter $p_{2c}$ is the probability of collision with second station only, without taking into account other stations. Factor $(1-p_{1c})$ takes into account the probability that conflict with the first station will not happen. For the case of $N$ stations in the network in general may be written

$$p_c \approx p_{1c} + p_{2c}(1-p_{1c}) + p_{3c}(1-p_{1c})(1-p_{2c}) +$$

$$+ ... + p_{(N-1)c}(1-p_{1c}) \cdot ... \cdot (1-p_{(N-2)c}). \quad (6)$$

If the collision probability with any other active station of the network will be the same and equal to $p_{ic} = 1/CW_1$, using (6) we shall obtain

$$p_c \approx \frac{1}{CW_1} + \frac{1}{CW_1}(1-\frac{1}{CW_1}) + \frac{1}{CW_1}(1-\frac{1}{CW_1})^2 + ... +$$

$$+ \frac{1}{CW_1}(1-\frac{1}{CW_1})^{N-2} = 1-(1-\frac{1}{CW_1})^{N-1} \quad (7)$$

The collision probability in homogeneous networks (all stations have the same probability of access to the network), calculated by using of ratio (7), is shown in Fig. 1.

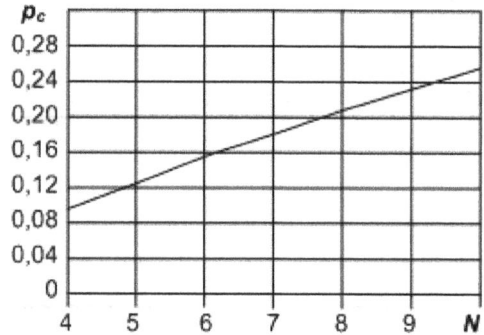

**Figure 1.** *The dependence of collision probability for one wireless network station when it try to transmit the data frame for the case of $CW_{min} = 31$*

## 4. The Wireless Channel Model of Unplanned Network Of 802.11 Standard

We propose to consider a wireless channel model based on the use of virtual contention window. First of all it necessary to find the value of virtual contention window VCW. We shall search it as the value of mathematical expectation for a binary exponential law of contention window CW increasing in case of collision, which is defined by basic specification of 802.11 standard

$$VCW = \frac{CW_1}{2}(1-p_c) + \frac{2CW_1}{2}(1-p_c)p_c +$$

$$+ \frac{4CW_1}{2}(1-p_c)p_c^2 + ... + \frac{2^m CW_1}{2}(1-p_c)p_c^m =$$

$$= \frac{CW_1 \cdot (1-p_c)}{2} \cdot \sum_{i=1}^{m+1}(2p_c)^{i-1}. \quad (8)$$

A graphic dependence of the virtual contention window as a function of the number of active stations in the network shows in Fig. 2. Graph constructed using ratios (7) and (8).

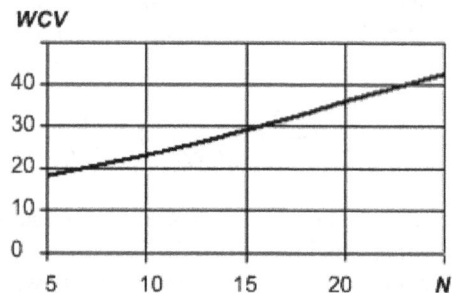

**Figure 2.** *Dependence of the virtual contention window VCW from the number of active stations on the network when used $CW_{min} = 31$*

From the graph it follows that with an increase in the number of active stations on the network value of the virtual contention window increases almost linearly with the speed of about 1.22 time slots for each station.

Let's define now the probability of successful transmission of a data frame each active station during a time of virtual contention window. We must define this probability as a total probability of a random process.

$$p_{sc} = (1-p_c) + (1-p_c)p_c + (1-p_c)p_c^2 +$$

$$+ \ldots + (1-p_c)p_c^m = (1-p_c)\frac{1-p_c^{m+1}}{1-p_c} = 1-p_c^{m+1}. \quad (9)$$

During virtual contention window each station of network transmits an average of about one frame of data. More precisely, this number is determined by the ratio (9). $N$ stations transmit $N$ frames.

Based on the concept of virtual contention window, it's possible to determine the number of collisions $N_c$, which occurs on average when transmitting $N$ frames of data. Based on the assumption that each collision involved only two stations, the number of conflict pairs will be equal to $N/2$.

We shall find the number $N_c$ as the expected number of collisions during a time of virtual contention window:

$$N_c = \frac{N}{2}p_c + \frac{N}{2}p_c^2 + \frac{N}{2}p_c^3 + \ldots + \frac{N}{2}p_c^m =$$

$$= p_c \cdot \frac{N}{2} \cdot \frac{1-p_c^{m+1}}{1-p_c}. \quad (10)$$

The obtained equations can be used to determine the throughput depending on the network settings. We can calculate throughput as the ratio between the average size of payload transmitted by all stations of network to time duration of a virtual contention windows:

$$S = \frac{N \cdot E[PL_1]}{T_{VCW}}, \quad (11)$$

where - $E[PL_1]$ averaged payload of one data frame.

If the frame size will be constant the overall payload which transmitted during one virtual contention window will equal to $N \cdot PL_1$. The average duration of virtual contention window will be

$$T_{VCW} = N \cdot \overline{T}_{PL} + N_c \cdot \overline{T}_c + \overline{T}_{idle}. \quad (12)$$

To determine the values of time intervals $\overline{T}_{PL}$ and $\overline{T}_c$ we can use well-known relations, in particular, given in [2, 3]. $N_c$ must be defined from (10). The average duration of empty time interval $\overline{T}_{idle}$ that arise during the implementation of virtual contention window will be equal to

$$\overline{T}_{idle} = \sigma \cdot [VCW \cdot N \cdot (1-p_c^{m+1}) - N_c]. \quad (13)$$

By using (11) it possible to define the maximum meaning of throughput as a dependence of packet payload. The expression (13) gives ability to get an accurate meaning of

$\overline{T}_{idle}$ duration.

The throughput, per one active station of the network will be therefore determined by the ratio

$$S_1 = \frac{E[PL_1]}{T_{VCW}}. \quad (14)$$

Fig. 3 shows graphs of one station throughput as a function of number of active stations on the network. Charts are built using the ratio (14) for a network that operates accordingly to 802.11a standard at the speed of signal flow of 24 Mbit/s.

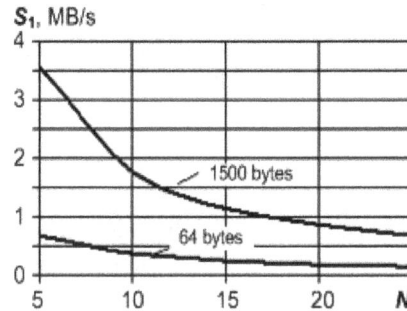

**Figure 3.** *Dependence of one station throughput (S₁) from the number (N) of active stations on the network*

## 5. Estimation of Quality of Service Parameters in Wireless Network

The ratios for some probabilities we got before give us ability to define some parameters of wireless network QoS. By using of some general expressions we can directly define such parameters as time delay of data packet transmission, the probability of data packet losing, jitter.

Time delay of data packet transmission over the wireless channel in accordance with the concept of virtual contention window will be equal to the duration of this virtual window,

$$\overline{\tau} = T_{VCW}. \quad (15)$$

The average delay time of the data frame transmitting as a function of payload size for 802.11a network in the case of signal transmitting velocity of 24 Mbit /s by using of (12) shown on Fig. 4 - 5.

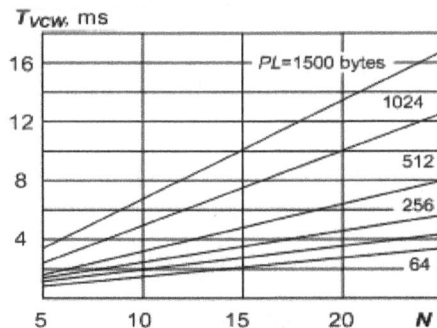

**Figure 4.** *The average time delay of the data frame transmitting in a networks with different number of active stations respectively with (15)*

$T_{VCW}$, ms

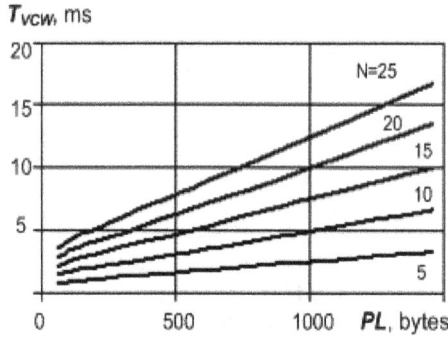

*Figure 5. The transmission time delay as a function of size of payload in the data frame*

The value of the probability of packet loss can be determined directly from the ratio for the probability of successful transmission (9)

$$P^{(rs)} = p_c^{m+1} . \qquad (16)$$

Values of packet loss probability on a network with $N$ stations calculated by using (16) is shown in Table 3.

**Table 3.**

| $N$ | 5 | 10 | 15 | 20 | 25 |
|-----|-----|-----|-----|-----|-----|
| $P^{(rs)}$ | 2,9E-06 | 2,4E-04 | 2,1E-03 | 8,6E-03 | 2,3E-02 |

To determine the jitter $\sigma^{(\tau)}$ make use of dispersion relations for the delay of data frame

$$D(\tau) = \frac{1}{N^{(b)}} \sum_i^{N^{(b)}} (\tau_i - \bar{\tau})^2 , \qquad (17)$$

where $N^{(b)}$ – the total number of transmitted data frames and $\tau_i$ – transmission delay of a single data frame.

Based on the concept of virtual contention window, in order to define time dispersion we shall take into account the average value of transmission delay during each of the possible attempts. In this case, expression (17) takes an other form, namely

$$D(\tau) = \frac{1}{N^{(b)}} \sum_{j=1}^{m+1} N_j \cdot (\tau_j^* - \bar{\tau})^2 , \qquad (18)$$

where $N_j$ and $\tau_j^*$ – the number of data frames and average transmission delay during the $j$-th attempt respectively.

$$\tau_j^* = N \cdot p_c^{i-1} \cdot (1 - p_c) \cdot \bar{T}_{PL} + N \cdot p_c^i + \sigma \cdot \frac{2^{i-1} CW_1}{2} \cdot p_c^{i-1} ,$$

$$N_j = N \cdot p_c^{i-1} \cdot (1 - p_c) . \qquad (19)$$

Jitter expression we shall find by using of general formula and taking into account (18) and (19)

$$\sigma^{(\tau)} = \tau^{(max)} - \tau^{(min)} = 2\sqrt{D(\tau)} , \qquad (20)$$

where $\tau^{(max)} = \bar{\tau} + \sqrt{D(\tau)}$, $\tau^{(min)} = \bar{\tau} - \sqrt{D(\tau)}$ ·

Graphs of $\sigma^{(\tau)} = f(PL)$ and $\sigma^{(\tau)} = f(N)$ dependences for 802.11a network with signal stream at 24 Mbit/s, in which all the stations use $CW_{min} = 31$, calculated respectively with (20) is shown in Fig. 6-7.

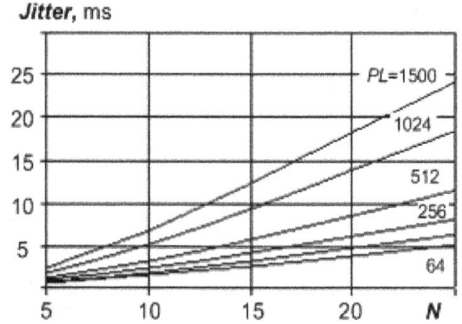

*Figure 6. Uneven of transmission delay of the data frame (jitter) on a network with different number of active stations (N)*

Due to graphs in Fig. 6, 7 we can conclude that the process of data frames transmitting over a wireless network radio channel is characterized by considerable irregularity in time. This inequality essentially depends on the number of stations in the network and the volume of payload in the data frame. Thus, for networks with $N = 5$, if the payload of a frame transmitted by the protocol UDP, is 64 bytes, jitter is 670 microseconds, and for the frame payload of 1024 bytes – 1742 microseconds.

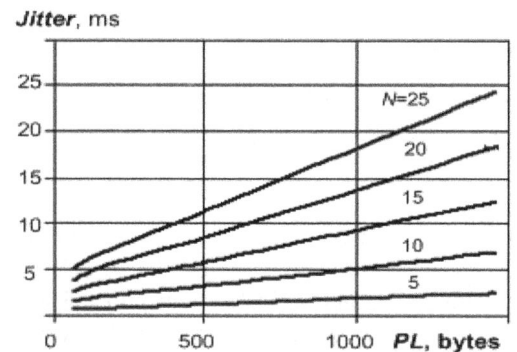

*Figure 7. Jitter as a dependence of payload in the data frame*

For a network with $N = 25$ jitter is equal to 5521 and 17116 microseconds when the loads are 64 bytes and 1024 bytes respectively.

## 6. Conclusions

The concept of virtual contention window give's ability to form a mathematical model of the wireless channel of the unplanned network 802.11 standard. This model provide a direct functional relationship between the probability characteristics of the channel and deterministic parameters of wireless network.

The proposed mathematical model of the wireless channel makes quantitative estimation of network operating parameters and quality of service characteristics. If it known the number of active stations in the network, the minimum value of the contention window, size of data frame, the maximum value of the contention window and permissible number of retries to transfer data frame in the event of collisions we can get all probable parameters of wireless network channel.

The model obtained for a saturated mode of unplanned 802.11 network may be applied to network with moderate payload by specification of collisions probability.

# References

[1]   IEEE Std 802.11, 2007 Edition, Wireless LAN Medium Access Control (MAC) and Physical Layer (PHY) specification. – 3 Park Avenue, New York, NY 10016-5997, USA, June 2007. – 1232 p.

[2]   Giuseppe Bianchi, Performance Analysis of the IEEE 802.11 Distributed Coordination Function/ Giuseppe Bianchi // IEEE Journal on Selected Areas in Communications. – 2000. – vol. 18 – № 3. – p. 1055 – 1067.

[3]   В.М.Вишневский. Широкополосные беспроводные сети передачи информации/ В.М.Вишневский, А.И.Ляхов, С.Л.Портной, И.В.Шахнович. – Москва: Техносфера, 2005, – 592с.

[4]   Emerging Technologies in Wireless LANs. Theory, Design, and Deployment/ Edited by BENNY BING. – Georgia Institute of Technology, Cambridge University Press 2008. – 897p.

[5]   V. M. Vishnevsky. 802.11 LANs: Saturation Throughput in the Presence of Noise/ V. M. Vishnevsky, A. I. Lyakhov  // in Proc. IFIP Networking'02. –  Italy, Pisa. – 2002.

[6]   Z. Hadzi-Velkov and B. Spasenovski, "Saturation Throughput – Delay Analysis of IEEE 802.11 DCF in Fading Channels," in Proc. IEEE ICC'03, Anchorage, Alaska.  – 2003, vol. 1. – pp. 121-126.

[7]   Javier del Prado. Link Adaptation Strategy for IEEE 802.11 WLAN via Received Signal Strength Measurement / Javier del Prado, Sunghyun Choi // In Proc. IEEE ICC'03. – USA, Alaska, Anchorage. – 2003

[8]   Daji Qiao. Goodput Analysis and Link Adaptation for IEEE 802.11a Wireless LANs / Daji Qiao, Sunghyun Choi,Kang G. Shin // IEEE Trans. on Mobile Computing (TMC). – 2002. - vol. 1, № 4. – pp. 278-292

[9]   Jean-Lien C. Wu. An Adaptive Multirate IEEE 802.11 Wireless LAN / Jean-Lien C. Wu, Hunh-Huan Liu, Yi-Jen Lung // in Proc. 15th International Conference on Information Networking – 2001. – pp. 411-418.

[10]  P. Chatimisios. Throughput and Delay Analysis of IEEE 802.11 Protocol, / P. Chatimisios, V. Vitsas, A. C. Boucouvalas // in Proc. the 5th IEEE International Workshop on Network Appliances (IWNA). – UK, Liverpool. – 2002.

# 28

# RSS-based near-collinear anchor aided positioning algorithm for Ill-conditioned scenario

**Senka Hadzic**[1,2]**, Du Yang**[1]**, Manuel Violas**[1,2]**, Jonathan Rodriguez**[1]

[1]Instituto de Telecomunicações, Aveiro, Portugal
[2]Universityof Aveiro, Aveiro, Portugal

**Email address:**

senka@av.it.pt (S. Hadzic), duyang@av.it.pt (D. Yang), mviolas@ua.pt (M. Violas), jonathan@av.it.pt (J. Rodriguez)

**Abstract:** The conventional Received Signal Strength (RSS) based positioning algorithms such as Least Square (LS) and Weighted LS (WLS) produce significant estimation errors when the anchor nodes positions approach a collinear scenario. In this paper, we propose the CAP (Collinear Anchor aided Positioning) algorithm to provide robust positioning performance under ill-conditioned matrix conditions, whilst contributing toward overall low computational complexity. The CAP algorithm outperforms traditional approaches such as the maximum likelihood algorithm, LS and WLS among others.

**Keywords:** Ill-Conditioned, Collinear Anchors, Unbiased Estimate, CRLB

## 1. Introduction

Wireless position estimation (locating a node using measurements from surrounding location-known anchors) is increasingly becoming a prominent feature for intelligent services and applications. There are mainly three popular estimation approaches: lateration, angulation, and mapping. Lateration techniques compute the position of an object by measuring its distances from multiple anchors. The distances could be obtained from received signal parameters such as Time of Arrival (ToA) [1]-[2] or Received Signal Strength (RSS) [3]. The target node position can be computed using estimation algorithms such as Least Square (LS) [4], Weighted Least Square (WLS) [5], Maximum Likelihood (ML) [3], etc. Another popular technique is angulation, which uses the measured Angle of Arrival (AoA) [6] to determine the target node position. The third one is mapping, which matches the measured signal parameter (e.g RSS) with a database consisting of the same type of signal parameters at known positions [7] so as to determine the position. The RSS-based lateration technique is the most practical in terms of implementation since it does not require any synchronization, antenna arrays or external database. In this paper, without loss of generality we will focus on RSS-based lateration technique in a 2-dimensional space. The geometric method can be applied to ToA based distance estimates, without any restrictions.

One limitation of lateration technique is that it requires distance measurements from at least 3 non-collinear anchors for calculating an object's position in 2-dimensional space. The accuracy of above-mentioned LS and WLS significantly degrades when the anchors are approaching collinear. The main reason for this degradation is that these two positioning algorithms involve matrix inversions which results in significant error injection when the matrix is ill-conditioned. The accuracy of high-complexity ML algorithm also degrades as well.

In this paper, we consider an ill-conditioned scenario, in which the anchor nodes are near collinear, and the target does not lie on the same line as the anchors. Furthermore, we assume an indoor scenario where GPS does not work, and the positioning procedure relies exclusively on available anchor nodes. Although its probability decreases with the increase of number of anchors, having near collinear anchor nodes is possible in practice. For example, it is possible that in public safety scenarios such as fire prevention, most of the well-planned indoor location sensors may be destroyed in a fire with only a few near collinear collocated sensors intact. Based on this scenario, we propose a Ner-Collinear Anchors-aided Positioning (CAP) algorithm, which provides significantly better localization results compared to the conventional LS/WLS algorithms. Its localization accuracy is comparable with ML, but with significantly reduced complexity.

The remainder of the paper is organized as follows: we present the related work in Section 2and our motivation to investigate the problem of collinear anchors; in Section 3 we describe the target scenario, RSS distance estimation, and some conventional positioning algorithms; Section 4 explains our proposed algorithm, namely the collinear anchors aided positioning algorithm; Section 5 evaluates the performance through simulations and finally, Section 6 concludes the paper.

## 2. Related Work

The problem of anchor placement is a well-known subject in localization literature [12-14]. Whether the localization algorithm is statistical and uses the Cramer-Rao lower bound as performance metric [13-14], or it introduces a new confidence metric for geometrical localization such as trilateration [12], the conclusions are similar - the best performance is given when anchor nodes are well separated around the target. One of the anchor constellations that have the biggest negative impact on localization performance is the collinear case, and several works have adopted methods to identify and discard such setups. It has been referred to as the 'pathological' case [8], and the goal is to avoid it. Specific lower bounds on thedegree of collinearity of anchor nodes sufficient to achieve optimal localization results have been proposed in [15]. The metric to measure collinearity is the height of the anchor triangle. The impact of anchor placement has also been studied in [16].

To our knowledge, there has not been work done in actually exploiting the near collinear case for localization purposes.

## 3. Preliminaries

### 3.1. Target Scenario

The target scenario we consider in this paper is illustrated in Fig. 1 that depicts the target node position in a 2-dimensional space with $N$near collinear anchors.

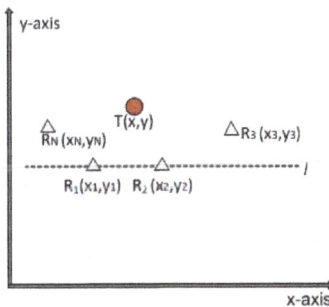

**Figure 1.**Target scenario.

The target node $T$ is located in a 2-dimensional space with unknown coefficients $(x,y)$. There are $N$ fixed near collinear located anchors. In order to quantify the degree of collinear between anchor nodes, we introduce the matrix $\mathbf{A}$ as in [8], which is formulated as:

$$\mathbf{A} = \begin{vmatrix} x_1 - x_N & y_1 - y_N \\ x_2 - x_N & y_2 - y_N \\ \vdots & \vdots \\ x_{N-1} - x_N & y_{N-1} - y_N \end{vmatrix} \quad (1)$$

This matrix contains information about geometrical configuration of anchor nodes. We use the condition number of matrix $cond(A)$ as an indicator of colinearity: in the extreme case when all three anchors are collinear, the matrix is singular and the condition number is infinite. If the condition number is too large, the matrix (in our case the scenario) is said to be ill-conditioned.

### 3.2. RSS-Based Distance Estimation

There are several methods for distance estimation, such as RSS [4], ToA [2], etc. Even though in practice it has been shown that the RSS ranging performs well in addition to location maps (fingerprinting), we adopt RSS method in this paper due to its simplicity. It is still most easy-to-apply method for practical systems, because there is no need for any additional hardware, neither for synchronization.

Supposed that the anchors $R_i$ transmit a signal, and the long-term averaged received signal strength at reference distance $d_0$ is $P_0$ (dBm), the long-term averaged received signal strength $P_i$ at the target node $T$ is formulated as

$$P_i(dBm) = P_0(dBm) - 10\alpha \log \frac{d_i}{d_0} + N(0,\sigma_i^2) \quad (2)$$

Here $d_i$represents the distance between $R_i$ and $T$, $\alpha$ represents the path-loss constant, and $N(0, \sigma^2)$ represents log-normal shadow fading variance. Employing the unbiased estimator proposed in [9]-[10], and based on the results of [11], the unbiased estimate of $d_i^2$is:

$$\tilde{d}_i^2 = e^{-\frac{2r_i}{\alpha} - \frac{2\lambda_i^2}{\alpha^2}} \quad (3)$$

In the above equation, the element

$$r_i = 0.1\ln(10)(P_i - P_0) - \alpha \ln(d_0), \; \lambda_i^2 = 0.01(\ln(10))^2 \sigma_i^2.$$

Moreover, the variance of the estimation is formulated as:

$$\text{var}(\tilde{d}_i^2) = d_i^4 (e^{\frac{4\lambda_i^2}{\alpha^2}} - 1) \quad (4)$$

Equation (4) demonstrates that the variance grows exponentially with the shadowing parameter $\sigma^2$ [11]. Since we assume equal channel conditions for all links, the variance of distance estimates will be proportional to the distance itself. Even though the assumption of equal link parameters does not reflect realistic scenarios, it serves its purpose to illustrate the algorithm performance.

### 3.3. Conventional Positioning Algorithms

Having the estimated distances $\tilde{d}_i^2$ and the knowledge of anchors' locations, the target node is capable of estimating its own location coefficients $(x, y)$ by exploiting the relationship $(x_i - x)^2 + (y_i - y)^2 = \tilde{d}_i^2$. The most common estimation algorithm is (linear-) Least Square (LS) [6], and its improved versions including the Weighted Least Square (WLS) [7], iterative least square (ILS) [9] etc. However, the above-mentioned algorithms require at least 3 well-conditioned anchors. The Maximum-likelihood (ML) estimation algorithm [4] does not have this constraint, but it requires iterative operations, which results in high computational complexity. Motivated by the insufficiency of current estimation algorithms, we propose a new Collinear Anchor aided Positioning (CAP) estimation algorithm, which is described in the next section.

# 4. Collinear Anchor Aided Positioning Algorithm

In this section, we describe the proposed algorithm given a scenario with $N = 3$ anchor nodes. The proposed algorithm can be extended to $N > 3$ scenario by first selecting 3 anchors having the most reliable estimation of distance, meaning having the smallest estimation variance according to Equation (4).

Supposed there are three near collinear anchors $R_1, R_2$ and $R_3$, $\operatorname{var}(\tilde{d}_1^2) \le \operatorname{var}(\tilde{d}_2^2) \le \operatorname{var}(\tilde{d}_3^2)$, the proposed CAP algorithm is detailed as follows:

Step 1: Employ the typical LS algorithm; obtain an initial estimation of the target node's location, which is represented as $(\tilde{x}^{(1)}, \tilde{y}^{(1)})$.

Step 2: By choosing two first two anchors $R_i$, $i = \{1, 2\}$, we have the following two equations:

$$(x_1 - \tilde{x})^2 + (y_1 - \tilde{y})^2 = \tilde{d}_1^2 \atop (x_2 - \tilde{x})^2 + (y_2 - \tilde{y})^2 = \tilde{d}_2^2 \qquad (5)$$

where $\tilde{d}_i^2$ is the estimated distance between node $R_i$ and node $T$ using the unbiased algorithm of Equation (3). Coordinates $(\tilde{x}, \tilde{y})$ represent the estimated coordinates. Since the anchors lie on the line $l$, without loss of generality, we choose the $x$-axis in parallel with this line $l$ to simplify the notations as shown in Fig.1. This condition can be easily achieved using transformations as translation, rotation and reflection. Hence, we have $y_1 = y_2$. By subtracting those two equations, after simple rearrangement we get an estimation of $x$:

$$\tilde{x}^{(2)} = \frac{\tilde{d}_2^2 - \tilde{d}_1^2 - (x_2^2 - x_1^2)}{2(x_1 - x_2)} \qquad (6)$$

By substituting $\tilde{x}^{(2)}$ into Equation (5), we have:

$$\tilde{y}^{(2+)} = \begin{cases} y_1 + \sqrt{\tilde{d}_1^2 - (x_1 - \tilde{x}^{(2)})^2} & \text{if} \quad \tilde{d}_1^2 - (x_1 - \tilde{x}^{(2)})^2 \ge 0 \\ 0 & \text{if} \quad \tilde{d}_1^2 - (x_1 - \tilde{x}^{(2)})^2 < 0 \end{cases}$$

$$\tilde{y}^{(2-)} = \begin{cases} y_1 - \sqrt{\tilde{d}_1^2 - (x_1 - \tilde{x}^{(2)})^2} & \text{if} \quad \tilde{d}_1^2 - (x_1 - \tilde{x}^{(2)})^2 \ge 0 \\ 0 & \text{if} \quad \tilde{d}_1^2 - (x_1 - \tilde{x}^{(2)})^2 < 0 \end{cases} \qquad (7)$$

From (7) we conclude that two estimates $\tilde{y}^{(\pm)}$ can be calculated. In cases where the solution is an imaginary number, e.g., when there is no solution, we decided to assume the value $y = 0$. This situation appears in case of a high $\sigma^2$ value. We obtain two second-step estimations of $y$: $\tilde{y}^{(2+)}$ and $\tilde{y}^{(2-)}$.

Step 3: So far we have three estimation results: $(\tilde{x}^{(1)}, \tilde{y}^{(1)})$, $(\tilde{x}^{(2)}, \tilde{y}^{(2+)})$, and $(\tilde{x}^{(2)}, \tilde{y}^{(2-)})$. The final estimation is chosen as follows:

$$\tilde{x}^{(3)} = \tilde{x}^{(2)} \qquad (8)$$

$$\tilde{y}^{(3)} = \underset{\tilde{y}^{(2+)}, \tilde{y}^{(2-)}}{\arg\min} \left\{ \left| \tilde{y}^{(2+)} - \tilde{y}^{(1)} \right|, \left| \tilde{y}^{(2-)} - \tilde{y}^{(1)} \right| \right\} \qquad (9)$$

The advantage of the CAP algorithm is that it avoids ill-conditioned matrix inversion. Hence it outperforms traditional localization algorithms when having near collinear anchors as shown in the next section.

# 5. Analysis and Simulation Results

In order to evaluate the performance of our proposed algorithm, we perform simulations in MATLAB. We compare the performance of the CAP algorithm to the state-of-the-art algorithms, specifically LS, WLS and ML, and also include the Cramér-Rao lower bound (CRLB) as a performance indicator. We consider a scenario having the minimum number of anchors (three) and one target. The target $T$ is placed at fixed coordinates (50, 50), namely in the center of a 100 x 100 m square area. We assume that the path loss value is the same throughout all area, namely $\alpha = 3$. We compute the root mean square error (RMSE) for different algorithms by running 1000 simulation runs, for shadowing variance values between -20 and 20 dB.

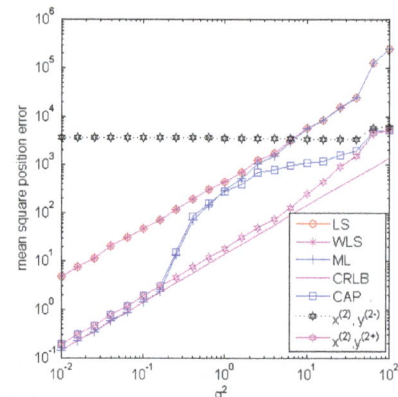

***Figure 2.*** *Position estimation RMSE versus shadowing variance for different positioning algorithms.*

In the first simulation, three anchors are fixed at coordinates (20, 20) and (70, 20), and (90, 30), results in matrix condition number $cond$(A)=10.9. The position estimation RMSE versus shadowing variance using different positioning algorithms is illustrated in Fig. 2. It shows that at low shadowing scenario $\sigma^2 = 0.1$, the RMSE using LS and WLS is 6.75 meter, whilst using the proposed CAP the RMSE is 1.37 meter. Note that Fig. 2 has logarithmic scale. The proposed algorithm is about 5 times more accurate than the LS and WLS algorithms. Moreover, at medium scenario $\sigma^2 = 1$, the performance of the proposed CAP algorithm is reduced, but still showing better performance than the LS and WLS algorithm. Furthermore, at high shadowing cases $\sigma^2 = 10$, the estimation error using CAP becomes again significantly better than the conventional LS and WLS. The performance of CAP algorithm is comparable with the performance of the ML algorithm in low and medium shadowing conditions. The ML estimator is usually implemented using the expectation-maximization (EM) iterative algorithm, which may converge to a local maximum depending on starting values. The relatively poor performance of ML in high shadowing scenario shown in Fig. 2 is due to using the EM algorithm, and using LS estimates as its starting value.

As illustrated in Fig. 2, the mean square error using ML is close to the CRLB when the shadowing variance is low, which means the distance estimation is relatively accurate. It increases as the shadowing variance increases, and finally converge to the curve using LS estimation. To better illustrate this phenomena, we demonstrate the spatial distribution of the estimated results using LS and ML for an ill-conditional scenario in Fig. 3. The three anchor nodes are located at fixed locations, namely (20, 20) and (70, 20), and (90, 30) (all in meters), while the target is in the center of the area, at coordinate (50,50).The lower the shadowing variance, the estimations are less scattered around the mean value.

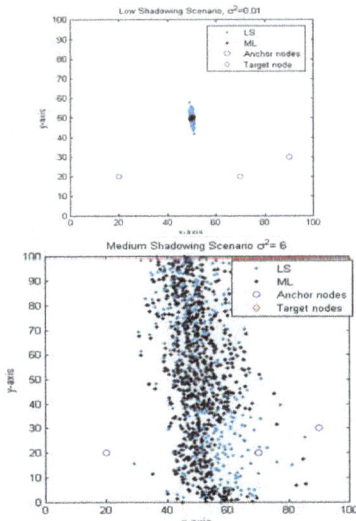

**Figure 3.** *Three anchors nodes, the target node, the estimates using LS, and the estimates using ML (implemented via EM iterative algorithm), when the shadowing variance is 0.01 (upper) and 6 (bottom).*

The ML estimates are capable of converging to the true target nodes when the shadowing value is 0.01.However, the ML estimates become similar to the LS estimates in a scenario with higher shadowing. Although randomly choosing multiple starting values can improve this drawback of EM, it still cannot guarantee the convergence to the global optimum, while it significantly increases the complexity. To sum up, ML should outperform our proposed algorithm in However, using the practical EM algorithm, ML becomes worse than our proposed algorithm in practice. In a high shadowing scenario, the performance of the ML algorithm converges to the LS performance since it employs the LS estimate as its initial value before iteration. Having this implementation in mind, our proposed CAP algorithm slightly outperforms the ML solution at high values of shadowing variance.

On the other hand, the proposed CAP algorithm is much more computationally efficient. Since the ML involves computationally demanding mathematical operations such as matrix inversion and matrix multiplication coupled with the iterative procedure, its complexity will be in the order of $N_{iter} * O(N^3)$, where $N_{iter}$ is the number of iterations (in our simulations we used $N_{iter} = 20$), and $N$ the number of anchor nodes. On the other hand, the CAP algorithm only involves simplest algebraic operations, such as addition, subtraction and division. Therefore its complexity is in the order of $O(N)$.

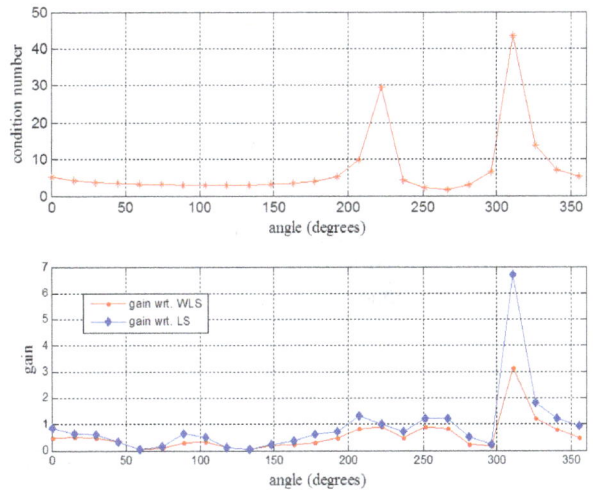

**Figure 4.** *Condition number as metric for ill-conditioned scenario (top), and the gain of CAP vs. LS and WLS(bottom).*

In the second simulation, the first two anchors remain at (20, 20) and (70, 20), while the third anchor is moved along a circle with radius 45m in steps of $15^0$. In Fig.4 (top) we show the plot of the condition number versus the angle of the third anchor with respect to line $l$ for a full $360^0$ circle. Furthermore, we compute the *gain* as the ratio between the RMSE using LS/ WLS and the RMSE using CAP. Hence, if *gain*$\leq$ 1, the proposed CAP algorithm provides less or equal estimation accuracy. If *gain*> 1, the proposed CAP algorithm provides higher estimation accuracy compared to the

LS/WLS algorithm. As we can see from Fig. 3, by setting $\sigma^2 = 0.1$, our CAP algorithm outperforms both the LS and WLS for ill-conditioned scenarios having large matrix condition value.

The gain over LS is larger than the gain over WLS, which indicates that the WLS is more robust to the ill-conditioned scenario than LS algorithm. We can also conclude that the condition number is a good indicator of accuracy.

## 6. Conclusion

We have proposed the CAP algorithm that performs well under ill-conditioned scenarios where the anchor nodes are almost collinear; which can be likely in public safety scenarios. The proposed algorithm has been shown to provide up to seven times more accuracy than the traditional LS and three times more than WLS algorithms, whilst showing almost three orders of magnitude less in terms of complexity.

In our future work we intend to apply different channel parameters for each link, since the identical channel model does not represent the realistic indoor channel conditions in the best way. However, our assumption serves well for the evaluation of the proposed method. We also aim at analyzing different simulation setups, for various geometric configurations of anchor nodes and location-unaware nodes.

## Acknowledgment

This work has been supported in part by the Portuguese Foundation for Science and Technology (Fundação para aCiência e Tecnologia - FCT) under grant number SFRH / BD / 61023 / 2009.

The research leading to these results has received funding from FEDER through Programa Operacional Factores de Competitividade - COMPETE and from National funds from the Portuguese Foundation for Science and Technology (Fundação para a Ciência e Tecnologia – FCT) under the project PTDC/EEA-TEL/119228/2010 – SMARTVISION.

## References

[1]    N. A. Alsindi, K. Pahlavan, and B. Alavi, "An Error Propagation Aware Algorithm for Precise Cooperative Indoor Localization", in Proceedings of IEEE Military Communications Conference MILCOM 2006, pp. 1-7, Washington, DC, USA, October 2006.

[2]    X. Wang, Z. Wang and B. O'Dea, "A TOA-based location algorithm reducing the errors due to non-line-of-sight (NLOS) propagation", in IEEE Transactions on Vehicular Technology, vol.52, issue 1, pp.112-116, Jan. 2003.

[3]    N. Patwari, A.O. Hero, M. Perkins, N.S.Correal, R.J. O'Dea, "Relative location estimation in wireless sensor networks," in IEEE Transactions on Signal Processing, vol. 51, no. 8, pp. 2137-2148, Aug. 2003.

[4]    I. Guvenc, S. Gezici, and Z. Sahinoglu, "Fundamental limits and improved algorithms for linear least-squares wireless position estimation," in Wiley Wireless Communications and Mobile Computing, Sep. 2010.

[5]    P. Tarrio, A.M. Bernardos, J.A. Besada and J.R. Casar, "A new positioning technique for RSS-Based localization based on a weighted least squares estimator," in Proc of IEEE International Symposium on Wireless Communication Systems, Reykjavik, Iceland, pp.633-637, Oct. 2008.

[6]    P. Rong and M. L. Sichitiu, "Angle of Arrival Localization for Wireless Sensor Networks ", in Proceedings of 3rd Annual IEEE Communications Society on Sensor and Ad Hoc Communications and Networks (SECON '06), vol.1, pp.374-382, Sept. 2006.

[7]    P. Bahl, and V. N. Padmanabhan, "RADAR: An in-building RF-based user location and tracking system," in InfoCom 2000, Tel Aviv, Israel, pp.775-784, March 2000.

[8]    J. Liu, Y. Zhang and F. Zhao, "Robust Distributed Node Localization with Error Management", in Proc. of the 7th ACM international symposium on Mobile ad hoc networking and computing (MobiHoc '06), pp.250-261, New York, USA, 2006.

[9]    H.C.So and L.Lin, "Linear least squares approach for accurate received signal strength based source localization," IEEE Transactions on Signal Processing, vol.59, no.8, pp.4035-4040, August 2011.

[10]   L.Lin and H.C.So, "Best linear unbiased estimator algorithm for received signal strength based localization," Proc. 2010 European Signal Processing Conference, Barcelona, Spain, pp.1989-1993, Aug. 2011.

[11]   S.D. Chitte, S. Dasgupta and Z. Ding, "Distance estimation from received signal strength under log-normal shadowing: bias and variance," IEEE Signal Processing Letters, vol.16, no.3, pp.216-218, Mar. 2009.

[12]   Zheng Yang; Yunhao Liu, "Quality of Trilateration: Confidence-Based Iterative Localization," IEEE Transactions on Parallel and Distributed Systems, vol.21, no.5, pp.631,640, May 2010.

[13]   Stefan O. Dulman, AlineBaggio, Paul J.M. Havinga, and Koen G. Langendoen, " A geometrical perspective on localization," Proceedings of the first ACM international workshop on Mobile entity localization and tracking in GPS-less environments (MELT '08). ACM, New York, NY, USA, 85-90, 2008.

[14]   Salman, N.; Maheshwari, H.K.; Kemp, A.H.; Ghogho, M., "Effects of anchor placement on mean-CRB for localization," Ad Hoc Networking Workshop (Med-Hoc-Net), 2011 The 10th IFIP Annual Mediterranean , vol., no., pp.115,118, 12-15 June 2011.

[15]   Kunz, T.; Tatham, B., "Localization in Wireless Sensor Networks and Anchor Placement," J. Sens. Actuator Netw. 1, no. 1, pp. 36-58, 2012.

[16]   Tatham, B.; Kunz, T., "Anchor node placement for localization in wireless sensor networks," Wireless and Mobile Computing, Networking and Communications (WiMob), 2011 IEEE 7th International Conference on , vol., no., pp.180,187, 10-12 Oct. 2011.

# 29

# A CPW-fed triangular monopole antenna with staircase ground for UWB applications

**Madhuri K. Kulkarni**[*], **Veeresh G. Kasabegoudar**

Post Graduate Department, Mahatma Basveshwar Education Society's, College of Engineering, Ambajogai, India, 431 517

**Email address:**

kulkarni3388@gmail.com (M. K. Kulkarni), veereshgk2002@rediffmail.com (V. G. Kasabegoudar)

**Abstract:** In this paper, a monopole antenna with triangular shaped patch based on coplanar waveguide (CPW)-feed is presented. The proposed antenna comprises a planar triangular patch element with a staircase ground which offers ultra wide bandwidth. The impedance bandwidth can be tuned by changing staircase shaped ground parameters (number of steps, step length, and/or its width). The overall size of the antenna is 28mm×26mm×1.6mm including finite ground CPW feeding mechanism. The antenna operates in the frequency range from 3.2-12GHz covering FCC defined UWB band with more than 100% impedance bandwidth. Stable omni-directional radiation patterns in the desired frequency band have been obtained. Measured data fairly agree with the simulated results.

**Keywords:** Microstrip Antenna, Coplanar Waveguide, and Monopole Antenna

## 1. Introduction

Due to increasing demands of wireless communication systems it is necessary to design antenna with compact size, wide bandwidth, and low profile [1]. One of the best choices for this is planar monopole antennas. These are low profile, etched on a dielectric substrate and can provide feature of broadband and multiband. Planar monopole antennas can be used for UWB applications due to their wideband characteristics [2, 3].

The FCC authorization for the unlicensed use of UWB in the frequency range of 3.1 to 10.6 GHz [4] encourages the designers to go for UWB antennas. It is well known that the CPW feed offers several advantages over microstrip feeds such as ease of mounting electronic components due to coplanar nature, no need to drill through substrate, and easy transition to slot line [5]. Hence, these feeds are preferred for applications where size is the constraint. There are several CPW fed antennas presented in the literature for UWB applications [6-14]. Antenna reported in [6] is CPW- fed fractal slot antenna which is having omni-directional radiation patterns over the operating frequency band. Another CPW-fed ultra-wideband (UWB) antenna with dual band-notched characteristics presented in [7] has little interference among WLAN and WiMAX. A compact CPW fed UWB antenna reported in [8] is designed using copper tape as radiating element and denim textile as substrate.

However, the radiation patterns of this geometry are not omni-directional over the entire operating frequency band. CPW-fed tapered ring slot antenna reported in [9] is limited by distorted radiation patterns. Most of the antennas presented in literature have complex geometry or large ground planes [10-12]. In this work, a coplanar waveguide fed triangular monopole antenna is presented. The antenna presented here is simple in structure and has a compact size. The proposed antenna consists of an inverted triangular strip with staircase shaped ground. Also, by modifying the ground geometry dual band characteristics can be obtained. The antenna is simulated using Ansoft's High Frequency Structure Simulator (HFSS) v.10 [15]. More details on this geometry are discussed in subsequent sections. Section 2 presents the geometry structure of the proposed antenna. Design and optimization procedure of the proposed antenna is presented in Section 3. Section 4 presents the validation of the fabricated prototype and discussions on the measured results are also presented there. Finally, conclusions of this study are presented in Section 5.

## 2. Antenna Geometry

Figure 1(a) shows the basic geometry of proposed triangular monopole antenna for UWB operation. The antenna is symmetrical with respect to the longitudinal direction. Substrate used for the design is FR4 with thickness of 1.6mm, dielectric constant of 4.4, and loss tangent equal to 0.02. Two equal finite ground planes, each with dimensions of length $L_g$ and width $W_g$ are placed symmetrically on either side of the CPW feed-line. It is well known that defected ground structure (DGS) helps in reducing the overall size of antenna [16], a staircased profile DGS has been introduced in the ground plane of proposed antenna. Besides reducing the size, DGS also alters the reactive part of the antenna's input impedance [17] and hence impedance bandwidth of the antenna may be tuned. Detailed analysis of modification of ground structure and its effect is discussed in Section 3.1. Triangular shaped patch with rectangular base is fed by signal conductor. The detailed optimization procedure of the proposed antenna and its optimum dimensions, and characteristics are presented in Section 3. All the parameters of the geometry are indicated in Figure 1(a).

# 3. Geometry Optimization and Discussions

In this section parametric study is conducted to optimize the proposed antenna. The key design parameters used for the optimization are number of steps in the staircase shaped ground plane, length and width of the staircase profile, and rectangular base dimensions. The detailed analysis of these parameters is investigated in the following paragraphs of this section.

## 3.1. Effect of Ground Geometry

As showed in Figure 2, ground plane of the geometry is modified to see its effect on the performance of antenna. For this, ground plane is changed to staircase shape. The number of steps in the staircase shaped geometry is changed from zero to three. In order to understand the effect of number steps in the staircase profile on radiation performance of the geometry, surface currents distributions on the geometry have been investigated (Figure 2). From the currents distribution on the geometry it may be noted that staircase ground with two steps exhibits better radiation than others. Furthermore, sharp edges of the staircase have been changed to smooth edges. Except slight improvement in the impedance bandwidth no significant change in the performance of the geometry was observed. However, smooth transitions in the geometry may be adopted if beam tilting is the major issue i.e., to avoid the beam tilting [18].

Return loss characteristics of this study are presented in Figure 3. From Figure 3 it may be noted that with two steps in the ground offers wide bandwidth whereas for one step, the return loss characteristics split in the middle. The optimum height and width for each step is achieved in two steps. In the first attempt height of the staircase is varied in steps of 0.5mm by keeping its width and all other parameters constant. After optimizing the height, width of the staircase is optimized. Figures 4 and 5 show return loss characteristics plots of this study. From these figures it may be noted that the wide bandwidth can be obtained for $h$=2.5mm and $w$=2.5mm. The percentage bandwidths obtained from these parametric studies are presented in Tables 1 & 2.

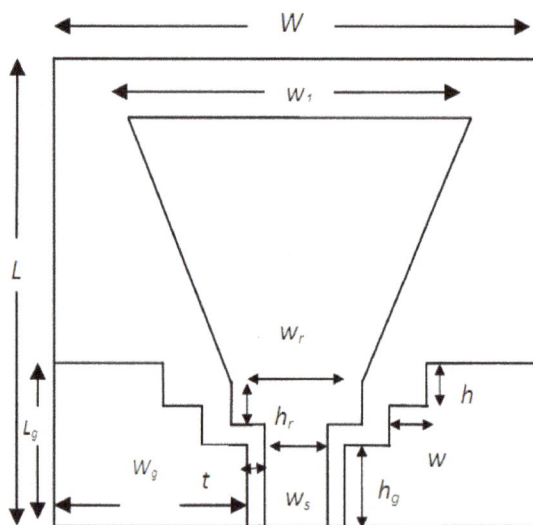

*(a) Geometry*

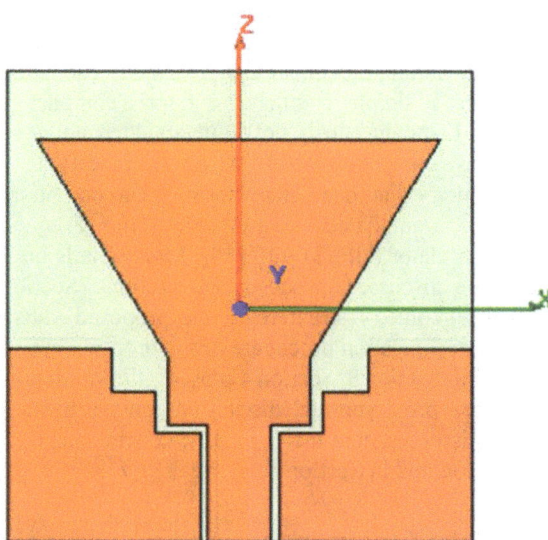

*(b) Simulation setup in HFSS*

**Figure 1.** *Geometry of proposed CPW-fed monopole antenna.*

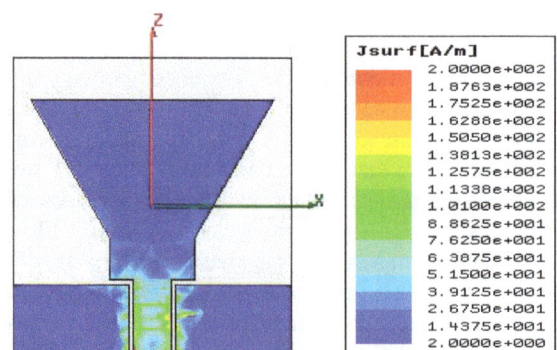

*a) No step*

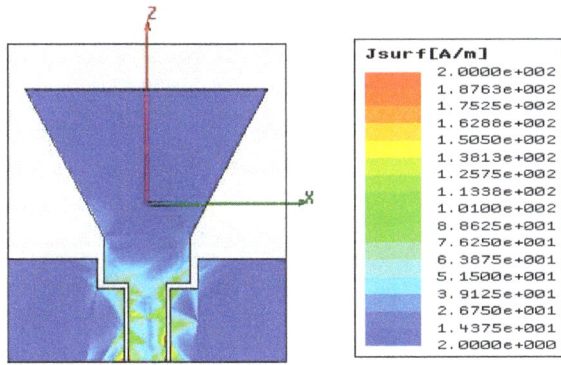

b) One step

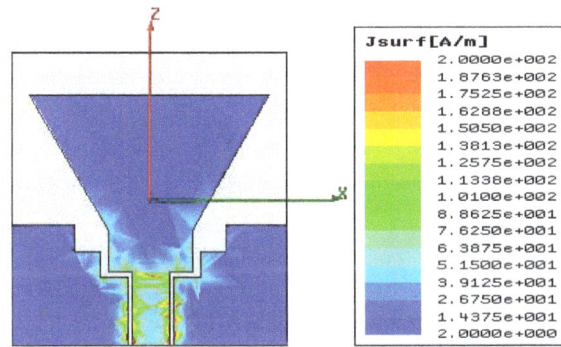

c) Two steps

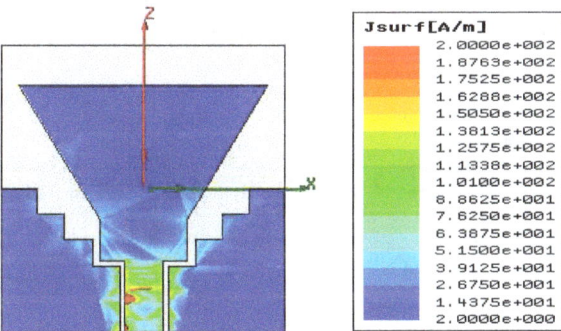

d) Three steps

**Figure 2.** Variation in number of steps in the staircase profile of ground and current distributions.

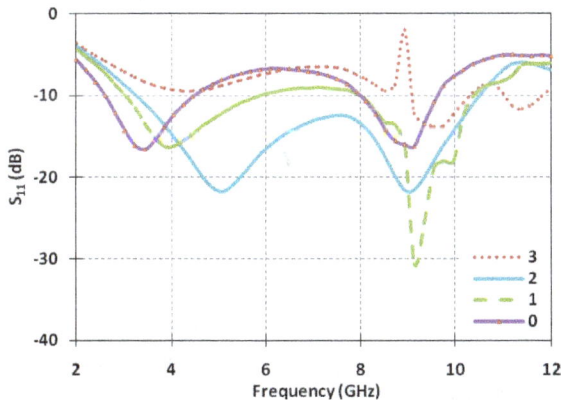

**Figure 3.** Return loss vs. frequency plot for variation in number of steps as indicated in Figure 2.

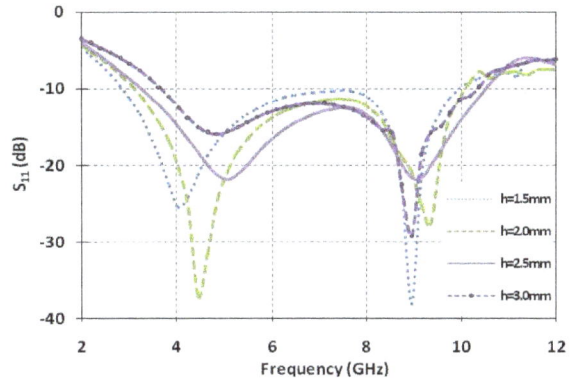

**Figure 4.** Return loss vs. frequency plot for variable step height (h).

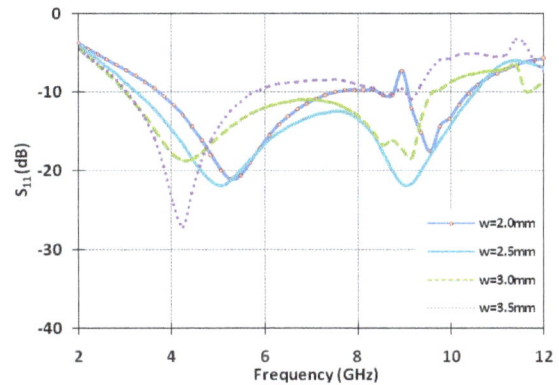

**Figure 5.** Return loss vs. frequency plot for variable step width (w).

**Table 1.** Effect of variation of step height on bandwidth of proposed antenna.

| Step height ($h$)(mm) | Frequency range (GHz) | Bandwidth (%) |
|:---:|:---:|:---:|
| 1.5 | 2.81-7.51 | 91.08 |
| 2.0 | 3.22-10.16 | 103.73 |
| 2.5 | 3.22-10.57 | 106.67 |
| 3.0 | 3.63-10.36 | 96.28 |

**Table 2.** Effect of variation of step width on bandwidth of proposed antenna with height=2.5mm.

| Step width (w)(mm) | Frequency range (GHz) | Bandwidth (%) |
|:---:|:---:|:---:|
| 2.0 | 3.83-7.51 | 64.90 |
| 2.5 | 3.22-10.57 | 106.67 |
| 3.0 | 3.02-9.55 | 103.98 |
| 3.5 | 3.02-5.87 | 64.18 |

### 3.2. Effect of Rectangular Base Dimensions

To study the effect of rectangular base dimensions on the antenna performance, its height ($h_r$) and width ($w_r$) are varied. Figures 6 (a) and (b) depict the current distributions on triangular monopole antenna without and with rectangular base. From the current distribution it is clear that better radiation effect can be observed in triangular monopole antenna mounted on rectangular base. This is mainly due to additional loading of reactance part to antenna's input impedance [17] and hence input impedance of the whole geometry can be fine tuned to get optimum performance. As in the case of staircase profile, edges of rectangular base was also made smooth but no significant improvement was obtained.

Initially, the height of the rectangle ($h_r$) is varied from 3mm to 4.5mm in steps of 0.5mm keeping width of the rectangular base constant (8mm). These effects on return loss characteristics are presented in Figure 7. From Figure 7, it may be noted that the lower cut-off frequency remains nearly constant whereas upper cut-off frequency varies slightly i.e., impedance bandwidth varies with respect to this parameter ($h_r$). From Table 3 it is found that $h_r = 4$mm offers the maximum bandwidth of 106.67%. With this optimum height ($h_r$), width of the base ($w_r$) was varied from 6.5mm to 8.5mm in steps of 0.5mm. Effect of width on return loss characteristics of the antenna are presented in Figure 8. Tables 3 & 4 illustrate the effect of $h_r$ and $w_r$ on percentage bandwidth. Accordingly, the optimum values of $h_r$ and $w_r$ are found to be 4mm and 8mm, respectively.

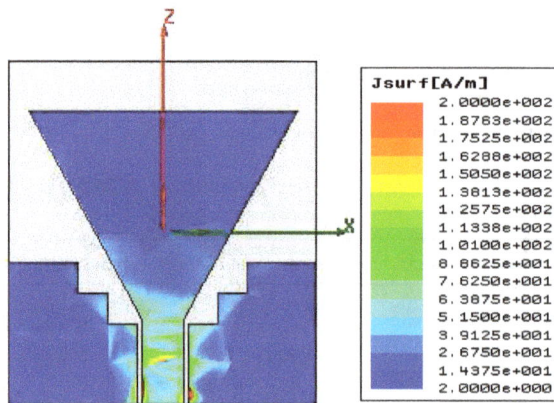

*(a) Geometry without rectangular base*

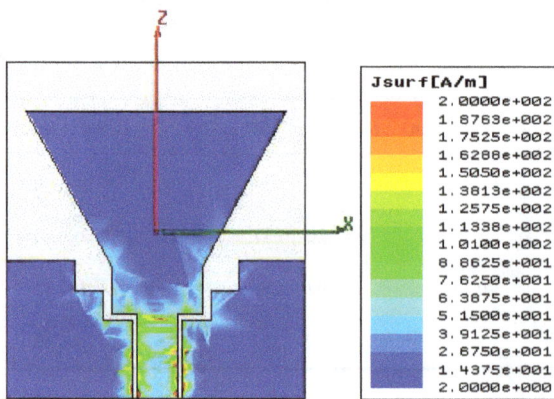

*(b) Geometry with rectangular base*

**Figure 6.** *Current distribution to show effect of rectangular base below triangular patch.*

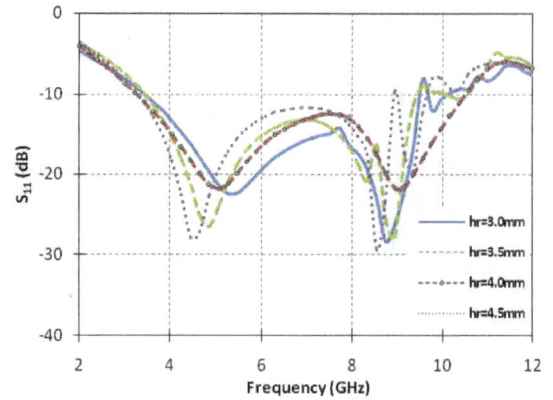

**Figure 7.** *Return loss vs. frequency plot with variable rectangle height ($h_r$).*

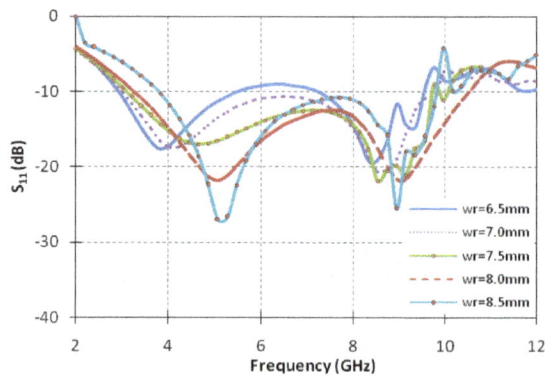

**Figure 8:** *Return loss vs. frequency plot with variable rectangle height ($w_r$).*

**Table 3.** *Effect of variation of rectangular base's height on bandwidth of proposed antenna.*

| Step height ($h_r$)(mm) | Frequency range (GHz) | Bandwidth (%) |
|---|---|---|
| 3.0 | 3.42-9.55 | 94.59 |
| 3.5 | 3.42-9.55 | 94.59 |
| 4.0 | 3.22-10.57 | 106.67 |
| 4.5 | 3.22-9.75 | 100.77 |

**Table 4.** *Effect of variation of rectangular base's width on bandwidth of proposed antenna.*

| Step height ($w_r$)(mm) | Frequency range (GHz) | Bandwidth (%) |
|---|---|---|
| 6.5 | 2.81-5.46 | 64.16 |
| 7.0 | 3.02-9.75 | 105.48 |
| 7.5 | 3.22-9.75 | 100.77 |
| 8.0 | 3.22-10.57 | 106.67 |
| 8.5 | 3.63-9.95 | 93.07 |

The proposed geometry is simulated with HFSS software (pl. ref. its setup in Figure 1(b)) which is FDTD based electromagnetic (EM) software. Optimized parameters of the proposed geometry are listed in Table 5.

**Table 5.** *Optimized parameters of geometry shown in Figure 1.*

| Parameter | Value(mm) |
|-----------|-----------|
| W | 26 |
| $w_1$ | 22.7 |
| L | 28 |
| $L_g$ | 11.5 |
| $W_g$ | 10.8 |
| w | 2.5 |
| h | 2.5 |
| t | 0.4 |
| $w_r$ | 8.0 |
| $h_r$ | 4.0 |
| $w_s$ | 3.6 |
| $h_g$ | 6.5 |

## 4. Experimental Validation of the Geometry and Discussions

The geometry shown in Figure 1 with its optimized dimensions presented in Table 5 was fabricated and tested. The substrate used for the fabrication is the FR4 glass epoxy with dielectric constant of 4.4, loss tangent (tan ($\delta$)) = 0.02, and thickness of 1.6mm. A photograph of the fabricated prototype is shown in Figure 9(a) and its $S_{11}$ measurement setup is shown in Figure 9(b). Return loss comparisons of measured and simulated values are compared in Figure 10. The measured results fairly agree with the simulated values.

From Figure 10 it may be noted that the proposed antenna is having the operating frequency range from 3.2GHz to 12GHz. This corresponds to an impedance bandwidth of about 115%. Radiation patterns of the geometry are presented at various frequencies in the band of operation (Figure 11). These patterns are symmetrical at the start and middle frequencies of the band of operation with omni-directional shape in the H-plane. However, at the end frequencies of operating band these patterns degrade. This degradation is mainly because of the change in the effective area over wide range of frequencies especially at high frequency end of operating band [18].

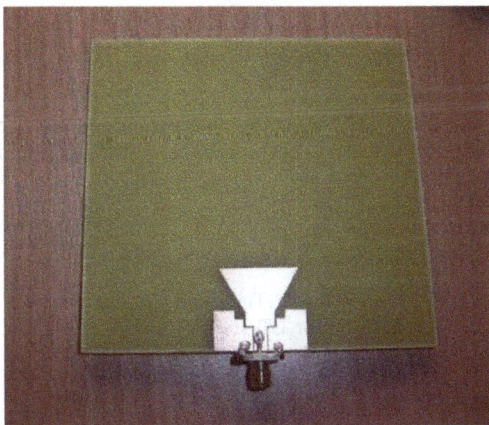

*b) $S_{11}$ measurement setup*

**Figure 9.** *Photographs of the fabricated antenna and its measurement setup.*

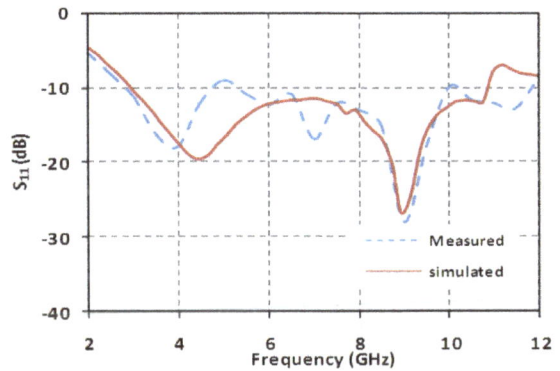

**Figure 10.** *Return loss characteristic plot of the proposed antenna shown in Figure 1.*

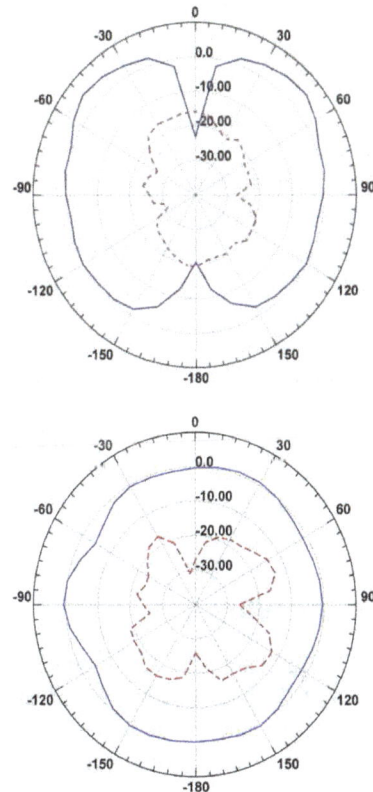

*a) Fabricated prototype*

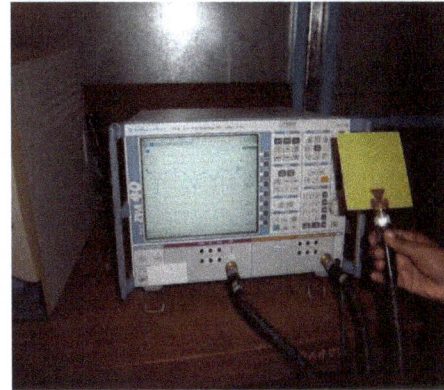

*(a) E- and H-plane patterns at 3.0 GHz*

*(b) E-and H-plane patterns at 6.7 GHz*

*(c) E- and H-plane patterns at 10.5 GHz*

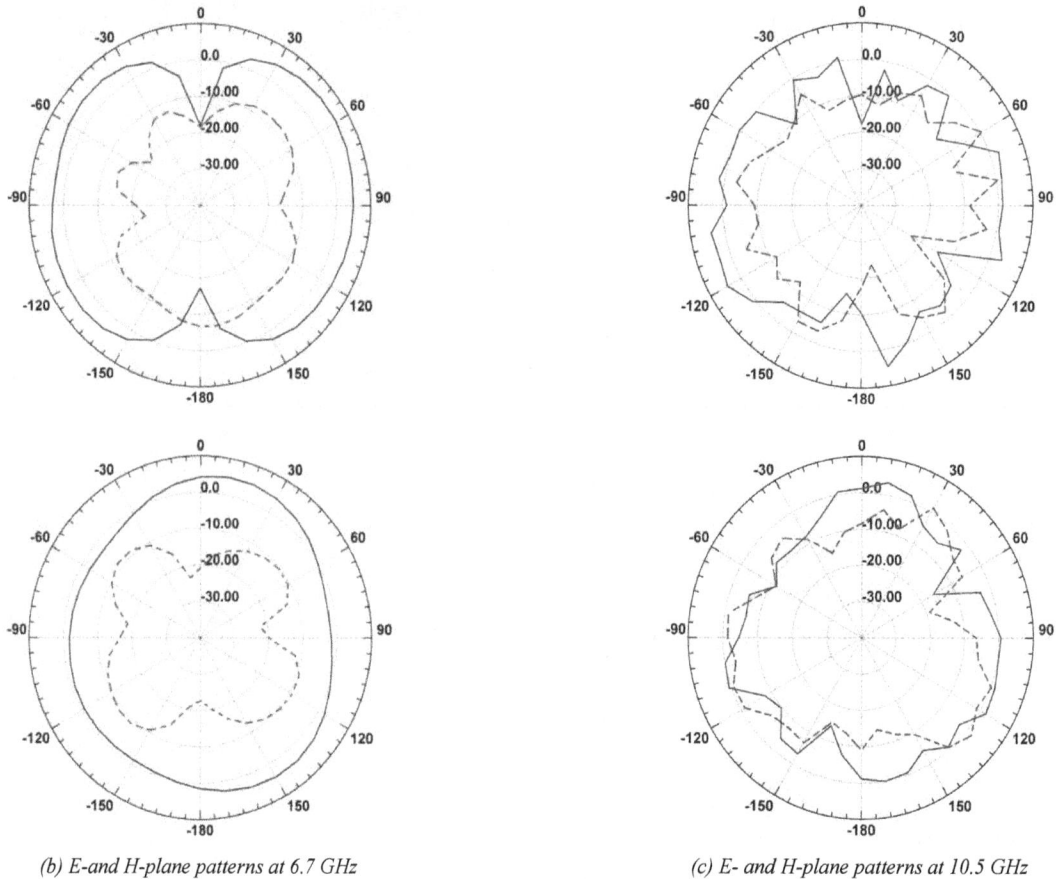

**Figure 11.** *E- and H- plane radiation patterns at various frequencies throughout the band of operation (Upper: E-plane patterns; Lower: H-plane patterns; Solid lines: Co-polarization patterns; Dashed lines: Cross-polarization patterns).*

## 5. Conclusions

A triangular shaped monopole antenna mounted on rectangular base with finite ground has been presented. Staircase shaped ground geometry was varied (number of steps, step size, and step width) to tune antenna's input impedance. The antenna presented here offers an impedance bandwidth of more than 100% in the frequency range of 3.2GHz to 12GHz which essentially covers the FCC defined ultra wideband (UWB) frequency range. Omni-directional radiation patterns were obtained throughout the band of operation. Also, it was observed that for one step in the staircase shaped ground, the geometry offers dual band operation with better gain in the desired bands. Therefore, the antenna presented here is a suitable candidate for the FCC defined UWB band of operation.

## Acknowledgements

We would like to thank Dr. C. Bhattacharya and his research students of Defence Institute of Advanced Studies (DIAT-DU), Girinagar, Pune, India for allowing and helping in measurements of fabricated prototypes.

## References

[1]   X. Y. Teng, X. M. Zhang, Z. X. Yang, Y. Wang, Y. Li, Q. F. Dai, and Z. Zhang, "A compact CPW-fed omni-directional monopole antenna for WLAN and RFID applications," *Progress In Electromagnetics Research Letters,* vol. 32, pp. 91-99, 2012.

[2]   H. D. Chen and H. T. Chen, "A CPW-fed dual-frequency monopole antenna," *IEEE Transactions on Antennas and Propagation*, vol. 52, no. 4, pp. 978-982, 2004.

[3]   G. Kumar and K. P. Ray, *Broadband Microstrip Antennas,* Artech House Boston, London.

[4]   First Report and Order, Revision of part 15 of the commission's rule regarding ultra-wideband transmission systems FCC 02-48, Federal Communications Commission, 2002.

[5]   K. C. gupta, R. Gerg, I. Bahl and P. Bhartia, *Microstrip Lines and Slotline,* Artech House, Boston, London.

[6]   H. Zhang, H. Y. Xu, B. Tian, and X. F. Zeng, "CPW-fed fractal slot antenna for UWB application," *Int. J. Antennas and Propagat.,* pp. 1-4, vol. 2012 (Article ID 129852).

[7]   Z.-A. Zheng and Q.-X. Chu, "Compact CPW-fed UWB antenna with dual band notched characteristics," *Progress In Electromagnetics Research Letters,* vol. 11, pp. 83-91, 2009.

[8]   M. E. Jalil, M. K. Rahim, M. A. Abdullah, and O. Ayop, "Compact CPW-fed ultra-wideband (UWB) antenna using denim textile material," *Proceedings of ISAP2012*, Nagoya, Japan.

[9]   T.-G. Ma and C.-H. Tseng, "An ultra wideband coplanar waveguide-fed tapered ring slot antenna," *IEEE Trans. Antennas Propagat.*, vol. 54, no. 4, pp. 1105-1110, 2006.

[10]  W.-P. Lin and C.-H. Huang, "Coplanar waveguide-fed rectangular antenna with an inverted-L stub for ultra wideband communications," *IEEE Antennas and Wireless Propagation Letters*, vol. 8, pp. 228-231, 2009.

[11]  A. C. Shagar and R. S. D. Wahidabanu, "New design of CPW-fed rectangular slot antenna for ultra wideband applications," International Journal of Electronics Engineering, 2(1), pp. 69-73, 2010.

[12]  R. V. Ram Krishna and R. Kumar, "Design of temple shape slot antenna for ultra wideband applications," *Progress In Electromagnetics Research B*, vol. 47, pp. 405-421, 2013.

[13]  A. K. Gautam, S. Yadav, and B. K. Kanaujia, "A CPW-Fed compact UWB microstrip antenna," *IEEE Antennas and Wireless Propagation Letters*, vol. 12, pp. 151-154, 2013.

[14]  J. Pourahmadazar, Ch. Ghobadi, J. Nourinia, N. Felegari, and H. Shirzad, "Broadband CPW-fed circularly polarized square slot antenna with inverted-L strips for UWB applications," *IEEE Antennas and Wireless Propagation Letters*, vol. 10, pp. 369-372, 2011.

[15]  HFSS10.0 User's Manual, Ansoft Corporation, Pittsburgh.

[16]  M. N. Moghadasi, R. Sadeghzadeh, L. Asadpor, and B. S. Virdee, "A small dual-band CPW fed monopole antenna for GSM and WLAN applications," *IEEE Antennas and Propoagat.* Lett., vol. 12, pp. 508-511, 2013.

[17]  L. H. Weng, Y. C. Guo, X. W. Shi, and X. Q. Chen, "An overview on defected ground structure," *Progress In Electromagnetics Research B*, vol. 7, pp. 173-189, 2008.

[18]  X. -C. Yin, C. -L. Ruan, C. -Y. Ding, and J. -H. Chu, "A planar U type monopole antenna for UWB applications," *Progress In Electromagnetics Research Letters*, vol. 2, pp. 1-10, 2008.

# 30

# Coplanar capacitive coupled compact microstrip antenna for wireless communication

**Swati Dhondiram Jadhav**[*], **Veeresh Gangappa Kasabegoudar**

Post Graduate Department, Mahatma Basveshwar Education Society's, College of Engineering, Ambajogai, India, 431 517

**Email address:**

swatijadhav5@gmail.com(S. D. Jadhav), veereshgk2002@rediffmail.com.com(V. G. Kasabegoudar)

**Abstract:** In this paper, a coplanar capacitive coupled suspended microstrip antenna with reduced air gap (compact) for wireless applications is presented. Suspended microstrip antennas offer wide bandwidth due to the reduced effective dielectric constant and surface waves. However, air gap in the suspended configuration is the prime constraint in compact applications. Therefore, in this paper it is demonstrated that the air gap can be significantly reduced by utilizing the previously reported approaches and modifying them would result in significant reduction compared to the results reported earlier for the similar antenna geometries. The designs presented here exhibit the fractional impedance bandwidth of nearly 25% for all cases studied in the frequency range of 2-10 GHz.

**Keywords:** Microstrip Antennas, Suspended Configuration, Capacitive Coupling, Compact Antennas

## 1. Introduction

Microstrip patch antennas are rapidly developing in modern wireless communications, satellite and missile applications due to their attractive features such as low profile, light weight, and ease of fabrication. However, they have inherently narrow impedance bandwidth. There are various methods to overcome this serious drawback to achieve ultra-wideband (UWB) performance. Since the frequency range of UWB wireless standard has been allocated from 3.1GHz to 10.6 GHz [1], nowadays obtaining a compact wideband antenna is exclusively noticeable in commercial and military systems. The antennas proposed here operate in L-, S-, C-, and X-bands, which are suitable for UWB applications such as wireless personal area network (WPAN), microwave imaging for detecting tumors and cancers, medical monitoring, and home networking applications. There are several techniques for impedance bandwidth enhancements which are reported by many researchers such as by using impedance matching network [2], stacked patches [3], using edge coupled parasitic patches [2-4], by changing the shape of geometry [5], by cutting slots into basic stapes [6]. Broad banding techniques can also be achieved by modifying the probe shape [7-9]. At the time of the fabrication and assembly some bandwidth enhancement techniques pose problems due to the use of multiple metal or dielectric layers in the geometries.

There are several microstrip patch antennas available in literature which utilize single layer. For example, single layer coplanar capacitive coupled probe fed suspended microstrip antennas [9-15] are simple to realize and offer high impedance bandwidth of nearly 50%. It is well known that bandwidth enhancement can be achieved by increasing the overall height of the composite air-dielectric medium [10-14]. However, the use of air gap increases the height/volume of the antenna which is undesired in several compact applications [15-18].

In this paper we have taken the compact geometry reported in [15] for the further investigation and tried to reduce the air gap further by more than 50% which is a significant reduction and useful in several compact applications. Section 2 presents the basic geometry of the antenna. The procedure of air gap reduction by selecting any center frequency is presented in Section 3. Parametric studies for optimization of air gap are presented in Section 4 followed by conclusions in Section 5.

## 2. Basic Geometry

Basic geometry of the antenna is shown in Figure 1 and its optimized dimensions are listed in Table 1 [15]. It consists of radiating patch and a rectangular feed strip which are etched on the same plane. The substrate is placed above the ground plane at a height equal to air-gap. This height of air gap plays a vital role for broad banding. The substrate used for the antenna design is a Rogers' make RO3003 with dielectric constant=3.0, loss tangent (tan ($\delta$)) = 0.0013 and thickness ($h$) = 1.56 mm.

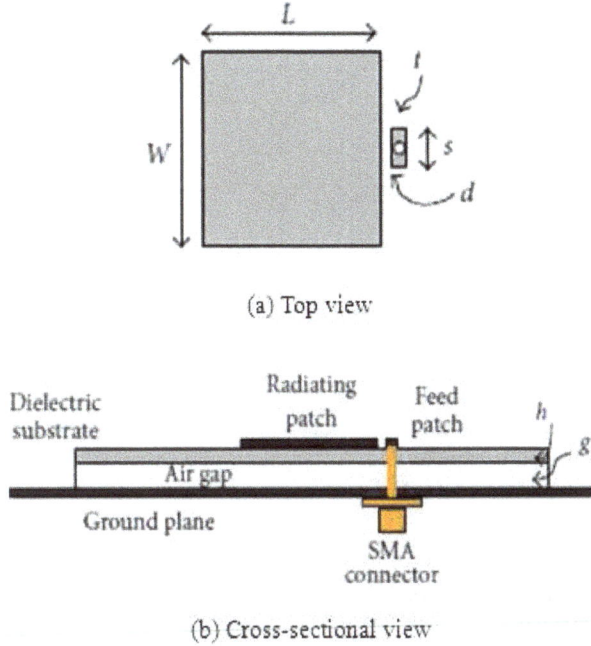

(a) Top view

(b) Cross-sectional view

**Figure 1.** *Geometry of a rectangular patch antenna with a small capacitive feed arrangement [10].*

**Table 1.** *Dimensions for the antenna designed for 10 GHz.*

| Parameter | Value |
|---|---|
| Length of the radiator patch ($L$) | 7.8 mm |
| Width of the radiator patch ($W$) | 10.6mm |
| Length of the feed strip($t$) | 2 mm |
| Width of the feed strip ($s$) | 5.2 mm |
| Separation of feed strip from the patch ($d$) | 0.5 mm |
| Air gap between substrates ($g$) | 0.6 mm |
| Relative dielectric constant ($\varepsilon_r$) | 3.0 |
| Thickness of substrate ($h$) | 1.56 mm |

For the optimization of all parameters IE3D is used which is method of moments (MoM) based electromagnetic (EM) software. The basic design of patch is started by selecting the center frequency as suggested in [10]. It has been shown that the impedance bandwidth of the antenna can be maximized by using the air gap value given in [10].

$$g \cong 0.16\lambda_0 - h\sqrt{\varepsilon_r} \qquad (1)$$

Where, g is the height of substrate above the ground, h is the thickness of the substrates, and $\varepsilon_r$ is dielectric constant of the substrate. However, equation (1) is used to calculate the initial value. Final optimized values would be within ± 5% of this value [10]. The parameters considered for

optimization are the height of substrate above the ground ($g$), the thickness of the substrates ($h$) separation between feed strip and patch ($d$) and feed strip dimensions (length ($t$), and width ($s$)). The complete optimization procedure of these key design parameters is presented in [10]. In this work optimization procedure is adopted for further reduction in air gap and is presented in detail in Section 3.

## 3. Parametric Studies and Discussion

As stated in Section 2, the geometry shown in Figure 1 was simulated using IE3D software. All key design parameters (length ($t$), width ($s$) and distance between feed strip and patch ($d$)) have been investigated to analyze the effect on antenna performance and are discussed in the following subsections

The air gap is reduced by keeping the impedance bandwidth high for compact applications as mentioned earlier. In order to reduce size of the proposed antenna geometry, an air gap was reduced in steps of 0.1 mm and feed strip parameters (length ($t$), width ($s$) and distance between feed strip and patch ($d$)) were re-optimized in each case. Variations in the parameters are stopped wherever the optimum bandwidth is obtained. The relation between reduction in air gap value and change in feed strip dimension are suggested in [15].

$$g_{modiried} = g_{opt} + \Delta g \qquad (2)$$

$$s_{midified} = s_{opt} + \Delta s \qquad (3)$$

In (2) and (3), $g_{opt}$ and $s_{opt}$ are the optimized air gap and feed strip width, respectively, as defined in [10]. And,

$$\Delta s \approx 1.5 \, \Delta g \; (\Delta g \leq g_{opt}/2) \qquad (4)$$

$$\Delta s \approx 1.75 \, \Delta g \; (\Delta g > g_{opt}/2) \qquad (5)$$

It should be noted that $s_{modified}$< W (width of the radiator patch) [15]. As suggested in [10]-[13] the parameter '$s$' controls the resistive part and '$d$' changes the reactive component of the input impedance. We have considered the optimized values as mentioned in [15] as a reference case and varied length of the feed strip '$t$' and stopped where the maximum impedance bandwidth is obtained. This procedure is repeated for remaining parameters, and finally considered all optimized values to obtain the high impedance bandwidth.

In this work besides procedure suggested in [15] for the reduction of air gap, other parameters have also been varied to get the further reduction in air gap value. As an example, 10 GHz design reported in [15] is reconsidered for further optimization of air gap value. As reported in [15], the best possible minimum value of air gap is 0.6mm. In this we have demonstrated that this value can be further reduced to 0.1mm keeping antenna's input impedance reasonably high (38%) and is suitable for FCC defined UWB applications. The optimization procedure is started from the minimum value of air gap reported in [15]. The optimization procedure for *g*= 0.5mm is presented in the following paragraphs and similar procedure has been adopted for other values of air gap values.

### 3.1. Effect of Distance between Feed Strip and Patch (d)

In order to reduce the air gap, dimensions of the feed strip and distance between feed strip and patch have been optimized. Width of the feed strip (*s*) is calculated using equation (3). The air gap is reduced in steps of 0.1 mm, and for each value of air gap (*g*) other parameters (*s*, *t*, and *d*) have re-optimized. As an example, for air gap value of 0.5 mm, optimization procedure is presented. The distance between feed strip and patch (*d*) dimensions are varied from 0.1 mm to 0.55 mm by keeping *t* = 2 mm and *s* = 5.35 mm constant. Optimized value obtained for this case is *d* = 0.5 mm (pl. ref. Figure 2).

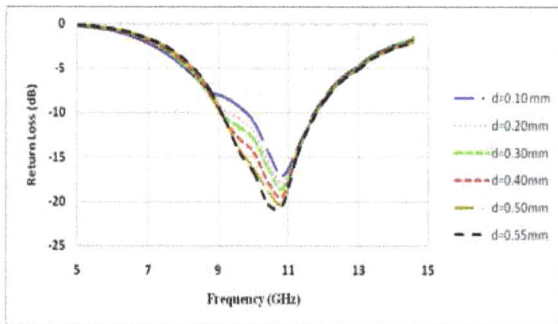

**Figure 2.** *Return loss characteristics with the variation in separation between radiator and feed strip (d).*

### 3.2. Effect of Feed Strip Length (t)

The length of the feed strip (*t*) is varied from 1.9 mm to 2.3 mm in steps of 0.1 mm by keeping the distance between feed strip and patch *d* = 0.5 mm (as obtained in Section 3.1) and width of the feed strip *s* = 5.35 mm. The return loss characteristic for this case is shown in Figure 3. From the Figure 3 it may be noted the maximum bandwidth is achieved for feed strip length *t* = 2.2mm.

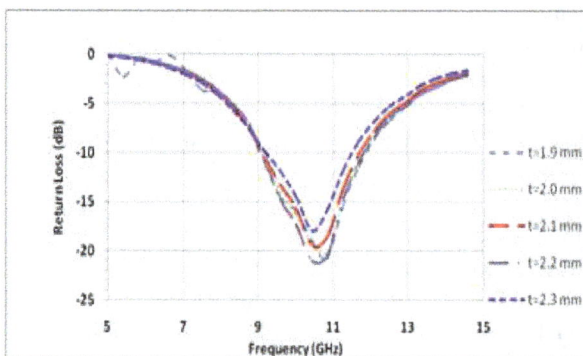

**Figure 3.** *Return loss characteristics with the variation in feed strip length (t).*

### 3.3. Effect of Feed Strip Width (s)

Finally, the width of feed strip (*s*) is varied from 4.35 mm

to 5.85 mm in a steps of 0.5 mm by keeping the optimum values obtained in previous sub sections *t* = 2.2 mm and *d*=0.5mm. The return loss characteristics for this case are presented in Figure 4.From Figure 4 it may be noted that the optimized value for this case is *s* = 5.35 mm.

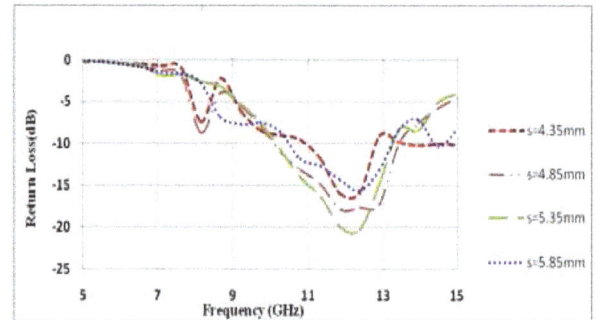

**Figure 4.** *Return loss characteristics with the variation in feed strip width (s).*

It may be noted that similar procedure is followed for air gap values of 0.3 mm, 0.2 mm, and for 0.1 mm. The optimized parameters for different air gap values of 10 GHz are shown in Table 2.

**Table 2.** *Re-optimized values of feed strip length, width and % BW for reduced air gap values of 10 GHz.*

| Air gap (mm) | 0.5 | 0.4 | 0.3 | 0.2 | 0.1 |
|---|---|---|---|---|---|
| *t, s* (mm) | 2.2, 4.85 | 1.2, 5.5 | 1.2, 5.65 | 1.2, 4.3 | 2, 4.45 |
| BW % | 40.0 | 39.8 | 41.0 | 32.9 | 38.8 |

The optimized air gap value obtained from the above studies is 0.1 mm with impedance bandwidth of about 38%. Obtained return loss characteristics for 10 GHz with different air gap values are shown in Figure 5. It may be noted that center frequency is shifted as the air gap reduces shown in Figure 5. However, the shift in the center frequency may be re-tuned to some extent by loading such as placing a slot on it [16].

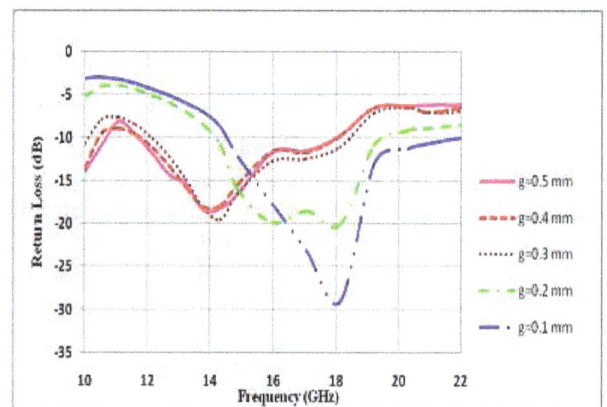

**Figure 5.** *Simulated return loss characteristics of geometry of 10 GHz for different air gap values.*

It may be noted that the reduction in the air gap at higher operating frequencies results in the shift in center frequency.

The similar procedure has been followed for the different center frequencies 2 GHz, 4.5 GHz, 5.9 GHz, and 8 GHz. For the center frequencies of 8 and 10 GHz, 90.00 % and 83.33 % air gap reductions have been achieved with the percentage bandwidths of 28.8% and 38.8% respectively. The obtained results for these cases are presented in Table 3.

*Table 3. Amount of air gap reduction and % bandwidth obtained for different bands of frequencie s*

| Center freq. ($f_c$) (GHz) | Radiator patch dimensions L,W (mm) | $t, s$ (mm) [15] | $t, s$ (mm) (this work) | Opt. air gap $g_{opt}$ (mm)[15] | Optimum air gap (mm) (this work) | BW (%) [15 ] | BW (%) (this work) | Air gap reduction (%) |
|---|---|---|---|---|---|---|---|---|
| 2.0 | 42.8, 53.0 | 3.0, 13 | 2.5, 14.5 | 8.50 | 7.50 | 34.5 | 29.02 | 11.76 |
| 4.5 | 21.0, 26.5 | 2.0, 10 | 2, 10.75 | 3.50 | 3.00 | 32.9 | 34.16 | 14.28 |
| 5.9 | 15.5, 16.4 | 1.2, 10 | 1.4, 9.8 | 2.25 | 1.75 | 34.5 | 34.58 | 22.22 |
| 8.0 | 10.0, 13.2 | 2.0, 5.2 | 2, 6.85 | 1.00 | 0.10 | 48.5 | 28.83 | 90.00 |
| 10.0 | 7.8, 10.6 | 2.0,5.2 | 2, 4.45 | 0.60 | 0.10 | 48.3 | 38.88 | 83.33 |

# 4. Conclusions

A procedure for reduction of air gap of coplanar capacitively coupled suspended microstrip antenna has been presented. An effort has been made (in all cases presented) to reduce the air gap significantly keeping all other parameters (of antenna geometry) optimum especially impedance bandwidth of nearly 25% which is suitable for most of the FCC defined UWB applications. It has been shown that more than 80% air gap reduction can be achieved at 8 GHz, and 10 GHz, respectively, compared to previously published works for similar antenna geometries. The antenna presented here proves to be the best candidate for the FCC defined compact wireless applications (UWB).

# References

[1] J. A. Wei, "Applications on Ultra Wideband," M. S. Report, University of Texas at Arlington, 2006.

[2] P. Bhartia, K. Rao, and R. Tomar, *Millimeterwave Micro strip and Printed Circuit Antennas*, Artech House, Canton, MA, 1991.

[3] D. M. Pozar and D. H. Schaubert, *Microstrip Antennas, the Analysis and Design of Microstrip Antennas and Arrays*, IEEE Press, New York, 1995.

[4] G. Kumar and K. Gupta, "Directly coupled multiple resonator wide-band microstrip antenna," *IEEE Transactions and Propagations*, vol. 33. no. 6, 1985.

[5] B. L. Ooi, and I. Ang, "Broadband semicircle fed flower-shaped microstrip patch antenna," *Electron. Lett.*, vol. 41, no. 17, 2005.

[6] K. M. Luk, K. F. Lee, and W. L. Tam, "Circular U-slot patch with dielectric substrate," *Electronics Lett.*, vol.33, no.12, pp. 1001-1002, 1997.

[7] C. L. Mak, K. F. Lee and K. M. Luk, "A novel broadband patch antenna with T-shaped probe," *Proc. Inst. Elect. Engg Microw., Antennas Propagat.*, vol. 147, pp. 73-76, 2000.

[8] H. W. Lai and K. M. Luk, "Wideband stacked patch fed by meandering probe," *Electron. Lett.*, vol. 41, no. 6, 2005.

[9] Mathew-Ridgers, G., J.W. Odondaal, and J. Joubert, "Single layer capacitive feed for wideband probe-fed microstrip antenna elements," *IEEE Trans. Antennas Propagat.*, vol. 51, pp. 1405-1407, 2003.

[10] V. G. Kasabegoudar, D. S. Upadhyay, and K. J. Vinoy, "Design studies of ultra wideband microstrip antennas with a small capacity feed," *Int. J. Antennas Propagat.*, pp. 1-8, 2007.

[11] V. G. Kasabegoudar and K. J. Vinoy, "A wideband microstrip antenna with symmetric radiation patterns," *Microw. Opt. Technol. Lett.*, vol. 50, no. 8, pp. 1991-1995, 2008.

[12] V. G. Kasabegoudar and K. J. Vinoy, "Coplanar capacitively coupled probe fed microstrip antennas for wideband applications," *IEEE Trans. Antennas Propagat.*," vol. 58, no. 10, pp. 3131-3138, 2010.

[13] V. G. Kasabegoudar and K. J. Vinoy, "A broadband suspended microstrip antenna for circular polarization," *Progress in Electromagnetics Research*, vol. 90, pp. 353-368, 2009.

[14] V. G. Kasabegoudar, "Dual frequency ring antenna with capacitive feed," *Progress in Electromagnetics Research C*, vol. 23, pp. 27-39, 2011.

[15] V. G. Kasabegoudar, "Low profile suspended microstrip antennas for wideband applications," *Journal of Electromagnetic Waves and Applications*, vol. 25, no. 13, pp. 1795- 1806, 2011.

[16] V. G. Kasabegoudar and K. J. Vinoy, "A coplanar capacitively coupled probe fed microstrip antenna for wireless applications," *Int. Symposium on Antennas and Propagation*, pp. 297-300, 2009.

[17] V. G. Kasabegoudar and K. J. Vinoy, "Input impedance modelling of capacitively coupled wideband microstrip antenna," *Int. Symposium on Antennas and Propagat.*, pp. 268-271, 2008.

[18] V. G. Kasabegoudar and A. Kumar, "Dual band capacitive coupled microstrip antennas with and without air gap for wireless applications," *Progress in Electromagnetics Research C*, vol. 36, pp. 105-117, 2013.

[19] "New public safety applications and broadband internet access among uses envisioned by FCC authorization of ultra wideband technology." First report and order (FCC 02-48), action by the Commission, February, 2002.

# 31

# Penetration loss of Walls and data rate of IEEE802.16m WiMAX

**Hala BahyEldeen Nafea, Fayez W. Zaki, Hossam E. S. Moustafa**

Dept. of Electronics and Communications Eng, Faculty of Engineering, Mansoura University, Egypt

**Email address:**
eg_hala2007@yahoo.com (H. B. Nafea)

**Abstract:** In this paper, the data rate for the downlink (DL) of OFDMA-based IEEE802.16m WiMAX system and the available DL throughput as a function of distance to the Base Station (BS) are estimated for a number of propagation scenarios (OUTDOOR; INDOOR 1, INDOOR 2 and INDOOR 3). Moreover, Walls penetration loss is also considered. Adaptive modulation and Coding (AMC) schemes will be assumed in the present study for 5 MHz and 20 MHz channel bandwidth.

**Keywords:** WiMAX, Broadband Wireless, Adaptive Modulation and Coding, Propagation Analysis

## 1. Introduction

One of the newest technologies, that satisfy the ongoing demand for faster data rates with longer transmission ranges and that are thus suitable for new applications is mobile WiMAX. Mobile WiMAX will compete with cellular, Wi-Fi, and last-mile Internet access technologies such as DSL and cable.

The next generation of mobile WiMAX is IEEE 802.16m amends the IEEE 802.16e, j specification to provide an advanced air interface for operation in licensed bands. It is a recommended candidate for 4G. Unlike other wireless standards, WiMAX allows data transport over multiple broad frequency ranges. This lets the technology avoid frequencies that would interfere with other wireless applications.

Pushed by the increasing market demand for wireless wideband services, strong industry support and a competitive edge over deployed 3.5G systems Orthogonal Frequency Division Multiple Access (OFDMA) based Mobile WiMAX is on the verge of becoming a reality all over the globe [1].

In this paper, a preliminary analysis of the data rate as a function of the distance of the subscriber station (SS) to the BS is developed, considering different propagation environments and 3.5 GHz carrier frequency.

## 2. System Parameters

### 2.1. Frequency Bands

Considering the 3.3-3.8 GHz spectrum, the WiMAX Forum™ Mobile System Profile [2] specifies the channel bandwidth combinations, Fast Fourier Transform (FFT) sizes and duplexing modes as shown in Table 1, for the possible frequency range configurations.

*Table 1. Possible WiMAX configurations for the 3.3-3.8 GHz band [2]*

| Frequency Range (GHz) | Channel Bandwidth (MHz) | FFT Size | Duplexing Mode |
|---|---|---|---|
| | 5 | 512 | TDD |
| 3.3-3.4 | 7 | 1024 | TDD |
| | 10 | 1024 | TDD |
| | 5 | 512 | TDD |
| 3.4-3.8 | 7 | 1024 | TDD |
| | 10 | 1024 | TDD |
| | 5 | 512 | TDD |
| 3.4-3.6 | 7 | 1024 | TDD |
| | 10 | 1024 | TDD |
| | 5 | 512 | TDD |
| 3.6-3.8 | 7 | 1024 | TDD |
| | 10 | 1024 | TDD |

### 2.2. OFDMA Parameters

OFDMA is a multiple access technique which divides

the total Fast Fourier Transform (FFT) space into a number of sub-channels (set of sub-carriers that are assigned for data exchange) whereas the time resource is divided into time slots (i.e. in WiMAX OFDMA PHY [3], the minimum frequency time unit of sub-channel is one slot, which is equivalent to 48 sub-carriers) and a frame is constructed from number of slots. Define the size of FFT as NFFT which denotes the total number of sub-carriers of all types (pilots, guard and data). Let Ndata denote the total number of data sub-carriers after reserving the pilot and guard sub-carriers. Ndata is divided into L groups, each with K = Ndata /L data sub-carriers.

For the possible channel bandwidth, Table 2 summarizes the standard IEEE 802.16 OFDMA parameters.

*Table 2. OFDMA Parameters [11]*

| Parameter | Fixed WiMAX OFDM-PHY | Mobile WiMAX Scalable OFDMA-PHY | | | |
|---|---|---|---|---|---|
| FFT Size | 256 | 128 | 512 | 1024 | 2048 |
| Number of used data subcarriers | 192 | 72 | 360 | 720 | 1440 |
| Number of pilot data subcarriers | 8 | 12 | 60 | 120 | 240 |
| Number of null/guard band data subcarriers | 56 | 44 | 92 | 184 | 368 |
| Cyclic prefix of guard time (Tg/Tb) | 1/32, 1/16, 1/8, 1/4 | | | | |
| Oversampling rate (Fs/BW) | Depends on bandwidth:7/6 for 256 OFDM, 8/7 for multiples of 1.75MHz, and 28/25 for multiples of 1.25 MHz, 1.5 MHz, 2 MHz ,or 2.75 MHz | | | | |
| Channel bandwidth (MHz) | 3.5 | 1.25 | 5 | 10 | 20 |
| Subcarrier frequency spacing (KHz) | 15.625 | 10.94 | | | |
| Useful symbol time (μs) | 64 | 91.4 | | | |
| Guard time assuming 12.5% (μs) | 8 | 11.4 | | | |
| OFDM symbol duration (μs) | 72 | 102.9 | | | |
| Number of OFDM symbols in 5 ms frame | 69 | 48.0 | | | |

In the present study a frame duration of 5 ms has been assumed, since, at least initially all WiMAX equipments will only support this duration.

### 2.3. Frame and Subchannel Structure

The 802.16e PHY [3] supports TDD, FDD; however the initial release of Mobile WiMAX certification profiles is includes only TDD. With ongoing releases, FDD profiles is considered by the WiMAX Forum to address specific market opportunities where local spectrum regulatory requirements either prohibit TDD or are more suitable for FDD deployments. To counter interference issues, TDD does require system-wide synchronization; nevertheless.

One should note that the first system profiles released by the WiMAX Forum only contemplate time division duplexing (TDD) modes, due to a number of advantages over frequency division duplexing (FDD). TDD is the preferred duplexing mode for the following reasons:

i.    TDD enables adjustment of the DL/ UL ratio to efficiently support asymmetric downlink/uplink traffic, while with FDD, downlink and uplink always have fixed (and most times equal) and generally, equal DL and UL bandwidths.

ii.   TDD assures channel reciprocity because the DL and UL frames are sent in the same band, for better support of link adaptation, MIMO, and other closed loop advanced antenna technologies (AAS).

iii.  Unlike FDD, which requires a pair of channels, TDD only requires a single channel for both downlink and uplink providing greater flexibility for adaptation to varied global spectrum allocations.

iv.   Transceivers design for TDD implementations are less complex and therefore less expensive.

Figure 1 illustrates the OFDM frame structure for a Time Division Duplex (TDD) implementation. Each frame is divided into DL and UL sub-frames separated by Transmit/Receive and Receive/Transmit Transition Gaps (TTG and RTG, respectively) to prevent DL and UL transmission collisions. In a frame, the following control information is used to ensure optimal system operation:

1)    Preamble: The preamble, used for synchronization, is the first OFDM symbol of the frame.

2)    Frame Control Head (FCH): The FCH follows the preamble. It provides the frame configuration information such as MAP message length, coding scheme and usable sub-channels.

3)    DL-MAP and UL-MAP: The DL-MAP and UL-MAP provide sub-channel allocation and other control information for the DL and UL sub-frames respectively.

4)    UL Ranging: The UL ranging sub-channel is allocated for mobile stations (MS) to perform closed-loop time, frequency, and power adjustment as well as bandwidth Requests.

5)    UL CQICH: The UL CQICH channel is allocated for the MS to feedback channel state information.

6) UL ACK: The UL ACK is allocated for the MS to feedback DL HARQ acknowledgement.

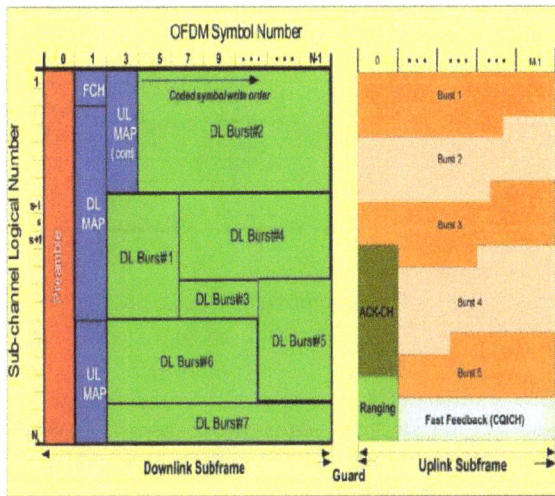

**Figure 1.** *WiMAX OFDMA TDD Frame. (Extracted from [4])*

The overheads in Figure1 have variable size depending on the type of traffic carried. In this analysis Full Buffer FTP traffic will be assumed, and a corresponding typical distribution of OFDMA symbols in the frame is shown in Table 3.

**Table 3.** *TDD Frame configurations used [4].*

| Parameter | Values | | |
|---|---|---|---|
| **Channel bandwidth (MHz)** | 5 | 10 | 7 |
| **Number of OFDMA Symbols/Frame** | 48 | 48 | 34 |
| **Total Number of OFDMA Overhead Symbols** | 10 | 10 | 7 |
| **Number of OFDMA Symbols for TTG (guard)** | 1 | 1 | 1 |
| **Total Number of OFDMA Data Symbols** | 37 | 37 | 26 |
| **DL:UL 3:1** DL OFDMA Data Symbols | 28 | 28 | 19 |
| UL OFDMA Data Symbols | 9 | 9 | 7 |

This option is somehow conservative, since most applications have a traffic which is bursty in nature and can operate efficiently with less overhead.

Moreover, as previously referred, TDD allows for flexible DL/UL ratios to cope with different traffic profiles. In this study a 3:1 DL: UL ratio will be analyzed, the respective data symbols distribution is also shown in Table 3.

IEEE 802.16 also allows different subcarrier permutations schemes, although the initial WiMAX system profile [5] only includes for the DL, Downlink Partial Usage of Subcarriers (DL PUSC) and Band Adaptive Modulation and Coding (Band AMC), with only the first being mandatory. Therefore, DL PUSC is considered as the subchannel permutation scheme. Table 4, shows the distribution of subcarriers for DL PUSC and UL PUSC mode.

**Table 4.** *PUSC Parameters [11].*

| Parameter | Downlink | UPlink | Downlink | UPlink |
|---|---|---|---|---|
| System bandwidth (MHz) | 5 | | 10 | |
| FFT Size | 512 | | 1024 | |
| Null Sub-Carriers | 92 | 104 | 184 | 184 |
| Pilot Sub-Carriers | 60 | 136 | 120 | 280 |
| Data Sub-Carriers | 360 | 272 | 720 | 560 |
| Sub-Channels | 15 | 17 | 30 | 35 |
| Symbols Period, Ts | 102.9 microseconds | | | |
| Frame Duration | 5 milliseconds | | | |
| OFDM Symbols/ Frame | 48 | | | |
| Data OFDM Symbols | 44 | | | |

## 2.4. Modulation and Coding Modes (Burst Profile)

Adaptive modulation and coding (AMC), Hybrid Automatic Repeat Request (HARQ) and Fast Channel Feedback (CQICH) were introduced with Mobile WiMAX to enhance coverage and capacity for WiMAX in mobile applications.

Support for QPSK, 16QAM and 64QAM are mandatory in the DL with Mobile WiMAX. In the UL, 64QAM is optional. Both Convolutional Code (CC) and Convolutional Turbo Code (CTC) with variable code rate and repetition coding are supported. Block Turbo Code and Low Density Parity Check Code (LDPC) are supported as optional features.

Table 5, summarizes the coding and modulation schemes supported in the Mobile WiMAX profile.

**Table 5.** *Supported Code and Modulations [11]*

| Modulation | | DL QPSK,16QAM,64 QAM | UL QPSK, 16QAM, 64QAM |
|---|---|---|---|
| **Code Rate** | CC | 1/2, 2/3, 3/4, 5/6 | 1/2, 2/3, 5/6 |
| | CTC | 1/2, 2/3, 3/4, 5/6 | 1/2, 2/3, 5/6 |
| | Repeteition | X2, X4, X6 | X2, X4, X6 |

From several burst profiles allowed by IEEE 802.16, the six listed in Table 6, along with the minimum required SNR, have been considered.

**Table 6.** *SNR required for considered burst profiles. (CTC – Convolution Turbo Codes)*

| Burst Profile | SNR Required (d B) |
|---|---|
| QPSK CTC 1/2 | 3.5 |
| QPSK CTC 3/4 | 6.5 |
| 16- QAM CTC 1/2 | 9.0 |
| 16- QAM CTC 3/4 | 12.5 |
| 64- QAM CTC 2/3 | 16.5 |
| 64- QAM CTC 3/4 | 18.5 |

## 2.5. Propagation Environments

### 2.5.1. Propagation Model
Propagation models are used to estimate the Path loss (PL) value in wireless communications and to predict the level of SNR at the receiver. The PL value is used to determine the coverage of the base station and mobile station's cell-range.

The propagation model COST 231 Hata [6] has been adopted. Although this model is based on empirical data obtained at 2 GHz, [7] it is also valid model at 3.5 GHz.

### 2.5.2. Penetration Loss
Penetration will be modeled as an excess loss introduced by the penetrated walls, using the model suggested in [8]

$$L_{ex} = L_{wi} k \left[ \frac{k+1.5}{k+1} - b \right] [dB] \qquad (1)$$

$$b = -0.064 + 0.0705 L_{wi} - 0.0018 L^2_{wi} \qquad (2)$$

Where $L_{wi}$ is the average excess attenuation per wall and k is the number of penetrated walls. Table (7-a) shows the penetration loss ($L_{wi}$) as a function of frequency for thin board dividing between rooms and thick walls made of reinforced concrete.

Table (7-a). Penetration loss as a function of frequency for two types of walls [9].

| Frequency [GHz] | Loss for thin Walls[d B] | Loss for thick Walls[d B] |
|---|---|---|
| 2 | 3.3 | 10.9 |
| 3.5 | 3.4 | 11.4 |
| 5 | 3.4 | 11.8 |

Three indoor scenarios have then been considered. These are listed in Table (7-b) along with the respective attenuations calculated by equations (1) and (2) for a 3.5 GHz frequency. The chosen Indoor scenarios try to represent possible limiting situations on propagation. The analysis was limited to two walls, since when the number of walls increases, other propagation mechanisms become dominant.

Table (7-b). Penetration Loss Parameters.

| Parameter | Indoor1 Thick Wall | Indoor 2 Thick Wall + Thin Wall | Indoor 3 2 Thick Walls |
|---|---|---|---|
| Total Penetration Loss [d B] | 11.4 | 12.9 | 18.0 |

### 2.5.3. Fading
The diverse fading components due to the propagation environment will be taken into account in the form of propagation margins.

Table 8, illustrates the adopted margins and the total margin for the different considered scenarios. WiMAX Forum™ reference studies have provided the guideline for this parameterization [10].

Table 8. Fading Margins Adopted.

| Margin | |
|---|---|
| Log Normal Fade Margin | 5.56 dB |
| Fast Fading Margin | 2.0 dB |
| Interference Margin | 2.0 dB |

The value of 5.56 dB for the shadow fade margin is based on a log-normal shadowing standard deviation of 8 dB assuring a 75% coverage probability at the cell edge and 90% coverage probability over the entire area. The interference margin assumes a cellular reuse pattern of 1 with 3 sectors per site.

### 2.6. Station Parameters

Tables 9 presents the parameters used for the link budget calculations for the BS and SS, again based on WiMAX Forum™ analysis [10].

Table 9. BS and SS the parameters

| Base Station Parameters | | | Subscriber Station Parameters | | |
|---|---|---|---|---|---|
| BS Height | $h_b$ | 32 m | Subscriber Station Height | $h_m$ | 1.5 m |
| $T_x$ Power per Antenna Element | $P_E$ | 10 W | Number of $R_x$ Antenna Elements | | 2 |
| Number of $T_x$ Antenna Elements | | 2 | Antenna Diversity Gain | $G_{DW}$ | 3 dB |
| Cyclic Combining Gain | $G_{CYC}$ | 3dB | $R_x$ Antenna Gain( Hand held Outdoor) | $G_R$ | -1 dBi |
| $T_x$ Antenna Gain | $G_E$ | 15 dBi | $R_x$ Antenna Gain(Fixed in door) | $G_R$ | 6 dBi |
| Pilot Power Boosting Attenuation | $A_{PILOT}$ | -0.7dB | Noise Figure | | 7 dB |

# 3. Mathematical Analysis and Computer Simulation

In order for the system to work correctly, the data rate in WiMAX can be calculated as:

$$R = \frac{N_{used} b_m c_r}{T_S} \qquad (3)$$

Where: R is the data rate in a WiMAX OFDM physical layer, $b_m$ is the number of bits per modulation symbol and is 1 for BPSK, 2 for QPSK, 4 for 16-QAM and in general if M is the modulation level in a M-QAM constellation, $M = 2^{\wedge} b_m$. The $c_r$ is the coding rate that can be found in [10, 11] for each different burst profile. The symbol duration $T_S$, $T_b$ is the useful symbol time, $T_g$ is the guard time according to Fig. 2, expressed as:

$$T_S = T_g + T_b$$
$$T_S = [G+1]T_b \qquad (4)$$

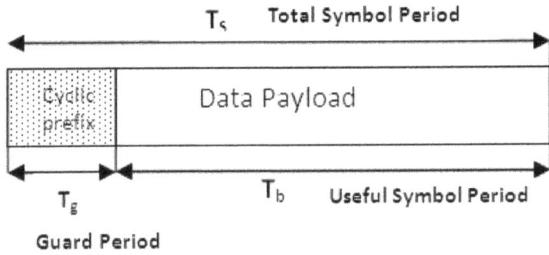

**Fig. 2.** *OFDM Symbol Structure with Cyclic Prefix.*

Where G is the ratio $Tg/Tb$, this value can be: 1/4, 1/8, 1/16 or 1/32. And $Tb = 1/\Delta f$, with the sub-carrier spacing $\Delta f$ given as

$$\Delta f = \frac{F_S}{N_{FFT}} \qquad (5)$$

$$F_S = floor(nBW/8000)8000 \qquad (6)$$

Where $F_S$ is the sampling frequency, $n$ is the sampling factor, $BW$ is the nominal channel bandwidth. The sampling factor in conjunction with $BW$ value has changed from OFDMA 802.16-2004 Standard and is set to 8/7 as follows: for channel bandwidths that are a multiple of 1.75 MHz then $n = 8/7$ else for channel bandwidths that are a multiple of any of 1.25, 1.5, 2 or 2.75 MHz then $n = 28/25$ else for channel bandwidths not otherwise specified then $n = 8/7$.

Sensitivity or minimum received power $S_R$ (receiver sensitivity) is different for each modulation and is expressed as [12]:

$$S_R = -102 + SNR_{(Rx)} + 10\log\left(Fs \times \left(\frac{N_{data}}{N_{FFT}}\right) \times \left(\frac{N_{subchannels}}{16}\right)\right) \qquad (7\text{-a})$$

Equation (7-a) may be re-expressed as [11]:

$$S_R = -114 + SNR_{(Rx)} - 10\log R + 10\log\left(\frac{F_S N_{used}}{N_{FFT}}\right) + ImpLoss + NF \qquad (7\text{-b})$$

Where: $N_{FFT}$ is the number of points for FFT or total number of subcarriers. $N_{data}$ is number of used data subcarries, and $N_{subchannels}$ is the number of subchannels. $N_{used}$ (the active subcarriers = total subcarriers – null subcariers). ImpLoss is the implementation loss, which includes non-ideal receiver effects such as channel

estimation errors, tracking errors, quantization errors, and phase noise. The assumed value is 7 dB. NF is the receiver noise figure, referenced to the antenna port. The assumed value is 8 dB, and R is the data rate in a WiMAX OFDM physical layer [13].

Tables 10, 11 present the different values of SR and Channel Capacity Cmodulation [bps] for each modulation at 5 MHZ and 20 MHZ bandwidth using SNR required for considered burst profiles. (CTC– Convolutional Turbo Codes).

**Table 10.** *Receiver Sensitivity for Each Modulation Type at 5MHZ.*

| Parameters | SNR (Rx) [dB] | RESICEVER Sensitivity $S_R$ [dB] | Usefull Channel Capacity Cmodulation [bps] |
|---|---|---|---|
| QPSK CTC 1/2 | 3.5 | -98.6561 | 4.0816e+006 |
| QPSK CTC 3/4 | 6.5 | -97.4170 | 6.1224e+006 |
| 16-QAM CTC1/2 | 9.0 | -96.1664 | 8.1633e+006 |
| 16-QAM CTC 3/4 | 12.5 | -94.4273 | 1.2245e+007 |
| 64-QAM CTC 2/3 | 16.5 | -91.6767 | 1.6327e+007 |
| 64-QAM CTC 3/4 | 18.5 | -90.1883 | 1.8367e+007 |

The received power may be calculated using link budget equations given as:

$$P_R = P_T + G_T + G_R - L_S - PL \qquad (8)$$

Where: $P_R$ is the received power, $P_T$ is the transmitted power, $G_T$ is the transmit antenna gain, $G_R$ is the receiver antenna gain, $L_S$ is the system loss and $P_L$ is the path loss.

$$PATHLOSS = PT + GT + GR - PR - Lex \qquad (9)$$

**Table 11.** *Receiver Sensitivity for Each Modulation Type at 20MHZ [13].*

| Parameters | SNR (Rx) [dB] | RESICEVER Sensitivity $S_R$ [dB] | Usefull Channel Capacity Cmodulation [bps] |
|---|---|---|---|
| QPSK CTC 1/2 | 3.5 | -98.6561 | 1.6327e+007 |
| QPSK CTC 3/4 | 6.5 | -97.4170 | 2.4490e+007 |
| 16-QAM CTC 1/2 | 9.0 | -96.1664 | 3.2653e+007 |
| 16-QAM CTC 3/4 | 12.5 | -94.4273 | 4.8980e+007 |
| 64-QAM CTC 2/3 | 16.5 | -91.6767 | 6.5306e+007 |
| 64-QAM CTC 3/4 | 18.5 | -90.1883 | 7.3469e+007 |

### 3.1. Cost-231 Hata Model

In this model, five parameters are used for propagation loss estimation. These are frequency f MHz, distance from base station to mobile station d (Km), base station height hb (m), the height of the mobile hm (m), and standard deviation constant Cm (dB). The pass loss in Hata model is expressed as:

$$PL = 46.3 + 33.9\log_{10}(f) - 13.82\log_{10}(h_b) - ah_m - (44.9 - 6.55\log_{10}(h_b))\log_{10}d + C_m \qquad (10)$$

Where the parameters Cm and ahm are used to specify the environmental characteristics as given below:
*Urban:

$$C_m = 3dB \quad ah_m = 3.20(\log_{10}(11.75h_m))^2 - 4.97 \qquad (11)$$

*Suburban/Rural:

$$C_m = 0dB \quad ah_m = (1.11\log_{10}f - 0.7)h_m - (1.56\log_{10}f - 0.8) \qquad (12)$$

Using the above equations, the relationship between the distance d and the propagation loss may be formulated as:

$$d = 10^{((\text{PATHLOSS}-46.3-33.9*\log10(f)+13.82*\log10(hb)+ahm-cm)/(44.9-6.55*\log10(hb)))} \qquad (13)$$

Where the PATHLOSS is calculated using Equation (9).

The simulation parameters are shown in Tables 7, 8, 9. Four propagation scenarios (OUTDOOR; INDOOR 1, INDOOR 2 and INDOOR 3) are considered.

The SNR (Rx) [dB] and useful channel capacity (C modulation) for each modulation at 5 MHZ and 20 MHZ bandwidth shown in Tables 10, 11 are considered too.

The variation of the DL data rate as a function of the distance to the BS has been computed for the 5 MHz and 20 MHz bandwidth and are shown in Figure 3 and Figure 4 .

The maximum distance to BS for each modulation scheme in the four propagation scenarios (OUTDOOR; INDOOR 1, INDOOR 2 and INDOOR 3), for both urban and suburban at frequency band 3.5GHZ, are shown in Tables12-15.

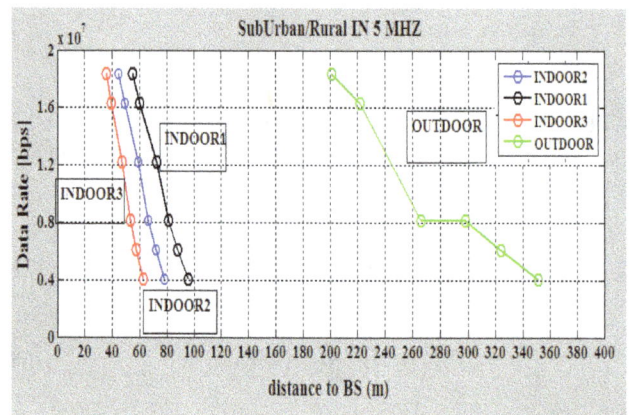

Figure 3 (b). R (Data Rate) with distance to the BS for COSTHATA SubUrban at (5MHZ)

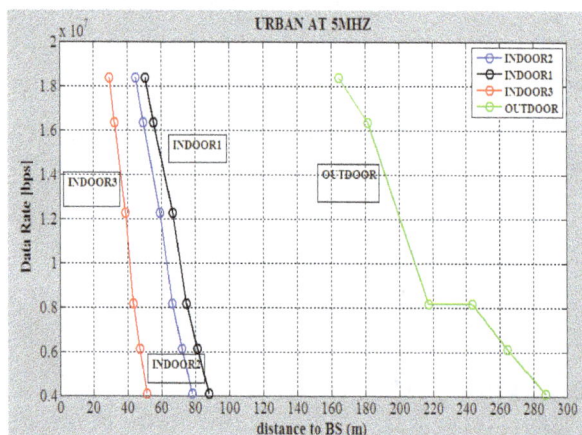

Figure 4 (a). R (Data Rate) with distance to the BS for COSTHATA Urban at (20MHZ)

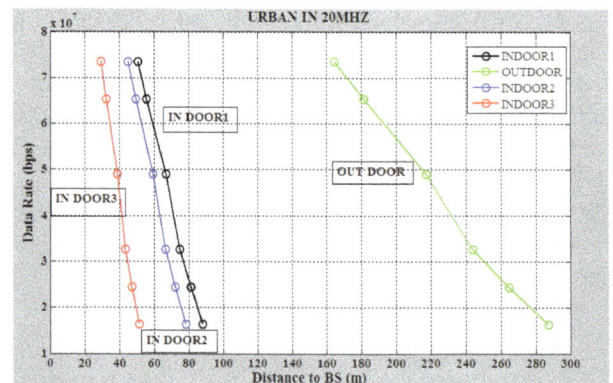

Figure 3 (a). R (Data Rate) with distance to the BS for COSTHATA Urban at (5MHZ)

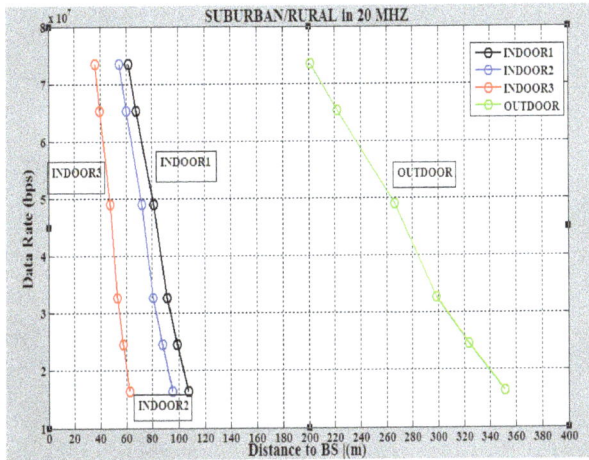

**Figure 4 (b).** *R (Data Rate) with distance to the BS for COSTHATA Suburban at (20MHZ)*

The results in figures 3 and 4 showed that the maximum distance to base station in the suburban case is increased by 22%, 22.3%, 22.5%, and 18.5% for INDOOR1, INDOOR2, INDOOR3, and OUTDOOR respectively as compared to the urban case.

Comparing OUTDOOR, INDOOR1, INDOOR 2 and INDOOR 3 scenarios, it is observed that the maximum distance to base station for INDOOR 2 is increased by 34.4 % more than INDOOR 3, whereas this distance is increase by 11.3% for INDOOR1 as compared to INDOOR2. On the other hand without any Penetration Loss in OUTDOOR scenario the distance increased by 69% as compared to INDOOR1.

**Table (12-a).** *Cost hata 231 model-3.5GHZ- Urban environment-Indoor1*

| Parameters | Maximum Distance To base station d[km] | Pathloss (d B) Indoor1 [Thick Wall] |
|---|---|---|
| QPSK CTC 1/2 | 0.0879 | 111.6320 |
| QPSK CTC 3/4 | 0.0810 | 110.3929 |
| 16-QAM CTC 1/2 | 0.0746 | 109.1423 |
| 16-QAM CTC 3/4 | 0.0665 | 107.4032 |
| 64-QAM CTC 2/3 | 0.0555 | 104.6526 |
| 64-QAM CTC 3/4 | 0.0504 | 103.1642 |

**Table (12-b).** *Cost hata 231 model -3.5GHZ- (SubUrban/Rural) environment- Indoor1*

| Parameters | Maximum Distance To base station d[km] | Pathloss(d B) Indoor1 [Thick Wall] |
|---|---|---|
| QPSK CTC 1/2 | 0.1075 | 111.6320 |
| QPSK CTC 3/4 | 0.0991 | 110.3929 |
| 16-QAM CTC1/2 | 0.0913 | 109.1423 |
| 16-QAM CTC 3/4 | 0.0814 | 107.4032 |
| 64-QAM CTC 2/3 | 0.0680 | 104.6526 |
| 64-QAM CTC 3/4 | 0.0616 | 103.1642 |

**Table (13-a).** *Cost hata 231 model-3.5GHZ- (Urban) environment-Indoor2*

| Parameters | Maximum Distance To base station d[km] | Pathloss(d B) Indoor2 [Thick Wall and Thin Wall] |
|---|---|---|
| QPSK CTC 1/2 | 0.0780 | 109.8200 |
| QPSK CTC 3/4 | 0.0719 | 108.5809 |
| 16-QAM CTC1/2 | 0.0662 | 107.3303 |
| 16-QAM CTC 3/4 | 0.0591 | 105.5912 |
| 64-QAM CTC 2/3 | 0.0493 | 102.8406 |
| 64-QAM CTC 3/4 | 0.0447 | 101.3522 |

**Table (13-b).** *Cost hata 231 model-3.5GHZ- (SubUrban/Rural) environment-Indoor2*

| Parameters | Maximum Distance To base station d[km] | Pathloss(d B) Indoor2 [Thick Wall and Thin Wall] |
|---|---|---|
| QPSK CTC 1/2 | 0.0954 | 109.8200 |
| QPSK CTC 3/4 | 0.0880 | 108.5809 |
| 16-QAM CTC1/2 | 0.0810 | 107.3303 |
| 16-QAM CTC 3/4 | 0.0723 | 105.5912 |
| 64-QAM CTC 2/3 | 0.0603 | 102.8406 |
| 64-QAM CTC 3/4 | 0.0547 | 101.3522 |

**Table (14-a)** *Cost hata 231 model-3.5GHZ- (Urban) environment-Indoor3*

| Parameters | Maximum Distance To base station d[km] | Pathloss(d B) Indoor3 [ 2 Thick Wall ] |
|---|---|---|
| QPSK CTC 1/2 | 0.0511 | 103.3962 |
| QPSK CTC 3/4 | 0.0471 | 102.1571 |
| 16-QAM CTC1/2 | 0.0434 | 100.9065 |
| 16-QAM CTC 3/4 | 0.0387 | 99.1674 |
| 64-QAM CTC 2/3 | 0.0323 | 96.4168 |
| 64-QAM CTC 3/4 | 0.0293 | 94.9284 |

**Table (14-b).** *Cost hata 231 model-5MHZ –3.5GHZ- (SubUrban/Rural) environment-Indoor3*

| Parameters | Maximum Distance To base station d[km] | Pathloss(d B) Indoor3 [ 2 Thick Wall ] |
|---|---|---|
| QPSK CTC 1/2 | 0.0626 | 103.3962 |
| QPSK CTC 3/4 | 0.0577 | 102.1571 |
| 16-QAM CTC1/2 | 0.0531 | 100.9065 |
| 16-QAM CTC3/4 | 0.0474 | 99.1674 |
| 64-QAM CTC2/3 | 0.0396 | 96.4168 |
| 64-QAM CTC3/4 | 0.0359 | 94.9284 |

*Table (15-a). Cost hata 231 model-3.5GHZ- Urban environment- outdoor*

| Parameters | Pathloss(d B) outdoor | Distance (km) |
|---|---|---|
| QPSK CTC 1/2 | 129.6561 | 0.2872 |
| QPSK CTC 3/4 | 128.4170 | 0.2647 |
| 16-QAM CTC1/2 | 127.1664 | 0.2438 |
| 16-QAM CTC 3/4 | 125.4273 | 0.2175 |
| 64-QAM CTC 2/3 | 122.6767 | 0.1815 |
| 64-QAM CTC 3/4 | 121.1883 | 0.1646 |

*Table (15-b). Cost hata 231 model-3.5GHZ- (SubUrban/Rural) environment- outdoor*

| Parameters | Pathloss(d B) outdoor | Distance (km) |
|---|---|---|
| QPSK CTC 1/2 | 129.6561 | 0.3514 |
| QPSK CTC 3/4 | 128.4170 | 0.3239 |
| 16-QAM CTC1/2 | 127.1664 | 0.2983 |
| 16-QAM CTC 3/4 | 125.4273 | 0.2661 |
| 64-QAM CTC 2/3 | 122.6767 | 0.2221 |
| 64-QAM CTC 3/4 | 121.1883 | 0.2014 |

The results in Tables 12, 13, 14, compars OUTDOOR, INDOOR1, INDOOR 2 and INDOOR 3 scenarios, it is observed that an increase in the maximum cell radius in OUTDOOR model causes increased path loss so, the path loss in QPSK CTC 1/2 for example is reached to 129.6561 dB, as compared to 103.3962 dB, 109.8200 dB and 111.6320 dB in INDOOR3, INDOOR 2, and INDOOR 1 respectively.

While the maximum cell radius in OUTDOOR model is reached to 0.3514 km in QPSK CTC 1/2 as compared to 0.0626 km, 0.0954 km, and 0.1075 km in INDOOR3, INDOOR 2, and INDOOR 1 respectively. As expected, if the system data rate is reduced, the cell radius is increased.

The variation of the Power received as a function of the distance to the BS has been computed using equations 8, 10, and the parameters given in Tables 7-a, 7-b, 8, 9, for the propagation scenarios (OUTDOOR; INDOOR 1, INDOOR 2 and INDOOR 3) in the cases of urban and suburban at frequency band 3.5GHZ, and the results are shown in figures 5- 8

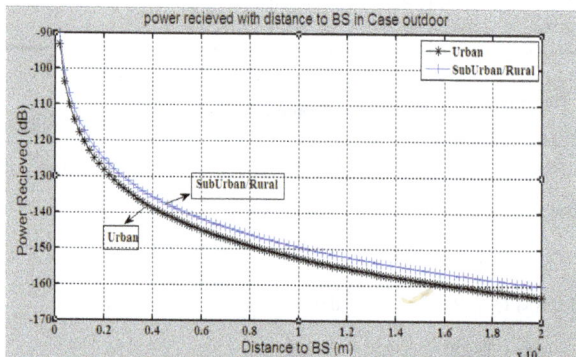

*Figure 5 Power received with distance to BS (COSTHATA) in case outdoor*

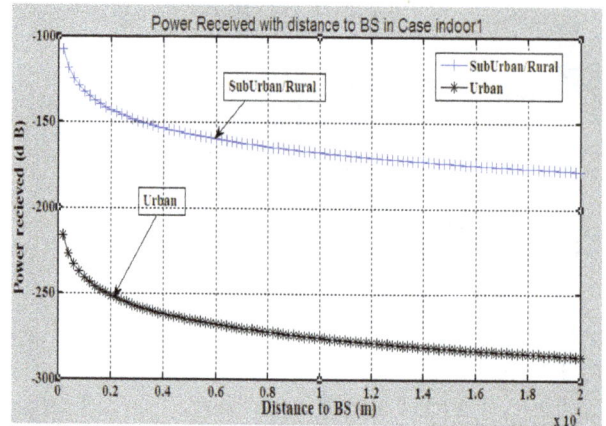

*Figure 6. Power received with distance to BS (COSTHATA) in case indoor1*

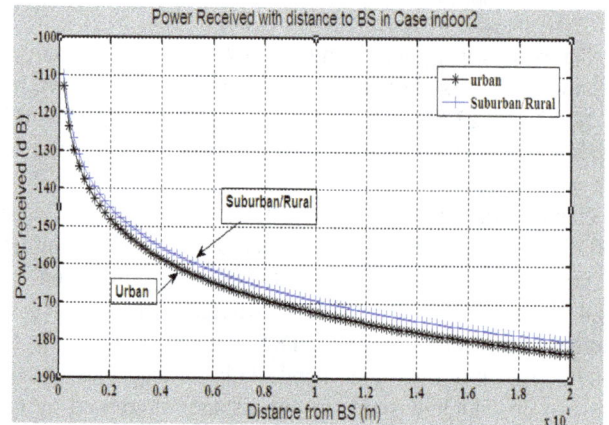

*Figure 7. Power received as a function of distance to BS (COSTHATA) in case indoor2*

The results in figures 5-8 showed that the maximum power received in the suburban case is increased by 100%, 3%, 2.5%, and 4% for INDOOR1, INDOOR2, INDOOR3, and OUTDOOR respectively as compared to the urban case.

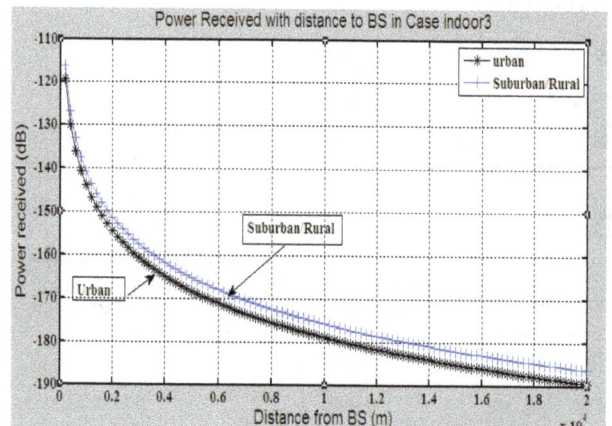

*Figure 8. Power received with distance to BS (COSTHATA) in case indoor3*

Comparing OUTDOOR, INDOOR1, INDOOR 2 and INDOOR 3 scenarios it is noticed that the increase in the maximum power received is -90 dB in OUTDOOR (suburban case), as compared to -100 dB, -110 dB, and -120 dB in INDOOR 1, INDOOR 2, and INDOOR 3 respectively.

While, in urban case the maximum power received is -92 dB in OUTDOOR, as compared to -220 dB, -110 dB, and -120 dB in INDOOR 1, INDOOR 2, and INDOOR 3 respectively.

# 4. Conclusions

Propagation models are used extensively in network planning, particularly for conducting feasibility studies and during initial deployment. They are also very useful for performing interference studies as the deployment proceeds. Knowledge on signal degradation enables RF designers to determine the required field strength for a reliable coverage in a specific area.

In this paper, the data rate for the downlink of OFDMA-based IEEE 802.16m WiMAX system and the available DL throughput as a function of distance to the Base Station (BS) are estimated for a number of propagation scenarios (OUTDOOR; INDOOR 1, INDOOR 2 and INDOOR 3). Moreover, Walls penetration loss is also considered. Adaptive modulation and Coding (AMC) schemes were assumed in the present study for 5 MHz and 20 MHz channel bandwidth.

Three indoor scenarios have then been considered. These are representing with the respective attenuations calculated by equations (1) and (2) for a 3.5 GHz frequency. The chosen Indoor scenarios try to represent possible limiting situations on propagation. The analysis was limited to two walls, since when the number of walls increases, other propagation mechanisms become dominant. The effects of the number of wall, and construction materials for each scenario were considered.

The results showed that the maximum distance to base station in the suburban case is increased by 22%, 22.3%, 22.5%, and 18.5% for INDOOR1, INDOOR2, INDOOR3, and OUTDOOR respectively as compared to the urban case.

In OUTDOOR case, the cell range increased as compared to INDOOR 1, INDOOR 2 and INDOOR 3. It is observed that the maximum distance to base station in INDOOR 2 increased by 34.4 % more than INDOOR 3, where in INDOOR1 this distance increase by 11.3% comparing INDOOR2, where without any Penetration Loss in OUTDOOR scenario this distance increase by 69% comparing INDOOR1.

While the maximum cell radius in OUTDOOR model is reached to 0.3514 km in QPSK CTC 1/2 as compared to 0.0626 km, 0.0954 km, and 0.1075 km in INDOOR3, INDOOR 2, and INDOOR 1 respectively. As expected, decreasing the system data rates, the cell radius is slightly increased.

Comparing OUTDOOR, INDOOR1, INDOOR 2 and INDOOR 3 scenarios it is observed that increase in the maximum power received is -90 dB in OUTDOOR (suburban case), as compared to -100 dB, -110 dB, and -120 dB in INDOOR 1, INDOOR 2, and INDOOR 3 respectively. While, in (urban case) the maximum power received is -92 dB in OUTDOOR, as compared to -220 dB, -110 dB, and -120 dB in INDOOR 1, INDOOR 2, and INDOOR 3 respectively.

In INDOOR 3 case the maximum distance to base station and the maximum power received is decreased as compared to INDOOR 2, INDOOR 1 due to the construction materials for each scenario.

At 20 MHz bandwidth one can observe an increasing in data rate as compared to the 5 MHZ bandwidth.

# References

[1] Garber, L, "Mobile WiMAX: The Next Wireless Battle Ground", IEEE Computer Society, Jun. 2008, vol. 41, No. 6, p p16-18.

[2] "WiMAX Forum Mobile System Profile, Release 1.0 approved specification, Revision 1.4.0", WiMAX Forum, 2007.

[3] H. Yaghoobi, "Scalalable OFDMA Physical Layer in IEEE802.16Wireless MAN", Intel Technology Journal, August 2004, Vol 08, pp. 201-212.

[4] J. G. Andrews, A. Ghosh, R. Muhamed, "Fundamentals of WiMAX", Prentice Hall, New York, 2007.).

[5] IEEE Computer Society & IEEE Microwave Theory and Techniques Society, "IEEE Std 802.16e™-2005: IEEE Standard for Local and metropolitan area networks – Part 16: Air Interface for Fixed and Mobile Broadband Wireless Access Systems; Amendment 2: Physical and Medium Access Control Layers for Combined Fixed and Mobile Operation in Licensed Bands", IEEE, 2005.

[6] COST 231, Digital mobile radio towards future generation systems, Final Report, COST Telecom Secretariat, European Commission, Brussels, Belgium, 1999.

[7] M. Hata, "Empirical formula for propagation loss in land mobile radio services", IEEE Transactions on Vehicular Technology, September 1981, vol. 29, pp. 317-325.

[8] L. M. Correia (Ed.), "Wireless Flexible Personalized Communications", Wiley, Chichester, 2001.

[9] P. Nobles, "A comparison of indoor pathloss measurements at 2 GHz, 5 GHz, 17 GHz and 60 GHz", COST 259, TD(99)100, Leidschendam, The Netherlands, September 1999.

[10] Doug, G., "Mobile WiMAX – part I: A technical overview and performance evaluation," WiMAX Forum, 2006.

[11] Ahmadzadeh, A. M. "Capacity and Cell-Range Estimation for Multitraffic Users in Mobile WiMAX" MSc. Dept. of Electrical ,Communication and Signal Processing Engineering , University College of Borås School of Engineering Sept. 2008.

[12] Koon Hoon Teo., Zhifeng Tao., and Jinyun Zrang. "The Mobile Broadband Standard" IEEE Signal Processing Magazine, September 2007.

[13] Hala. B. Nafea, Fayez W. Zaki," PERFORMANCE OF IEEE 802.16m WIMAX USING ADAPTIVE MODULATION AND CODING" The Mediterranean Journal of Electronics and Communications, Vol. 7, No. 2, 2011

# 32

# Energy-Efficient in wireless sensor networks using fuzzy C-Means clustering approach

**Mourad Hadjila[1], Hervé Guyennet[1], Mohammed Feham[2]**

[1]LIFC UFR ST, Besançon, France
[2]UABB STIC Laboratory, Tlemcen, Algeria

**Email address:**

mhadjila_2002@yahoo.fr (M. Hadjila), herve.guyennet@femto-st.fr (H. Guyennet), m_feham@mail.univ-tlemcen.dz (M. Feham)

**Abstract:** Extending the lifetime of a wireless sensor networks remains one of the prominent research topics in recent years. Clustering has been proven to be energy-efficient in sensor networks since data routing and relaying are only operated by cluster heads. The present paper focuses on proposing two algorithms. In the former nodes organize themselves into clusters using fuzzy c-means (FCM) mechanism then a randomly node chooses itself cluster head in each cluster since initially all nodes have the same amount of power. Then the node having the higher residual energy elects itself cluster head. All non-cluster head nodes transmit sensed data to the cluster head. This latter performs data aggregation and transmits the data directly to the remote base station. The second algorithm which is a improvement of the former uses the same principle in forming clusters and electing cluster heads but operates in multi-hop manner when it routes data from cluster heads to the base station. Simulation results show that the proposed algorithms improve energy consumption and consequently resulting in an extension of the network lifetime. In addition, the second algorithm proves its ability to be applied in large-scale wireless sensor networks.

**Keywords:** Wireless Sensor Networks, Fuzzy C-Means, Clustering, Lifetime

## 1. Introduction

In recent decades, the need to observe and monitor hostile environments has become essential for many military and scientific applications. The nodes used must be independent, has a miniature size and can be deployed in a random manner in the dense and monitored field. A special class of ad hoc networks called wireless sensor networks comes to the rescue. They have appeared thanks to technological developments such as miniaturization of electronic components, reducing manufacturing costs and increasing performance and storage capacity, and energy calculation.

A wireless sensor network consists of a massive number of small, inexpensive, self-powered devices that can sense, compute, and communicate with other devices for the purpose of gathering local information to make global decision about a physical environment [1]. Data gathered may be a variety of environment conditions such as temperature, humidity, pressure, movement, light, density of air pollutants, early fire detection, and so on [2].

The constraints imposed by these networks are very well known: very limited computation, communication, storage capabilities, and energy resources. This last aspect, which limits the lifetime of the network and therefore its utility, has received considerable attention by the research community over the last several years. The design of energy-aware protocols, algorithms, and mechanisms, with the goal of saving as much energy as possible, and therefore extending the lifetime of the network, has been the topic of many research studies [3].

The hierarchical organization of the sensors, grouping them and assigning those specific tasks into the group before transferring the information to higher levels, is one of the mechanisms proposed to deal with the sensors limitations and is commonly referred to as clustering [4-6]. In this context came our proposed algorithms named respectively proposed1 and proposed2. They consist of forming clusters applying fuzzy c-means mechanism followed by selecting cluster heads using residual energy of nodes in each cluster but they differ in their mode of transmitting data to the base station. The remainder of this paper is organized as follows: in the next section, related works about the clustering-based routing protocols have been presented. Section III describes in detail the proposed algorithms. Simulation results are discussed in section IV. Finally, section V concludes the paper.

## 2. Related Works

Several clustering-based routing in wireless sensor networks has been addressed by a number of researchers [5,6]. Heinzelman and al [7,8] proposed a typical clustering scheme called Low-Energy Adaptive Clustering Hierarchy (LEACH). It is an energy-efficient protocol designed for sensor networks with continuous data delivery mechanism and no mobility. In LEACH, nodes organize themselves into local clusters, with one node acting as the local cluster-head. LEACH includes randomized rotation of the high-energy cluster-head position such that it rotates among the various sensors in order to not drain the battery of a single sensor [9]. A Hybrid Energy-Efficient Distributed clustering approach (HEED) [10] introduces a variable known as the cluster radius, which defines the transmission power to be used for intra-cluster broadcast. The initial probability for each node to become a tentative cluster head depends on its residual energy, and final cluster heads are selected according to the intra-cluster communication cost. HEED relies on the assumption that cluster heads can communicate with each other and form a connected graph; realizing this assumption in practical deployments could be tricky.

In [11], the authors propose a new energy-efficient clustering approach (EECS) for single-hop wireless sensor networks, which is more suitable for the periodical data gathering applications. EECS extends LEACH algorithm by dynamic sizing of clusters based on cluster distance from the base station. D. C. Hoang & al in [12] apply an approach based on fuzzy C-Means for clustering calculation, cluster head selection and data transmission. Threshold sensitive Energy Efficient sensor Network protocol (TEEN) [13] is a hybrid of hierarchical clustering and data-centric protocols designed for time-critical applications. It is a responsive protocol to sudden changes of some of the attributes observed in the wireless sensor network such as temperature [6]. However, TEEN cannot be applied for sensor networks where periodic sensor readings should be delivered to the sink. An adaptive threshold sensitive Energy Efficient sensor Network protocol (APTEEN) [14] that is an extension of TEEN is used for both periodic and responsive data collection. We just mention this non-exhaustive list of clustering protocols. For more reading refer to [5,6].

## 3. Proposed Protocols

This manuscript proposes two energy-efficient algorithms for the extension of the lifetime of a wireless sensor networks based on fuzzy C-means clustering approach. Before explaining the principle of the proposed approaches, we briefly describe the fuzzy c-means algorithm used to accomplish clustering task.

### 3.1. Fuzzy C-Means Algorithm

FCM clustering protocol is centralized clustering algorithm, the base station computes and allocates sensor nodes into clusters according to the information of their location and the cluster head is assigned to the node having the largest residual energy [12]. Consider a network of N nodes and partitioned into c clusters. The objective function of FCM algorithm is to minimize the following equation [15]:

$$J_m = \sum_{i=1}^{c} \sum_{j=1}^{N} \mu_{ij}^m d_{ij}^2 \qquad (1)$$

where

$\mu_{ij}$ is node j's degree of belonging to cluster i.
$d_{ij}$ is the Euclidean distance between node j and the center of cluster i.
The $z_j$ centroid of the $j^{th}$ cluster, is obtained using (2).

$$z_j = \frac{\sum_{i=1}^{N} \mu_{ij}^m o_i}{\sum_{i=1}^{N} \mu_{ij}^m} \qquad (2)$$

The degree $\mu_{ij}$ of node j respected to cluster is calculated and fuzzified with the real parameter m>1 as below:

$$\mu_{ij} = \frac{1}{\sum_{k=1}^{c} \left(\frac{d_{ij}}{d_{kj}}\right)^{2/(m-1)}} \qquad (3)$$

The FCM algorithm is iterative and can be stated as follows [15]:

1- Select m   (m >1); initialize the membership function values $\mu_{ij}$, i = 1,2,..., n ;j = 1,2,..., c.

2- Compute the cluster centers $z_j$   , j =   1,2,..., c, according to (2).

3- Compute Euclidian distance $d_{ij}$ , i =   1,2,..., n ; j = 1,2,..., c

4- Update the membership function $\mu_{ij}$ , i =   1,2,..., n ; j =   1,2,..., c   according to (3).

5- If not converged, go to step 2.

### 3.2. First Proposed Algorithm

The first proposed algorithm mimics Leach protocol but the difference lies in the mode of clusters formation and cluster heads selection. It operates into three steps. In the first step clusters are formed by the Fuzzy C-Means method. Each clusters contains a set of a nodes and the number of nodes is not necessary equal in the clusters. In the second step, a cluster head is initially elected in each cluster. Since all nodes have the same amount of energy, a random number is generated between 1 and the number of nodes in each cluster and the node corresponding to this number elects itself cluster head. This is done only at the beginning but after a rotation mechanism based on the remaining energy is applied to select the next cluster head. Non-cluster head nodes send gathered data to the corresponding cluster head.

In the third step, cluster heads receive and aggregate data then communicate it in single hop manner to the remote base station.

The pseudo-code of the first proposed algorithm is described as follows:

Apply FCM algorithm to form clusters.

Each cluster c(i) contains a number of nodes, i=1, ..., nc

Initially all nodes have the same amount of energy.

//cluster heads selection

maxE=zeros(1,nc) ;

maxE is a row vector contains nc zeros

it_max: maximum number of iterations

ET: is the totalnetwork energy

while(it≤it_max || ET>0)

for i = 1 to nc do

if it==1

r=rand*length(c(i)) ;

ch(i)=c(i).r ;

else

for j=1 to length(c(i))   do

if maxE(i)<c(i).E(j)

maxE(i)=c(i).E(j)

end if

end for

ch(i)=maxE(i)

end if

end for

// non CHs send data to CHs

for i – 1 to nc do

for j=1 to length(c(i)) do

c(i).j send data to ch(i)

end for

end for

// CHs send data to BS

for i = 1 to nc do

ch(i) aggregates and transmits directly data to BS

end for

end while

Figure below illustrates the operating mode of the first proposed algorithm where cluster heads send data directly to the base station.

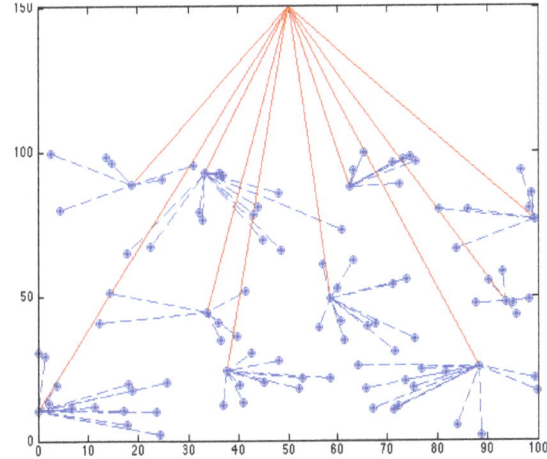

**Figure. 1.** *Single hop with clustering.*

### 3.3. Second Proposed Algorithm

The second proposed protocol uses the same principle of clusters formation and cluster head selection than the first protocol. The difference is in the data transmission from the CHs to the base station. Data transmission is made by multi-hop. Each cluster head sends aggregated data to the closest cluster head in the direction of the base station. This is repeated until it reaches the base station.

Since the mode of clusters formation and the cluster heads selection is the same as the first algorithm, we present below only the pseudo-code of the part concerning the data transmission from the cluster heads to the base station.

Calculation of distances between CHs and distances between CHs and BS

for i=1 to nc

for j=1 to nc do

$$d(i,j) = \sqrt{(x_i - x_j)^2 + (y_i - y_j)^2}$$

end for

$$dBS(i) = \sqrt{(x_i - x_{BS})^2 + (y_i - y_{BS})^2}$$

end for

for i=1 to nc do

if dBS(i)==mindBS

ch(i) sends directly to BS

else

for j=1 to nc do

```
if (i≠j && d(i,j)==min(i) && dBS(i)>dBS(j))

        ch(i) sends data to ch(j)

                end if

                end for

                end if

                end for
```

Figure below represents the principle of the second proposed algorithm.

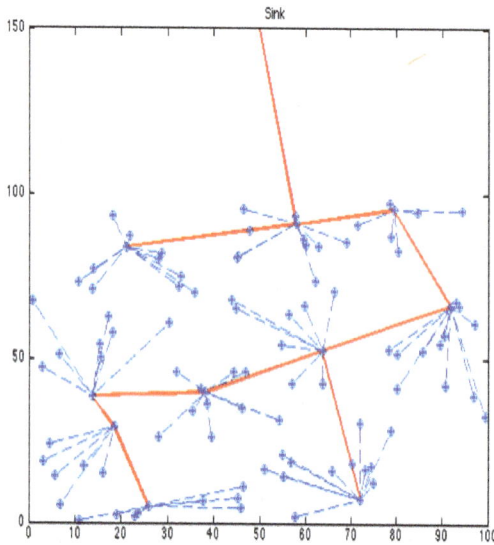

*Figure. 2.* *Multi-hop with clustering*

The blue dashed lines represent intra-cluster communication while the red bold lines represent the inter-cluster heads communication and the base station.

We emphasize that both proposed algorithms are centralized and controlled by the base station.

# 4. Evaluation

Before discussing the simulation part, we make some assumptions then we present the used radio model.

## 4.1. Assumptions

The following properties are assumed in regard to the sensor network being simulated:

Both sensor nodes and base station are stationary after being deployed in the field.

Base station is located outside the area of the sensor nodes.

The wireless sensor network consists of the homogeneous sensor nodes.

All sensor nodes have the same initial energy.

Base station is not limited in terms of energy, memory and computational power.

The radio channel is symmetric such that energy consumption of transmitting data from node X to node Y is the same as that of transmission from node Y to node X.

Each sensor nodes can operate either in sensing mode to monitor the environment parameters and transmit to the base station or cluster head mode to gather data, compress it and forward to the base station.

## 4.2. Radio Model

The radio model we have used is similar to [8]. There are two different radio models. The free space model and the multi-path fading channel model. When the distance between the transmitter and receiver is less than threshold value $d_0$, the algorithms adopt the free space model ($d^2$ power loss). Otherwise the algorithms adopt the multi-path fading channel model ($d^4$ power loss). So if the transmitter sends a k-bit message to the receiver up to a distance of d, the energy consumption of the receiver can be calculated by the following equations:

$$E_{TX} = \begin{cases} k.E_{elec} + kE_{fs}d^2 & if \quad d < d_0 \\ kE_{elec} + kE_{mp}d^4 & if \quad d \geq d_0 \end{cases} \quad (4)$$

Threshold value $d_0$ is given by:

$$d_0 = \sqrt{\frac{E_{fs}}{E_{mp}}} \quad (5)$$

Where

$E_{elec}$ is the amount of energy consumption per bit to run the transmitter or receiver.

$E_{fs}$ and $E_{mp}$ represent the energy consumption factor of amplification in the two radio models.

$E_{RX}$ is the amount of energy consumption in receiving a packet with k bits. This is given by (6).

$$E_{RX} = k \ . \ E_{elec} \quad (6)$$

## 4.3. Simulation Results and Analysis

The proposed algorithms are evaluated by simulating a network of 100 nodes. Sensor nodes are dispersed in an area of 100 x 100 m² shown in figure 1, the base station is located at (50,150), so it is at least 50m from the closest sensor node. The number of clusters is chosen equal 10, which is the square root of the total number of nodes. Each sensor node transmits a 1000 bit message. The initial energy supplies to each sensor node is 0.1J. Table 1 gives all simulations parameters.

For the evaluation of the algorithms two metrics have been chosen: energy consumption and number of alive nodes.

Figure 3 shows the simulation of the energy consumption of nodes in each round for both approaches comparing with the direct transmission algorithm. The total energy dissipated of proposed1 and proposed2 is better than direct transmission algorithm, and also the total energy dissipated of proposed2 is a few better than that of proposed1.

**Table 1.** *Simulation parameters*

| Parameters | Values |
|---|---|
| Network size | $(100 \times 100) \text{ m}^2$ |
| Number of nodes | 100 |
| Initial energy | 0.1 J |
| Data packet size | 1000 bits |
| $E_{elec}$ | $50 \times 10^{-9}$ |
| $E_{fs}$ | $10^{-11}$ |
| $E_{mp}$ | $1.3 \times 10^{-15}$ |
| EDA | $5 \times 10^{-9}$ |

*EDA represents energy data aggregation.*

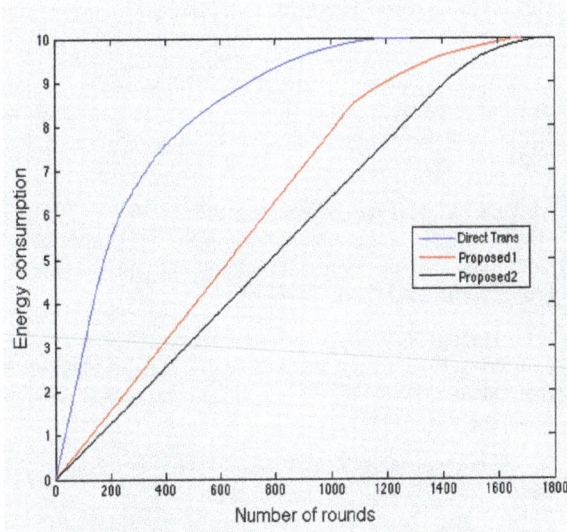

**Figure. 3.** *Energy consumption vs. number of rounds*

From the simulation result shown in figure 4 and given by table 2, we observe that the first node dies in direct transmission algorithm after 131 rounds while in proposed1 and proposed2 first node dies after 1077 and 1411 rounds respectively. We also observe that the last node dies in direct transmission algorithm after 1283 rounds while in proposed1 and proposed2 last node dies after 1681 and 1753 rounds respectively. Therefore, we note that proposed2 is about 4.11% more efficient in term of network lifetime than proposed1 and about 26.82% than direct transmission algorithm.

In the following part, we simulate proposed2 algorithm in five different network sizes. We assume that nodes are deployed in an area of 300 x 300 m²; the base station is located at (150,400). The two metrics mentioned above are used for evaluation.

Figure 5 represents the variation of energy consumption with respect to the number of rounds in different networks containing respectively 100, 300, 600, 900 and 1200 nodes.

**Table 2.** *Duration of first and last node die in the network*

| | Direct Transmission | Proposed1 | Proposed2 |
|---|---|---|---|
| First node dies | 131 | 1077 | 1411 |
| Last node dies | 1283 | 1681 | 1753 |

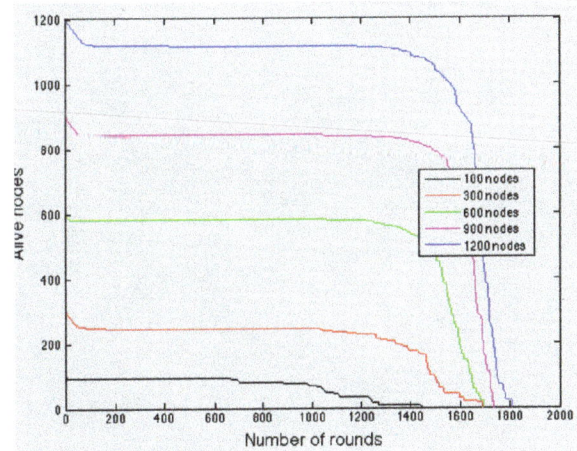

**Figure 4.** *Number of alive nodes vs. number of rounds*

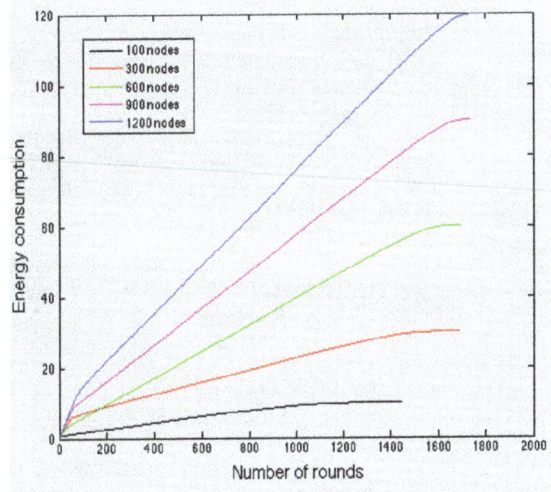

**Figure. 5.** *Energy consumption vs. number of rounds*

Figure 6 below shows the variation of alive nodes versus number of rounds for the precedent five network sizes.

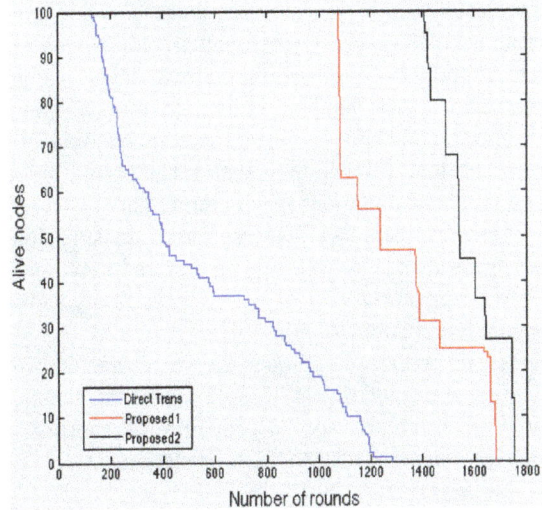

**Figure 6.** *Number of alive nodes vs. number of rounds*

# 5. Conclusion

In wireless sensor networks, it is recommended to apply a clustering scheme specially when you are confronted to the problem of power consumption. In this paper, we have proposed two approaches that consist to form clusters by FCM algorithm and choose cluster heads by residual energy of nodes but they differ in their data transmission mode to the base station. Simulations results show that ours schemes offer a better performance than the direct transmission algorithm in terms of energy consumption and network lifetime. Moreover, scalability of the proposed2 algorithm is also verified over simulation.

The second proposed approach can be further improved by adopting some intelligent algorithm such as genetic algorithms or ant colony specially to find the shortest path between the cluster heads and the base station.

# References

[1]   S. Olariu, "Information Assurance in Wireless sensor networks", Parallel and Distributed Processing Symposium. 19th IEEE International, April 2005.

[2]   G. Haosong and Y. Younghwan, "Distributed Bottleneck Node Detection in Wireless Sensor Networks", IEEE 10th International Conference on Computer and Information Technology (CIT), July 2010, pp. 218-224.

[3]   P. M. Wightmanl and M. A. Labrador, "Topology Maintenance: Extending the Lifetime of Wireless Sensor Networks", IEEE LATIN AMERICA TRANSACTIONS, Vol. 8, No. 4, Aug. 2010.

[4]   A. Boukerche, "Algorithms and Protocols for Wireless Sensor Networks", John Wiley & Sons, Inc, pp.161, 2009.

[5]   A. A. Abbasi and M. Younis, "A survey on clustering algorithms for wireless sensor networks", Elsevier, Computer Communications, 2007, pp. 2826–2841.

[6]   A. M, A. Boukerche and R. W. Nelem Pazzi, "A Taxonomy of Cluster-based Routing Protocols for Wireless Sensor Net-

works", IEEE International Symposium on Parallel Architectures, Algorithms, and Networks, pp. 247–253, 2008.

[7]   W. Heinzelman, A. Chadrakasan and H. Balakrishnan, "Energy efficient communication protocol for wireless microsensor networks", in Proceedings of the 33rd Annual HawaiII International Conference on System Sciences, Jan 4-7, 2000.

[8]   W.B. Heinzelman, A.P. Chandrakasan, H. Balakrishnan, "An application-specific protocol architecture for wireless microsensor networks", IEEE Transactions on Wireless Communications, Vol. 1, Issue 4, pp. 660-670, 2002.

[9]   Si-Ho Cha1 and Minho Jo, "An Energy-Efficient Clustering Algorithm for Large-Scale Wireless Sensor Networks", Advanced in Grid and Pervasive Computing, Vol. 4459, 2007, pp. 436–446.

[10]  O. Younis, S. Fahmy, "HEED: A hybrid, energy-efficient, distributed clustering approach for ad hoc sensor networks", IEEE Transactions on Mobile Computing, Vol. 3, Issue 4, 2004, pp. 366-379.

[11]  M. Ye, C. Li, G. Chen, J. Wu, "An energy efficient clustering scheme in wireless sensor networks", 24th International Conference on Performance, Computing and Communication, 7-9 April 2005, pp. 535-540.

[12]  D.C. Hoang, R. Kumar and S.K. Panda, "Fuzzy C-Means Clustering Protocol for wireless sensor networks", IEEE International Symposium on Idustrial Electronics (ISIE), July 2010, pp. 3477-3482.

[13]  A. Manjeshwar and D. P. Agarwal, "TEEN: a Routing Protocol for Enhanced Efficiency in Wireless Sensor Networks", Parallel and Distributed Processing Symposium., Proceedings 15th International, 2001, pp. 2009-2015.

[14]  A. Manjeshwar and D. P. Agarwal, "APTEEN: a hybrid protocol for efficient routing and comprehensive information retrieval in wireless sensor networks", Parallel and Distributed Processing Symposium, Proceedings International, 2002, pp. 195-202.

[15]  H. Izakian, A. Abraham and V. Snasel, "Fuzzy Clustering Using Hybrid Fuzzy c-means and Fuzzy Particle Swarm Optimization", IEEE World Congress on Nature & Biologically Inspired Computing, USA, 2009, pp.1690-1694

# 33

# Two aspect authentication system using secure mobile devices

S. Uvaraj[1], S. Suresh[2], N. KannaiyaRaja[3]

[1]Arulmigu Meenakshi Amman College of Engineering
[2]Sri Venkateswara College of Engineering
[3]Defence Engineering College, Ethiopia

**Email address:**

ujrj@rediffmail.com(S. Uvaraj), ss12oct92@gmail.com(S. Suresh), kanniya13@hotmail.co.in(N. KannaiyaRaja)

**Abstract:** Mobile devices are becoming more pervasive and more advanced with respect to their processing power and memory size. Relying on the personalized and trusted nature of such devices, security features can be deployed on them in order to uniquely identify a user to a service provider. In this paper, we present a strong authentication mechanism that exploits the use of mobile devices to provide a two-aspect authentication system. Our approach uses a combination of one-time passwords, as the first authentication aspect, and credentials stored on a mobile device, as the second aspect, to offer a strong and secure authentication approach. By Adding an SMS-based mechanism is implemented as both a backup mechanism for retrieving the password and as a possible mean of synchronization. We also present an analysis of the security and usability of this mechanism. The security protocol is analyzed against an adversary model; this evaluation proves that our method is safe against various attacks, most importantly key logging, shoulder surfing, and phishing attacks. Our simulation result evaluation shows that, although our technique does add a layer of indirectness that lessens usability; participants were willing to trade-off that usability for enhanced security once they became aware of the potential threats when using an untrusted computer.

**Keywords:** Computer Network Security, Mobile Handsets, One-Time Password, Smart Mobile Phones

## 1. Introduction

Today security concerns are on the rise in all areas such as banks, governmental applications, healthcare industry, military organization, educational institutions, etc. Government organizations are setting standards, passing laws and forcing organizations and agencies to comply with these standards with non-compliance being met with wide-ranging consequences. There are several issues when it comes to security concerns in these numerous and varying industries with one common weak link being passwords. Mobile-OTP was introduced in 2003. As of 2012 there are more than 40 independent implementations of the Mobile-OTP algorithm making it a de facto standard for strong mobile authentication. Most systems today rely on static passwords to verify the user's identity. However, such passwords come with major management security concerns. Users tend to use easy-to-guess passwords, use the same password in multiple accounts, write the passwords or store them on their machines, etc. Furthermore, hackers have the option of using many techniques to steal passwords such as shoulder surfing, snooping, sniffing, guessing, etc.

Several 'proper' strategies for using passwords have been proposed [1]. Some of which are very difficult to use and others might not meet the company's security concerns. Two aspect authentication using devices such as tokens and ATM cards has been proposed to solve the password problem and have shown to be difficult to hack. Two aspect authentication also have disadvantages which include the cost of purchasing, issuing, and managing the tokens or cards. From the customer's point of view, using more than one two- aspect authentication system requires carrying multiple tokens/cards which are likely to get lost or stolen.

Mobile phones have traditionally been regarded as a tool for making phone calls. But today, given the advances in hardware and software, mobile phones use have been expanded to send messages, check emails, store contacts, etc. Mobile connectivity options have also increased. After standard GSM connections, mobile phones now have infra-red, Bluetooth, 3G, and WLAN connectivity. Most of us, if not all of us, carry mobile phones for communication purpose. Several mobile banking services available take advantage of the improving capabilities of mobile devices. From being able to receive information on account balances in the form of SMS messages to using WAP and Java together with GPRS to allow fund transfers between accounts, stock trading, and confirmation of direct payments via the phone's micro browser [12].

Installing both vendor-specific and third party applications allow mobile phones to provide expanded new services other than communication. Consequently, using the mobile phone as a token will make it easier for the customer to deal with multiple two aspect authentication systems; in addition it will reduce the cost of manufacturing, distributing, and maintaining millions of tokens. We have developed a scheme that enables the use of mobile devices for authenticating users to a web service provider. The approach provides a two- aspect authentication mechanism by combining one-time passwords (OTPs), as the first authentication aspect, together with encrypted user credentials stored on a mobile phone as the second authentication aspect. In this approach, we treated the mobile device as a trusted digital wallet to securely encrypt and store users' long term credentials. These credentials (i.e., her username and password) are encrypted using the public key of the service provider, stored on the mobile device, and are transferred to the service provider when needed. The storage of long term credentials on the mobile device enables users to use stronger passwords for their accounts, as they don't need to remember and retype the passwords for each and every login time. One-time passwords further protect the stored credentials on the cell phone, if stolen or lost, by requiring additional information at the time of log in. In addition to a description of this authentication protocol, we present the results of security and usability evaluations of our two-aspect mobile authentication system. The security analysis evaluates the mobile authentication mechanism against an adversary model. Our analysis shows that the security of our devised method is improved over similar authentication approaches that use mobile devices [MvO07, MWL04, PKP06, WGM04], due to the addition of the OTP which leads to having a strong authentication mechanism. Furthermore, the results of our usability study show that our participants were willing to adopt the new technology once became aware of the potential threats to their passwords when using untrusted computers. Participants indicated they would accept a lower level of usability in return for the higher level of security of the mobile technology. However, for this new technology to be a complete replacement to conventional username/password based systems, it should be signicantly simpler.

In this paper, we propose and develop a complete two aspect authentication system using mobile phones instead of tokens or cards. The system consists of a server connected to a GSM modem and a mobile phone client running a J2ME application. Two modes of operation are available for the users based on their preference and constraints. The first is a stand-alone approach that is easy to use, secure, and cheap. The second approach is an SMS-based approach that is also easy to use and secure, but more expensive. The system has been implemented and tested.

In the next section we provide a related works about authentication aspect s and existing two aspect authentication systems. Section 3 & 4 describes the proposed system design, the OTP algorithm, client, server, and the database. Section 6 & 7 provides simulation setup and results of testing the system. Section 8 concludes the paper and provides future work.

## 2. Related Works

By definition, authentication is the use of one or more mechanisms to prove that you are who you claim to be. Once the identity of the human or machine is validated, access is granted.

Three universally recognized authentication aspect s exist today: what you know (e.g. passwords), what you have (e.g. ATM card or tokens), and what you are (e.g. biometrics). Recent work has been done in trying alternative aspects such as a fourth aspect, e.g. somebody you know, which is based on the notion of vouching [10]. Two aspect authentications [4] is a mechanism which implements two of the above mentioned aspects and is therefore considered stronger and more secure than the traditionally implemented one aspect authentication system. Withdrawing money from an ATM machine utilizes two aspect authentications; the user must possess the ATM card, i.e. what you have, and must know a unique personal identification number (PIN), i.e. what you know. Passwords are known to be one of the easiest targets of hackers. Therefore, most organizations are looking for more secure methods to protect their customers and employees. Biometrics are known to be very secure and are used in special organizations, but they are not used much in secure online transactions or ATM machines given the expensive hardware that is needed to identify the subject and the maintenance costs, etc. Instead, banks and companies are using tokens as a mean of two aspect authentication. A security token is a physical device that an authorized user of computer services is given to aid in authentication. It is also referred to as an authentication token or a cryptographic token. Tokens come in two formats: hardware and software. Hardware tokens are small devices which are small and can be conveniently carried. Some of these tokens store cryptographic keys or biometric data, while others display a PIN that changes with time. At any particular time when a user wishes to log-in, i.e. authenticate, he uses the PIN displayed on the token in addition to his normal account password. Software tokens are programs that run on computers and provide a PIN that change with time. Such programs implement a One Time Password (OTP) algorithm. OTP algorithms are critical to the security of systems employing them since unauthorized users should not be able to guess the next password in the sequence. The sequence should be random to the maximum possible extent, unpredictable, and irreversible. Aspects that can be used in OTP generation include names, time, seed, etc. Several commercial two aspect authentication systems exist today such as BestBuy's BesToken, RSA's SecurID [6], and Secure Computing's Safeword [2]. BesToken applies two-aspect authentication through a smart card chip integrated USB token. It has a great deal of functionality by being able to both generate and store users' information such as passwords, certificates and keys. One application is to use it to log into laptops. In this case, the user has to enter a password while the USB token is plugged to the laptop at the time of the login. A hacker must compromise both the USB and the user account password to log into the laptop.SecurID from RSA uses a token (which could be hardware or software) whose internal clock is synchronized with the main server. Each token has a unique seed which is used to generate a pseudo-random number. This seed is loaded into the server upon purchase of the token and used to identify the user. An OTP is generated using the token every 60 seconds. The same process occurs at the server side. A user uses the OTP along with a PIN which only he knows to authenticate and is validated at the server side. If the OTP and PIN match, the user is authenticated [8]. In services such as ecommerce, a great deal of time and money is put into countering possible threats and it has been pointed out that both the client and the server as well as the channel of communication between them is imperative [1]. Using tokens involves several steps including registration of users, token production and distribution, user and token authentication, and user and token revocation among others [6]. While tokens provide a much safer environment for users, it can be very costly for organizations. For example, a bank with a million customers will have to purchase, install, and maintain a million tokens. Furthermore, the bank has to provide continuous support for training customers on how to use the tokens. The banks have to also be ready to provide replacements if a token breaks or gets stolen. Replacing a token is a lot more expensive than replacing an ATM card or resetting a password. From the customer's prospective, having an account with more than one bank means the need to carry and maintain several tokens which constitute a big inconvenience and can lead to tokens being lost, stolen, or broken. In many cases, the customers are charged for each token. We propose a mobile-based software token that will save the organizations the cost of purchasing and maintaining the hardware tokens. Furthermore, will allow customers to install multiple software tokens on their mobile phones. Hence, they will only worry about their mobile phones instead of worrying about several hardware tokens.

# 3. System Design

In this paper, we propose a mobile-based software token system that is supposed to replace existing hardware and computer-based software tokens. The proposed system is secure and consists of three parts: (1) software installed on the client's mobile phone, (2) server software, and (3) a GSM modem connected to the server. The system will have two modes of operation:

• Connection-Less Authentication System: A onetime password (OTP) is generated without connecting the client to the server. The mobile phone will act as a token and use certain aspects unique to it among other aspects to generate a one-time password locally. The server will have all the required aspects including the ones unique to each mobile phone in order to generate the same password at the server side and compare it to the password submitted by the client. The client may submit the password online or through a device such as an ATM machine. A program will be installed on the client's mobile phone to generate the OTP.

• SMS-Based Authentication System: In case the first method fails to work, the password is rejected, or the client and server are out of sync, the mobile phone can request the one time password directly from the server without the need to generate the OTP locally on the mobile phone. In order for the server to verify the identity of the user, the mobile phone sends to the server, via an SMS message, information unique to the user. The server checks the SMS content and if correct, returns a randomly generated OTP to the mobile phone. The user will then have a given amount of time to use the OTP before it expires. Note that this method will require both the client and server to pay for the telecommunication charges of sending the SMS message.

## 3.1. OTP Algorithm

In order to secure the system, the generated OTP must be hard to guess, retrieve, or trace by hackers. Therefore, its very important to develop a secure OTP generating algorithm. Several aspects can be used by the OTP algorithm to generate a difficult-to-guess password. Users seem to be willing to use simple aspects such as their mobile number and a PIN for services such as authorizing mobile micropayments [9]. Note that these aspects must exist on both the mobile phone and server in order for both sides to generate the same password. In the proposed design, the following aspects were chosen:

• IMEI number: The term stands for International Mobile Equipment Identity which is unique to each mobile phone allowing each user to be identified by his device. This is accessible on the mobile phone and will be stored in the server's database for each client.

• IMSI number: The term stands for International Mobile Subscriber Identity which is a unique number associated with all GSM and Universal Mobile Telecommunications System (UMTS) network mobile phone users. It is stored in the Subscriber Identity Module (SIM) card in the mobile phone. This number will also be stored in the server's database for each client.

• Username: Although no longer required because the IMEI will uniquely identify the user anyway. This is used together with the PIN to protect the user in case the mobile phone is stolen.

- PIN: This is required to verify that no one other than the user is using the phone to generate the user's OTP. The PIN together with the username is data that only the user knows so even if the mobile phone is stolen the OTP cannot be generated correctly without knowing the user's PIN. Note that the username and the PIN are never stored in the mobile's memory. They are just used to generate the OTP and discarded immediately after that. In order for the PIN to be hard to guess or brute-forced by the hacker, a minimum of 8-characters long PIN is requested with a mixture of upper- and lower-case characters, digits, and symbols.

- Hour: This allows the OTP generated each hour to be unique.

- Minute: This would make the OTP generated each minute to be unique; hence the OTP would be valid for one minute only and might be inconvenient to the user. An alternative solution is to only use the first digit of the minute which will make the password valid for ten minutes and will be more convenient for the users, since some users need more than a minute to read and enter the OTP. Note that the software can modified to allow the administrators to select their preferred OTP validity interval.

- Day: Makes the OTP set unique to each day of the week.

- Year/Month/Date: Using the last two digits of the year and the date and month makes the OTP unique for that particular date. The time is retrieved by the client and server from the telecommunication company. This will ensure the correct time synchronization between both sides.

The above aspects are concatenated and the result is hashed using SHA-256 which returns a 256 bit message. The message is then XOR-ed with the PIN replicated to 256 characters. The result is then Base64 encoded which yields a 28 character message. The message is then shrunk to an administrator-specified length by breaking it into two halves and XOR-ing the two halves repeatedly. This process results in a password that is unique for a ten minute interval for a specific user. Keeping the password at 28 characters is more secure but more difficult to use by the client, since the user must enter all 28 characters to the online webpage or ATM machine. The shorter the OTP message the easier it is for the user, but also the easier it is to be hacked. The proposed system gives the administrator the advantage of selecting the password's length based on his preference and security needs.

### 3.2. Client Design

A J2ME program is developed and installed on the mobile phone to generate the OTP. The program has an *easy to-use* GUI that is developed using the NetBeans drag and drop interface. The program can run on any J2ME-enabled mobile phone. The OTP program has the option of (1) generating the OTP locally using the mobile credentials, e.g. IMEI and IMSI numbers, or (2) requesting the OTP from the server via an SMS message. The default option is

the first method which is cheaper since no SMS messages are exchanged between the client and the server. However, the user has the option to select the SMS-based method. In order for the user to run the OTP program, the user must enter his username and PIN and select the OTP generation method. The username, PIN, and generated OTP are *never* stored on the mobile phone. If the user selects the connection-less method the username and PIN are used to locally generate the OTP and are discarded thereafter. The username and PIN are stored on the server's side to generate the same OTP. If the user selects the SMS-based method, the username and PIN, in addition to the mobile identification aspects, are encrypted via a 256-bit symmetric key in the OTP algorithm and sent via an SMS to the server. The server decrypts the message via the same 256-bit symmetric key, extracts the identification aspects, compares the aspects to the ones stored in the database, generates an OTP and sends the OTP to the client's mobile phone if the aspects are valid. The advantage of encrypting the SMS message is to prohibit sniffing or man-in-the-middle attacks. The 256-bit key will be extremely hard to brute-force by the hacker. Note that each user will have a pre-defined unique 256-bit symmetric key that is stored on both the server and client at registration time.

### 3.3. Database Design

A database is needed on the server side to store the client's identification information such as the first name, last name, username, pin, password, mobile IMEI number, IMSI number, unique symmetric key, and the mobile telephone number for each user. The password field will store the hash of the 10 minute password. It will not store the password itself. Should the database be compromised the hashes cannot be reversed in order to get the passwords used to generate those hashes. Hence, the OTP algorithm will not be traced.

### 3.4. Server Design

A server is implemented to generate the OTP on the organization's side. The server consists of a database as described in Section 3.C and is connected to a GSM modem for SMS messages exchange. The server application is multithreaded. The first thread is responsible for initializing the database and SMS modem, and listening on the modem for client requests. The second thread is responsible for verifying the SMS information, and generating and sending the OTP. A third thread is used to compare the OTP to the one retrieved using the connection-less method. In order to setup the database, the client must register in person at the organization. The client's mobile phone/SIM card identification aspects, e.g. IMEI/IMSI, are retrieved and stored in the database, in addition to the username and PIN. The J2ME OTP generating software is installed on the mobile phone. The software is configured to connect to the server's GSM modem in case the SMS option is used. A unique symmetric key is also generated and installed on both the mobile phone and server. Both parties are ready to generate the OTP at that point.

# 4. Existing OTP Algorithm

Certain types of encryption in the networks, by their mathematical properties cannot be easily overcome by brute force. An example of this is the one-time password algorithm (OTP) [5], where every clear text bit has a corresponding key bit. One-time passwords rely on the ability to generate a truly random sequence of key bits. A brute force attack would eventually reveal the correct decoding, but also every other possible combination of bits, and would have no way of distinguishing one from the other. A small, 100-byte, one-time-password encoded string subjected to a brute force attack would eventually reveal every 100-byte string possible, including the correct answer, but possibly low chance. Here we have analyzed one-time password algorithm for a secure network [6] available today based on mobile authentication and we have listed the possible attacks [7] to the one-time password algorithm.

## 4.1. One-Time Password Algorithm

In existing one-time password algorithm, Java MIDlet is a client application and we assume that this runs in client mobile phones which can be able to receive one time passwords. A MIDlet is an application that uses the Mobile Information Device Profile (MIDP) of the Connected Limited Device Configuration (CLDC) for the Java ME environment. Typical applications include games running on mobile devices and cell phones which have small graphical displays, simple numeric keypad interfaces and limited network access over HTTP. This whole design describes the two main protocols used by Java MIDlet system. Initially, the user downloads the client (Java MIDlet) to his mobile phone. Then the client executes a protocol to register with both server and a service provider utilizing server system for user authentication. After the successful execution of the activation protocol the user can run the authentication protocol an unlimited number of times.

## 4.2. Activation Protocol

After the user has downloaded the client software from a service onto his mobile phone, he must activate the phone as an OTP receiver before it can be used for authentication to a web-based secure service [8]. The activation protocol takes place between five parties: the user, the user's mobile phone, the user's PC, the Server (SE), and the service provider (SP). The main steps of the protocol are summarized below.

1. The user authenticates himself to SP using credentials already known to SP.

2. When the user asks to activate his mobile phone as an OTP receiver, SP redirects the user's browser to SE with a URL that contains an activation request and a Secure Object [9].

3. SE verifies that the Secure Object comes from SP, and gets the user's phone number and other unique identification number like IMEI.

4. SE sends an activation code to the user's PC and an SMS message to the user's phone asking it to start the client software.

5. The mobile phone asks the user to enter the activation code, available on his PC, and transmits the code to SE.

6. SE verifies that the activation code is the same as the one sent to the PC, and sends a challenge to the mobile phone together with an encryption key K0 (The role of K0 is explained separately).

7. The user chooses a personal identification number (PIN) and enters it on the mobile phone, which generates a security code and a response. The response is the encryption of the challenge using the security code as key. The security code and response are sent to SE, and SE stores the security code.

8. SE verifies that the response and the security code correspond to the challenge, and if so, the user has activated the mobile phone as an OTP receiver for use with SP.

These steps should ensure that the PC and the mobile phone are in the same location, or at least that there exists a communication link between the person using the PC and the holder of the phone. Since the person using the PC is authenticated and has transferred the activation code to the phone, we can assume that this person really wants to activate the mobile phone as an OTP generator.

## 4.3. Authentication Protocol

The OTP-based authentication protocol takes place between five parties: the user, the user's mobile phone, the user's PC, the SE, and SP. The main steps of the protocol are described below.

1. The user enters the identity he shares with SP on its login page.

2. SP asks the user for an OTP, and sends a request to ES to generate an OTP for the user.

3. SE first sends an SMS to the user's mobile phone to start the client software. It then sends a challenge to the phone together with two encryption keys Ki and Ki+1 (The role of Ki and Ki+1 is explained separately).

4. The user enters his PIN on the phone, and the phone computes the same security code generated at the time of activation. The phone then encrypts the challenge with the security code as key and sends the cipher text as a response to SE.

5. ES verifies that the response from the mobile phone corresponds to the challenge, and sends an OTP to the phone.

6. The user enters the OTP on the SP's login page, and SP contacts ES to verify that the OTP is indeed the correct one for this user.

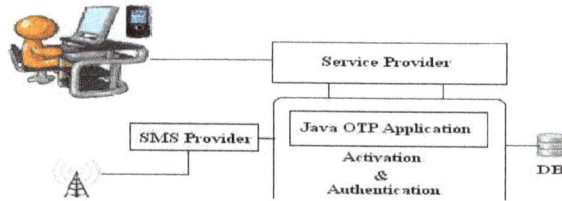

**Fig 1.** *Existing OTP Generation Mechanism*

Fig1 shows architecture of existing OTP mechanism. The authentication protocol's main goal is to ensure that only the legitimate user can obtain an OTP from SE. The goal is not achieved fully because the phone's response to SE challenge is the encryption of the challenge using the key (security code) made during activation. The answer for this challenge may be known to non-legitimate users also. The correct generation of this key requires the correct PIN and other unique information, which possibly other person who are not legitimate also is supposed to have. This person was in turn authenticated at the time of activation; hence we cannot be confident that he is the legitimate user.

# 5. Proposal

Here we designed a PassText based authentication scheme in order to produce a security code instead of using a challenge. Our proposed idea explores the usage of PassText [14] which is impossible to break with brute force attack and stay remains as unpredictable. Proposed works are listed below.

### 5.1. Proposed Activation Protocol

1. The user authenticates himself to SP using credentials already known to SP.

2. When the user asks to activate his mobile phone as an OTP receiver, SP redirects the user's browser to SE with a URL that contains an activation request and a Secure Object.

3. SE verifies that the Secure Object comes from SP, and gets the user's phone number and other unique identification number like IMEI. Here users have to specify PIN.

4. SE sends an activation code to the user's PC and an SMS message to the user's phone asking it to start the client software.

5. The mobile phone asks the user to enter the activation code, available on his/her PC, and transmits the code to SE.

6. SE verifies that the activation code is the same as the one sent to the PC, and sends a Passphrase to the mobile phone together with an encryption key K0.

7. The user chooses a personal identification number (PIN) and enters it on the mobile phone, which generates a security code and a PassText response. The response is the encryption of the PassText using the security code as key. PassText is known only to the legitimate user. The security code and response are sent to SE, and SE stores the security code.

8. SE verifies that the response and the security code correspond to the PassPhrase send, and if so, the user has activated the mobile phone as an OTP receiver for use with SP.

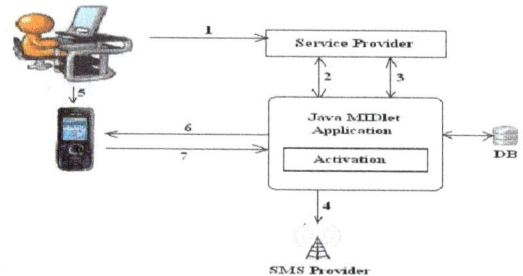

**Fig 2.** *Proposed Layout of Activation Protocol*

In the above proposed protocol, Passphrase is a simple passage given by either a user or SP and it sent from SP to user while activation process. User converts this Passphrase into PassText by some remember able changes. These changes are known to and done by only legitimate users. So this leads to maximum security and so security level for this scheme becomes unpredictable and proposed security level of this measure described later.

### 5.2. Proposed Authentication Protocol

The proposed protocol steps are as follows. This ensures the necessary authentication steps for recognizing legitimate user.

1. User requesting SP for an OTP at the first step, rather than user enters the secure login page.

2. SP verifies the mobile request with existing database and if it's approved request, SP request SE to generate an OTP for the user.

3. SE sends a PassPhrase to the phone together with two encryption keys $K_i$ and $K_{i+1}$.

4. The user enters his PIN and PassText from Passphrase on the phone, and so the mobile computer security code. The mobile then encrypts the PassText with the security code as key and sends the cipher text as a response to SE.

5. ES verifies that the response from the mobile phone corresponds to the passphrase, and sends a sms to the users mobile to login using the identity shares with SP.

6. Now session cookie is activated by SE and it will be send to user's PC after completing login session by the user.

7. If session cookie has arrived to user's PC, secure login page will be redirected automatically by session cookie after a successful logon process finished by a user. If user logins already without mobile authentication, secure page will not be redirected as there is no session cookie. This method completely avoids malicious user to login with the secure system.

8. Redirected URL will request OTP for a secure authentication. Now SE will send an OTP to user mobile through SP.

9. User can authenticate them by the OTP received through mobile. Successful users only can receive OTP

from SP after completing a strong mobile authentication.

This proposed protocol ensures that only legitimate users are accessing the secure system and it also avoids malware based attacks. It's proven that PassText is a string which is known neither only to the legitimate users nor to unauthorized. So brute force attack turns to be zero and its measures are discussed later.

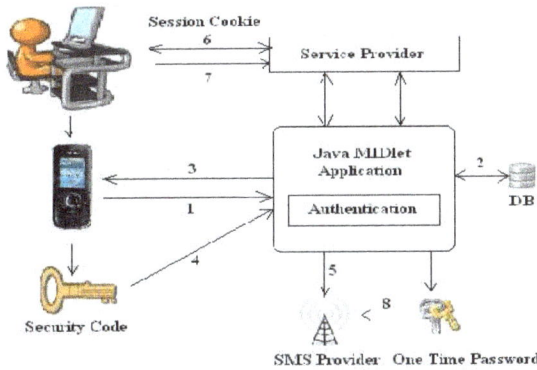

**Fig 3**. *Proposed Layout of Authentication Protocol*

### 5.3. Generation of Security code and Responses

The hash function SHA-1 and the encryption algorithm AES with a 16-byte key are used to generate the security code and the responses. Hashing is denoted by H () and encryption with key K is denoted by EK (). The security code, SC, is computed by the following hash, truncated to 16 bytes,

$$SC=H(PIN||IMEI||CR||SPID)16 \qquad (1)$$

where || denotes concatenation. PIN is a secret number with at least four digits entered by the user; IMEI is a 14 digit code uniquely identifying the mobile phone where the Java MIDlet client is installed; CR or client reference is a 40-byte random string generated on the mobile phone during the activation protocol; and SPID is a public value identifying the SP to whom the user wishes to authenticate. The client reference needs to be stored on the mobile phone for later use. It is only stored in encrypted form. During the activation protocol it is encrypted using AES with the key K0 which is sent by ES together with the PassPhrase. When the client reference is needed in the authentication protocol it is first decrypted using Ki, and when it goes back into storage it is encrypted using Ki+1. The keys Ki and Ki+1 are sent from SE together with the challenge in the authentication protocol. The generation of a 16-byte response, R, to a Passphrase PP, is defined by the expression,

$$R = ESC (PP) \qquad (2)$$

where SC is the 16-byte security code defined by (1) and PP is a 16-byte PassPhrase received from ES.

## 6. Simulation Setup

It has been analyzed that in order to thwart the possible attacks in a secure system, three countermeasures have to be considered. To protect against malware-based replay attack, the proposed activation protocol needs to be secured against the replay of old requests. SE must ensure that each secure object is only used once. Also it should not be allowed to activate another mobile phone as OTP generator for an account, if there already exists a mobile phone activated for the specific account. In order to protect against malware-based impersonation attack and phishing attack there is a need for a tighter control over the transition from old OTP generator to a new mobile phone. Simulation was done between a system and a mobile phone which is having the capacity of receiving and storing the secure object (Fig 4). Mobile phone can be replaced with any device which should have an enough capacity to manipulate a secure object. While analyzing the existing one-time password algorithm, obtained results shows that brute force attack is possible in most of the secure networks. Even though an OTP giving more reliability, according to the cryptographic policy, brute force attack will try all possible combinations of passwords until it finds the correct one. It is very difficult to defend against brute force. First when a user requests SP to provide an OTP, SP verifies that whether the mobile request has been registered already. If not, the request will be discarded immediately. If the requested is approved, SE sends a Passphrase which can be considered as a challenge to the user along with the keys. By the help of personal identification number and a passtext, mobile phone will generate a response to the challenge. Here we have designed a schema which will ensure that same challenge and response should not be given to different users. In this way a secure system can be protected safely from a number of user's. If the response is approved to ES, a message will be sent to user's mobile.

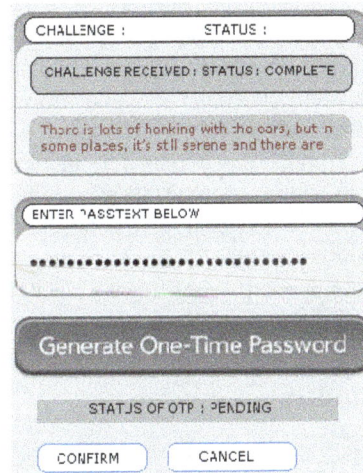

**Fig 4**. *Simulation setup (Making of OTP)*

After getting the confirmation message from ES, user will logon by having his/her own credentials. This mechanism completely avoids malware based impersonation attack since user will not advised to use

URL at the first step itself. Fig6. User logon process when the authenticated user logon into the system, SE will send a session cookie to the user's PC and this will redirect to the secure system. If the user have an OTP token, SE will never send session cookie and this will be stop the malicious users to access the secure system.

# 7. Simulation Results

It seems unfair to say that any set of alphanumeric characters are equally easy to commit to memory. For example "A7Jo0" and word "commit" are not both equal to six units of memory. We have compared password space with different password schemas we can identify the most secure approaches with respect to brute force attack. Table1 demonstrates comparison of password space and password length for popular user authentication schemas. Table1 shows that the approach presented by us is both more secured and the easiest to remember. At the same time, it is relatively fast to produce during an authentication procedure.

*Table 1. Password space comparison.*

| Authentication System | Alphabet | Password Length | Password Space Size |
|---|---|---|---|
| Password | 64 | 8 Char | 2.8 x 1014 |
| Pin Number | 10 | 4 Numbers | 1 x 104 |
| Text with Graphical Assistant | 10 spaces | 8 Char | 2 x 106 |

### 7.1. System Testing

The server was implemented using Java. A Siemens MC35i GSM modem was used for sending and receiving SMS messages on the server side. The smslib3.2.0 [17] library was used to send the messages and the SHA 4j [16] library was used to hash the password. An Oracle 10g was used as a database. The client was implemented using J2ME and installed on a Nokia E60 and Nokia E61 phone. Both methods, the connection-less and SMS-based, were tested. In the first method, fake user accounts were setup on both the mobile phone and server. The mobile phone was used to generate 5000 random OTPs at various times of the day and all 5000 generated OTPs matched the OTPs generated on the server side. The use of date and time from the telecommunication company helped solve the synchronization problem.

# 8. Conclusions

Today, single aspect authentication, e.g. passwords, is no longer considered secure in the internet and banking world. Easy-to-guess passwords, such as names and age, are easily discovered by automated password-collecting programs. Two aspect authentications has recently been introduced to meet the demand of organizations for providing stronger authentication options to its users. In most cases, a hardware token is given to each user for each account. The

increasing number of carried tokens and the cost the manufacturing and maintaining them is becoming a burden on both the client and organization. Since many clients carry a mobile phone today at all times, an alternative is to install all the software tokens on the mobile phone. This will help reduce the manufacturing costs and the number of devices carried by the client. This paper focuses on the implementation of two-aspect authentication methods using mobile phones. It provides the reader with an overview of the various parts of the system and the capabilities of the system. The proposed system has two option of running, either using a free and fast connection-less method or a slightly more expensive SMSbased method. Both methods have been successfully implemented and tested, and shown to be robust and secure. The system has several aspects that make it difficult to hack. Future developments include a more user friendly GUI and extending the algorithm to work on Blackberry, Palm, and Windows-based mobile phones. In addition to the use of Bluetooth and WLAN features on mobile phones for better security and cheaper OTP generation.

# References

[1] A. Jøsang and G. Sanderud, "Security in Mobile Communications: Challenges and Opportunities," in Proc. of the Australasian information security workshop conference on ACSW frontiers, 43-48, 2003.

[2] Aladdin Secure SafeWord 2008. Available at http://www.securecomputing.com/index.cfm?skey=1713

[3] A. Medrano, "Online Banking Security – Layers of Protection," Available at http://ezinearticles.com/?Online-Banking-Security---Layers-of-Protection&id=1353184

[4] B. Schneier, "Two-Aspect Authentication: Too Little, Too Late," in Inside Risks 178, Communications of the ACM, 48(4), April 2005.

[5] D. Ilett, "US Bank Gives Two-Aspect Authentication to Millions of Customers," 2005. Available at http://www.silicon.com/financialservices/0,3800010322,391 53981,00.htm

[6] D. de Borde, "Two-Aspect Authentication," Siemens Enterprise Communications UK- Security Solutions, 2008. Available at http://www.insight.co.uk/files/whitepapers/Twoaspect%20au thenticatio n%20(White%20paper).pdf

[7] A. Herzberg, "Payments and Banking with Mobile Personal Devices,"Communications of the ACM, 46(5), 53-58, May 2003.

[8] J. Brainard, A. Juels, R. L. Rivest, M. Szydlo and M. Yung, "Fourth-Aspect Authentication: Somebody You Know," ACM CCS, 168-78.2006.

[9] NBD Online Token. Available at http://www.nbd.com/NBD/NBD_CDA/CDA_Web_pages/In ternet_Banking /nbdonline_topbanner

[10] N. Mallat, M. Rossi, and V. Tuunainen, "Mobile Banking

Services,"Communications of the ACM, 47(8), 42-46, May 2004.

[11] "RSA Security Selected by National Bank of Abu Dhabi to Protect Online Banking Customers," 2005. Available at http://www.rsa.com/press_release.aspx?id =6092

[12] R. Groom, "Two Aspect Authentication Using BESTOKEN Pro USBToken." Available at http://bizsecurity.about.com/od/mobilesecurity/a/twoaspect.htm

[13] Sha4J. Available at http://www.softabar.com/home/content/view/46/68/

[14] SMSLib. Available at http://smslib.org

# Permissions

# List of Contributors

**YUE Xiangyu**
School of Management and Engineering, Nanjing University; Nanjing Jiangsu; 210008; PR China.

**Adinya John Odey and Li Daoliang**
College of Information and Electrical Engineering, China Agricultural University, Beijing, PRC.

**Uma Datta**
Electronics & Instrumentation Dept., CSIR-CMERI, Durgapur, India.

**Sumit Kundu**
Dept. of ECE, National Institute of Technology Durgapur, Durgapur, India.

**S. Uvaraj**
Arulmigu Meenakshi Amman College of Engineering, Kanchipuram, India.

**S. Suresh**
Sri Venkateswara College of Engineering, Chennai, India.

**E. Mohan**
Pallavan College of Engineering, Kanchipuram, India.

**Adam Mohammed Saliu and Mohammed Idris Kolo**
Department of Computer Science, Federal University of Technology, Minna, Nigeria.

**Mohammed Kudu Muhammad**
Academic Planning Unit, Federal University of Technology, Minna, Nigeria.

**Lukman Abiodun Nafiu**
Department of Mathematics and Statistics, Federal University of Technology, Minna, Nigeria.

**Sarita Maurya and R. L. Yadava**
Department of Electronics & Communication Engineering, Galgotia's College of Engineering and Technology, Greater Noida, India.

**R. K. Yadav**
Department of Electronics & Communication Engineering, I.T.S Engineering College, Greater Noida, India.

**Ana Régia de M. Neves**
Dept. of Electrical Engineering, University of Brasilia, Brazil.

**Humphrey C. Fonseca and Célia G. Ralha**
Dept. of Computer Science, University of Brasilia, Brazil.

**Elnaz Shafigh Fard and Mohammad H. Nadimi**
Faculty of Computer Engineering, Najafabad branch, Islamic Azad University, Isfahan, Iran.

**A. A. Ojugo., R. Abere. and B. C Orhionkpaiyo.**
Department of Mathematics/Computer, Federal University of Petroleum Resources Effurun, Delta State.

**E. R. Yoro**
Department of Computer, Delta State Polytechnic Ogwashi-Uku, Delta State.

**A. O. Eboka**
Dept. Of Computer Sci., Federal College of Edu, (Technical), Asaba, Delta State.

**Poonam Singh**
Electronics & Comm. Engg. Dept. National Institute of Technology Rourkela, India.

**Saswat Chakrabarti**
GSSST, Indian Institute of Technology, Kharagpur, India.

**Hussein Mohammed Salman**
College of Material Engineering, Babylon University, Babil, Iraq

Babylon University, Babil, Iraq.

**Miti Bharatkumar Sukhadia and Veeresh Gangappa Kasabegoudar**
Post Graduate Department, Mahatma Basaveshwar Education Society's, College of Engineering, Ambajogai, India.

**Rakesh Ranjan**
Department of Electronics and Communication Engineering, National Institute of Technology (NIT), Patna, 800005, India.

**Dipen Bepari and Debjani Mitra**
Department of Electronics Engineering, Indian School of Mines (ISM), Dhanbad, 826004, India.

**Jian Li, Yifeng He, Yun Tie and Ling Guan**
Department of Electrical and Computer Engineering, Ryerson University, Toronto, Canada.

**Diponkar Paul, Subrata Kumar Sarkar and Rajib Mondal**
World University of Bangladesh.

**Costas Daskalakis, Nikos Sakkas and Maria Kouveletsou**
Applied Industrial Technologies Ltd., Gerakas, Attiki, Greece.

**YUE Xiangyu**
School of Management and Engineering, Nanjing University; Nanjing Jiangsu; 210008; PR China.

**Basma A. Mahmoud, Esam A. A. Hagras and Mohamed A. Abo El-Dhab**
Department of Electronics & Communications, Arab Academy for Science, Technology & Maritime Transport, Cairo, Egypt.

**Jingcheng Zhang, Allan Huynh, Patrik Huss, Qin-Zhong Ye and Shaofang Gong**
Department of Science and Technology, Linköping University, Bredgatan 33, 60174, Norrköping, Sweden.

**AMALA GRACY**
Department of Information Technology, RMK Engineering College, Anna University, Chennai, INDIA.

**CHINNAPPAN JAYAKUMAR**
Department of Computer Science and Engineering, RMK Engineering College, Anna University, Chennai, INDIA.

**Rushikesh Dinkarrao Maknikar and Veeresh Gangappa Kasabegoudar**
Post Graduate Department, Mahatma Basaveshwar Education Society's, College of Engineering, Ambajogai, India.

**Abdullahi Ibrahim Abdu and Muhammed Salamah**
Computer Engineering Department, Eastern Mediterranean University, KKTC, Mersin 10, TURKEY.

**Md. Mainul Islam Mamun and Shaikh Enayet Ullah**
Dept. of Applied Physics and Electronic Engineering, Rajshahi University, Rajshahi, Bangladesh.

**Joarder Jafor Sadique**
Dept. of Electrical and Electronic Engineering (EEE), University of Information Technology and Sciences (UITS), Dhaka, Bangladesh.

**Amel Boufrioua**
Electronics Department, Technological Sciences Faculty, University Constantine, Ain El Bey Road, 25000, Constantine, Algeria.

**Arianit Maraj**
Faculty of Computer Science, Public University of Prizren, Prizren, Republic of Kosova.

**Ruzhdi Sefa**
Faculty of Electrical and Computer Engineering, University of Prishtina, Prishtina, Republic of Kosova.

**Anton V. Lazebnyy**
Scientific and Production Enterprise "Ukrservisbud" Ltd., Kyiv, Ukraine, system administrator, MA.

**Volodymyr S. Lazebnyy**
National Technical University of Ukraine "Kyiv Polytechnic Institute", Kyiv, Ukraine, Associate Professor, Ph.D. NTUU "KPI", Kyiv, Ukraine.

**Senka Hadzic, Du Yang and Jonathan Rodriguez**
Instituto de Telecomunicacoes, Aveiro, Portugal.

**Manuel Violas**
Instituto de Telecomunicacoes, Aveiro, Portugal Universityof Aveiro, Aveiro, Portugal.

**Madhuri K. Kulkarni and Veeresh G. Kasabegoudar**
Post Graduate Department, Mahatma Basveshwar Education Society's, College of Engineering, Ambajogai, India.

**Swati Dhondiram Jadhav and Veeresh Gangappa Kasabegoudar**
Post Graduate Department, Mahatma Basveshwar Education Society's, College of Engineering, Ambajogai, India.

**Hala BahyEldeen Nafea, Fayez W. Zaki andHossam E. S. Moustafa**
Dept. of Electronics and Communications Eng, Faculty of Engineering, Mansoura University, Egypt.

**Mourad Hadjila and Hervé Guyennet**
LIFC UFR ST, Besancon, France.

**Mohammed Feham**
UABB STIC Laboratory, Tlemcen, Algeria.

**S. Uvaraj**
Arulmigu Meenakshi Amman College of Engineering.

**S. Suresh**
Sri Venkateswara College of Engineering.

**N. KannaiyaRaja**
Defence Engineering College, Ethiopia.

# Index

www.ingramcontent.com/pod-product-compliance
Lightning Source LLC
Chambersburg PA
CBHW080249230326
41458CB00097B/4183